Sensors and Biosensors Related to Magnetic Nanoparticles

Sensors and Biosensors Related to Magnetic Nanoparticles

Editor

Galina V. Kurlyandskaya

MDPI • Basel • Beijing • Wuhan • Barcelona • Belgrade • Manchester • Tokyo • Cluj • Tianjin

Editor
Galina V. Kurlyandskaya
Euskal Herriko Unibertsitatea
Spain

Editorial Office
MDPI
St. Alban-Anlage 66
4052 Basel, Switzerland

This is a reprint of articles from the Special Issue published online in the open access journal *Sensors* (ISSN 1424-8220) (available at: https://www.mdpi.com/journal/sensors/special_issues/sens_biosens_mag_nano).

For citation purposes, cite each article independently as indicated on the article page online and as indicated below:

LastName, A.A.; LastName, B.B.; LastName, C.C. Article Title. *Journal Name* **Year**, *Volume Number*, Page Range.

ISBN 978-3-0365-4615-5 (Hbk)
ISBN 978-3-0365-4616-2 (PDF)

© 2022 by the authors. Articles in this book are Open Access and distributed under the Creative Commons Attribution (CC BY) license, which allows users to download, copy and build upon published articles, as long as the author and publisher are properly credited, which ensures maximum dissemination and a wider impact of our publications.

The book as a whole is distributed by MDPI under the terms and conditions of the Creative Commons license CC BY-NC-ND.

Contents

About the Editor vii

Preface to "Sensors and Biosensors Related to Magnetic Nanoparticles" ix

Mohammad Reza Zamani Kouhpanji and Bethanie J. H. Stadler
Magnetic Nanowires for Nanobarcoding and Beyond
Reprinted from: *Sensors* 2021, 21, 4573, doi:10.3390/s21134573 1

Gabriele Barrera, Federica Celegato, Matteo Cialone, Marco Coïsson, Paola Rizzi
and Paola Tiberto
Effect of the Substrate Crystallinity on Morphological and Magnetic Properties of $Fe_{70}Pd_{30}$
Nanoparticles Obtained by the Solid-State Dewetting
Reprinted from: *Sensors* 2021, 21, 7420, doi:10.3390/s21217420 17

Anastasia V. Artemova, Sergey S. Maklakov, Alexey V. Osipov, Dmitriy A. Petrov,
Artem O. Shiryaev, Konstantin N. Rozanov and Andrey N. Lagarkov
The Size Dependence of Microwave Permeability of Hollow Iron Particles
Reprinted from: *Sensors* 2022, 22, 3086, doi:10.3390/s22083086 31

Sergey N. Starostenko, Dmitriy A. Petrov, Konstantin N. Rozanov, Artem O. Shiryaev
and Svetlana F. Lomaeva
Effect of Temperature on Microwave Permeability of an Air-Stable Composite Filled with
Gadolinium Powder
Reprinted from: *Sensors* 2022, 22, 3005, doi:10.3390/s22083005 43

Arthur V. Dolmatov, Sergey S. Maklakov, Polina A. Zezyulina, Alexey V. Osipov,
Dmitry A. Petrov, Andrey S. Naboko, Viktor I. Polozov, Sergey A. Maklakov,
Sergey N. Starostenko and Andrey N. Lagarkov
Deposition of a SiO_2 Shell of Variable Thickness and Chemical Composition to Carbonyl Iron:
Synthesis and Microwave Measurements
Reprinted from: *Sensors* 2021, 21, 4624, doi:10.3390/s21144624 57

Tatyana V. Terziyan, Alexander P. Safronov, Igor V. Beketov, Anatoly I. Medvedev,
Sergio Fernandez Armas and Galina V. Kurlyandskaya
Adhesive and Magnetic Properties of Polyvinyl Butyral Composites with Embedded Metallic
Nanoparticles
Reprinted from: *Sensors* 2021, 21, 8311, doi:10.3390/s21248311 71

Stanislav O. Volchkov, Anna A. Pasynkova, Michael S. Derevyanko, Dmitry A. Bukreev,
Nikita V. Kozlov, Andrey V. Svalov and Alexander V. Semirov
Magnetoimpedance of CoFeCrSiB Ribbon-Based Sensitive Element with FeNi Covering:
Experiment and Modeling
Reprinted from: *Sensors* 2021, 21, 6728, doi:10.3390/s21206728 89

Arthur L. R. Souza, Matheus Gamino, Armando Ferreira, Alexandre B. de Oliveira,
Filipe Vaz, Felipe Bohn and Marcio A. Correa
Directional Field-Dependence of Magnetoimpedance Effect on Integrated YIG/Pt-Stripline
System
Reprinted from: *Sensors* 2021, 21, 6145, doi:10.3390/s21186145 103

Sergey V. Komogortsev, Irina G. Vazhenina, Sofya A. Kleshnina, Rauf S. Iskhakov, Vladimir N. Lepalovskij, Anna A. Pasynkova and Andrey V. Svalov
Advanced Characterization of FeNi-Based Films for the Development of Magnetic Field Sensors with Tailored Functional Parameters
Reprinted from: *Sensors* **2022**, *22*, 3324, doi:10.3390/s22093324 . . . . **115**

Tomoo Nakai
A Uniform Magnetic Field Generator Combined with a Thin-Film Magneto-Impedance Sensor Capable of Human Body Scans
Reprinted from: *Sensors* **2022**, *22*, 3120, doi:10.3390/s22093120 . . . . **127**

Nikita A. Buznikov and Galina V. Kurlyandskaya
A Model for the Magnetoimpedance Effect in Non-Symmetric Nanostructured Multilayered Films with Ferrogel Coverings
Reprinted from: *Sensors* **2021**, *21*, 5151, doi:10.3390/s21155151 . . . . **145**

Grigory Yu. Melnikov, Vladimir N. Lepalovskij, Andrey V. Svalov, Alexander P. Safronov and Galina V. Kurlyandskaya
Magnetoimpedance Thin Film Sensor for Detecting of Stray Fields of Magnetic Particles in Blood Vessel
Reprinted from: *Sensors* **2021**, *21*, 3621, doi:10.3390/s21113621 . . . . **159**

About the Editor

Galina V. Kurlyandskaya

Galina V. Kurlyandskaya, Senior Member of IEEE Magnetic Society, graduated from the Physics Department of Ural State University A.M. Gorky, Ekaterinburg, Russia, in 1983. She started her research work in 1983 at the Institute of Metal Physics UD RAS and obtained a Ph.D. in physics of magnetic phenomena in 1990. Her Doctorate of Science degree in 2007 from Ural State University A.M. Gorky. Prof. Kurlyandskaya received advanced training at the Institute of Applied Magnetism, University of Complutense, University of Oviedo, University of the Basque Country (Euskal Herriko Unibertsitatea UPV/EHU), University of Dusseldorf Heinrich Heine, ENS Cashan, University of Maryland, Rowan University, Immanuel Kant Baltic Federal University, Ural State University A.M. Gorky, Ural Federal University B.N. Yeltsin (Laboratory of Magnetic Sensors), Brazilian Research Center of Physics (CBPF), and University of Santa Maria. Her main research areas are fabrication and the magnetic and transport properties of nanostructured magnetic materials, magnetic domain structure, magnetoabsorption, magnetic sensors and biosensors, and biomedical applications of magnetic nanocomposites.

Previously Galina V. Kurlyandskaya served as Guest Editor of Special Issues of MDPI

"Biosensors with Magnetic Nanocomponents" (2020)—

https://www.mdpi.com/journal/sensors/special_issues/Magneticnano

"Magnetic Materials Based Biosensors" (2018)

https://www.mdpi.com/journal/sensors/special_issues/MagneticMaterials

Preface to "Sensors and Biosensors Related to Magnetic Nanoparticles"

Present-day sensor devices and sensor networks are representative of the multidisciplinary approach of science, technology, and biomedicine. The list of related research areas (nanosensors, biosensors, physical and chemical, intelligent, remote, optical, electronic sensors, sensor networks, communications, non-destructive testing, etc.) is rapidly growing, and public attention grows and transforms, becoming a more and more important area of attention. Therefore, sensor materials become more and more important, elaborate, and sophisticated, from multilayered structures to magnetic polymer composites, from magnetic nanoparticles and nanorods toward multifunctional structured soft matter. However, the most common image of modern sensor material can be described as a composite.

There are different types of magnetic effects capable of creating highly sensitive detectors for electronic applications, communications, automatic control, biology, medicine, etc. One of the main goals of their efficient development is to create a new generation of sensor material. In this Special Issue, we considered both materials and devices with magnetic nanoparticles and materials and devices for magnetic nanoparticle detection and evaluation of their concentration, distribution, and contribution to the other physical properties of composites.

This book describes interesting examples of magnetic materials with magnetic nanoparticles or compact devices using composites with nanoparticles, including new engineering solutions and theoretical contributions on the magnetic biosensing of soft matter composites. I would like to thank the authors from different countries who formed international team of experts, and the reviewers and Editorial members of MDPI for their special effort to ensure the high standard of this book and to make it interesting for wide audience.

Galina V. Kurlyandskaya
Editor

Magnetic Nanowires for Nanobarcoding and Beyond

Review

Mohammad Reza Zamani Kouhpanji [1,2] and Bethanie J. H. Stadler [1,*]

[1] Department of Electrical and Computer Engineering, University of Minnesota, Minneapolis, MN 55455, USA; zaman022@umn.edu
[2] Department of Biomedical Engineering, University of Minnesota, Minneapolis, MN 55455, USA
* Correspondence: stadler@umn.edu

Abstract: Multifunctional magnetic nanowires (MNWs) have been studied intensively over the last decades, in diverse applications. Numerous MNW-based systems have been introduced, initially for fundamental studies and later for sensing applications such as biolabeling and nanobarcoding. Remote sensing of MNWs for authentication and/or anti-counterfeiting is not only limited to engineering their properties, but also requires reliable sensing and decoding platforms. We review the latest progress in designing MNWs that have been, and are being, introduced as nanobarcodes, along with the pros and cons of the proposed sensing and decoding methods. Based on our review, we determine fundamental challenges and suggest future directions for research that will unleash the full potential of MNWs for nanobarcoding applications.

Keywords: magnetic nanowires; nanobarcodes; encoding; sensing and decoding

1. Introduction

Initially, barcodes were invented for the authentication of products in anti-counterfeiting, which is of the foremost importance due to the continuous growth of non-transparent trading [1–4]. Nanostructured materials are the backbone in barcoding applications, because their similar appearance hides them from the naked eye, while their physical and chemical properties are significantly different and are suitable for authentication [5,6]. As a result, the unmet demands for miniaturized barcodes led to the emergence of nanobarcodes, such as magnetic nanoparticles [7–9], magneto-optic nanoparticles [10,11], and photonic nanoparticles [12–14], in diverse applications, including nanomedicine and cell biology [15–18], as well as computing and cryptography [19–21]. Changing the composition and size of nanomaterials is probably the most convenient approach to generate numerous nanobarcodes with distinct codes [22]. However, generating nanobarcodes with unique codes does not necessarily guarantee that reliable sensing and decoding is also possible, especially when there is more than one nanobarcode at the scanner. This restriction obligates defining the essential merits for nanobarcodes and designing nanobarcodes that meet these merits [5,6].

In the big picture, there are the following three essential merits for the ideal nanobarcode: (1) expandable encoding, (2) secure sensing, and (3) reliable decoding, as shown in Figure 1. Simply, the codes are physical properties with high flexibility that can be easily tailored and measured. For many applications, the sensing must be done by non-destructive measurement techniques with high repeatability [23] that can be readily translated to the portable devices, suitable for daily applications. Therefore, the first two essential merits are strongly correlated, and they can be tackled by choosing nanomaterials/nanostructures with special properties, which can be readily engineered and measured. The first two merits may discard several proposed nanomaterials/nanostructures for nanobarcoding, but there is still a vast number of nanomaterials/nanostructures that meet these two merits and are deemed promising. To specialize this review, here, we only focus on one-dimensional magnetic nanoparticles, also known as magnetic nanowires (MNWs), with ferromagnetic properties, to deeply discuss their recent progress, particularly in nanobarcoding applications and how they potentially can transform the future of this field.

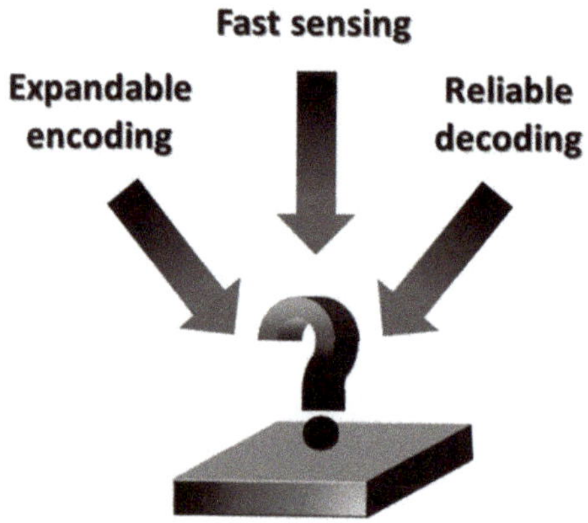

Figure 1. A flowchart rendering the essential merits for nanobarcodes.

2. Why Magnetic Nanowires for Nanobarcodes?

Recently, MNWs became the center of the research in nanobarcoding because of the revealed potential for making the next generation of nanobarcodes and/or biolabels [23–26], driven by the fact that the MNWs can be remotely and selectively detected [27–29]. Moreover, the MNWs are dominantly fabricated using electrodeposition techniques that are cheap, fast, and scalable for mass production [30–33]. More importantly, their magnetic response can be readily extracted from the background signals, leading to a high signal-to-noise ratio–suitable for miniaturizing the barcodes size [7]; this is because the majority of materials are diamagnetic or paramagnetic, which do not produce magnetic signals, such as irreversible switching, which is exclusively a property of ferromagnetic materials [34,35]. Thus, as opposed to optical- or radio-frequency barcodes, magnetic signals are not contaminated by the background noise [36–39]. Aside from these advanced benefits of MNWs for nanobarcoding, they also meet the aforementioned merits of expandable encoding, fast sensing, and reliable decoding, which we discuss in detail in the following sections. First, due to the strong correlation between the encoding merit and the sensing merit, we discuss these merits together. We next discuss the current state-of-the-art for the reliable decoding of multiple MNW-based nanobarcodes at the readout, because its progression currently substantially lags behind the other merits' progression.

2.1. Encoding and Sensing of Magnetic Nanowire (MNW)-Based Nanobarcodes

Each magnetic nanowire (MNW)-based nanobarcode is made of a collection of MNWs, where the magnetic properties of each MNW and the intra-magnetic interactions can be used for encoding [40,41]. The most favorable magnetic signatures for nanobarcoding are those that can be rapidly measured, with high repeatability, to fit the daily applications as expected for nanobarcodes. This requirement limits the number of magnetic measurements to a few applicable measurements, which can be categorized into the following two groups: (1) DC measurements, and (2) AC measurements. The DC measurements usually need simpler equipment, and they have been widely used for the magnetic characterization of MNWs. As a result, there has been much progress in the development of instruments for fast and repeatable DC measurements. The DC measurements include hysteresis loop measurements, the first-order reversal curve (FORC) method [42–44], remanence curve method [45,46], and, most recently, the projection method and the backward remanence method [34,47]. The AC measurements are magnetic particle spectroscopy and ferromagnetic resonance spectroscopy, as the most well-established and common methods that might be transferrable to daily applications [23,48,49].

2.1.1. DC Measurements

The hysteresis loop measurement is the most popular and fastest method for magnetic property extraction, and it provides the saturation magnetization and the coercivity of any sample. Figure 2a schematically illustrates the hysteresis loop method. The saturation magnetization is a function of the MNW composition, as shown in Figure 3a–e. Making MNWs as alloys of one magnetic (such as cobalt or iron) and one non-magnetic component allows the tailoring of the saturation magnetization from zero to the saturation magnetization of the magnetic component [50]. The common magnetic and non-magnetic components that can be easily co-electrodeposited are (1) iron with gold [31,51] or copper [52], (2) nickel with gold [53] or copper [54,55], or (3) cobalt with gold or copper [56].

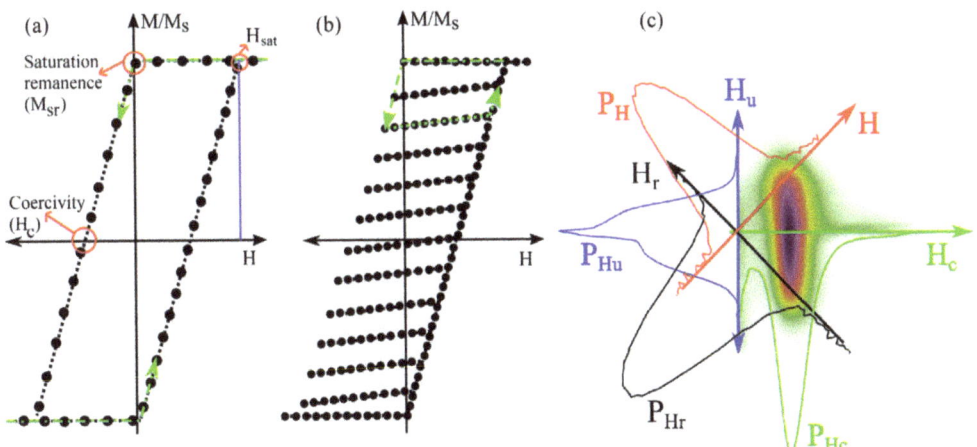

Figure 2. Schematically depicting the hysteresis loop method (**a**) and the FORC method (**b**,**c**), where (**b**) is the FORC data collection and (**c**) is a FORC heat-map. In subfigures (**a**,**b**), the green arrows show the data acquisition direction. In subfigure (**c**), the red distribution is the projection of the FORC heat-map on the applied field (H), the blue distribution is the projection of the FORC heat-map on the interaction field axis (interaction field distribution), the black distribution is the projection of the FORC heat-map on the reversal field (Hr), and the green distribution is the projection of the FORC heat-map on the coercivity axis (coercivity distribution). Figure adapted from [34,35].

Tailoring the saturation magnetization is best to be conducted using magnetic components with high saturation magnetization, such as iron, if the other components are non-magnetic [57–59]. In this case, the saturation magnetization can be tailored over a wider range, from nearly zero up to the saturation magnetization of the magnetic component. Therefore, among all of the compositions, alloys containing iron might be more favorable as they have high saturation magnetization, leading to a wider achievable range of saturation magnetizations. Note that the alloyed MNWs were also made of both magnetic components, such as iron–nickel [60,61], iron–cobalt [62,63], nickel–cobalt [64,65], or iron–cobalt–nickel [66]. When both components are magnetic, the saturation magnetization range will be limited to the minimum and maximum saturation magnetization of the components, except iron–cobalt MNWs with a 2:1 atomic ratio that leads to higher saturation magnetization [63,67,68], as shown in Figure 3f. Generally speaking, having both magnetic components does not provide much flexibility to tailor the saturation magnetization as the encoding parameters. This is also valid when the MNWs are made of three components, such as iron, nickel, and cobalt, as trinary [69–71]. However, these cases are very useful to tailor other magnetic properties, such as coercivity, where, as an example, Permalloy (iron–nickel with a 1:4 atomic ratio) is one of the most popular compositions [61,67,72,73].

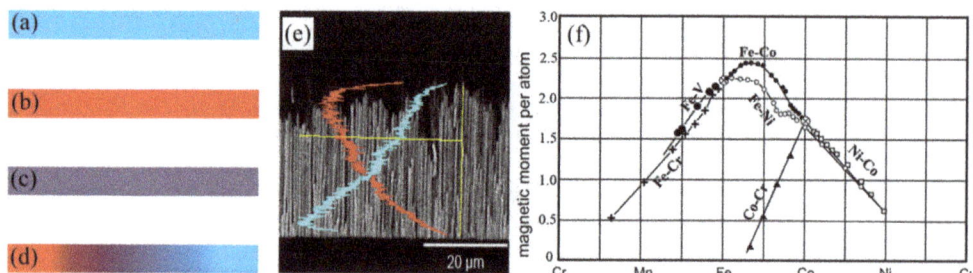

Figure 3. Demonstrating different approaches for tailoring the magnetization saturation and the coercivity of MNWs for encoding. (**a**,**b**) Single component MNWs; (**c**) multi-component or alloyed MNWs; (**d**) MNWs with modulated composition; (**e**) SEM image of modulated composition MNWs, adapted from [74]; and (**f**) represents a Slater–Pauling curve illustrating the dependency of the magnetic moment on the composition, adapted from [58].

Coercivity is another important magnetic property that determines how resilient the spins are against changing their direction. Simply, coercivity is the required magnetic field to rotate the magnetization 90 degrees, or for very anisotropic samples, it is the field required to produce equal "up" and "down" spins. The simplest way to tailor the coercivity is probably to vary the MNW sizes [75–77], as shown in Figure 4a,b. For MNWs with a very large diameter, the spins switch via the nucleation and propagation of a magnetic domain wall, which usually requires lower energies, or equivalently, smaller coercivities. As the MNWs diameter increases, the nucleation and propagation of the magnetic domain walls become easier, which causes the coercivity to decrease [78]. This is because MNWs with large diameters hold multiple magnetic domains, leading to the presence of exchange coupling between the magnetic domains [50]. The exchange coupling contributes to the coercivity that is proportional to the inverse square of the MNWs diameter. Thus, for diameters larger than the critical diameter (the diameter in which the MNWs are a single domain), the coercivity decreases as the diameter increases [79]. Note, once the MNWs diameter becomes smaller than the critical diameter, all spins rotate simultaneously. In this case, which is also known as coherent rotation, the coercivity significantly increases to large values. It should be mentioned that the MNWs length can also impact the reversal mechanism of the spins (i.e. transverse well mode instead of the coherent mode if the length is very long) [80,81]. However, the effects of the length on coercivity are usually taken out, because the definition of MNW obligates a much longer length compared to the diameter. Note, when the length is much longer than the diameter, the shape anisotropy is constant. Thus, the coercivity becomes independent of the length.

Inducing any chemical or physical changes/mismatches that facilitate or hinder the switching direction of the spins would result in tailoring the coercivity [72,82]. An example for the chemical approach is to synthesize MNWs with different compositions or crystal structures [83–86]. For instance, it was shown that by varying the pH of the electrolyte during electrodeposition, one can manipulate the crystal structure of cobalt MNWs from hexagonal close-packed (hcp), to a mixture of hcp and face-center-cubic (fcc), to a purely fcc crystal structure [85,87]. Over the last few years, the physical approaches for tailoring the coercivity have been intensively studied. The basic for physical approaches is to pin magnetic domain walls by inducing a discontinuity; a few examples are depicted in Figure 4. The examples for physical approaches are diameter modulation (an MNW with multiple diameters along its length) [88,89], multi-segmented [90,91], inducing notches [92,93], and interconnecting MNWs [29,94]. Figure 4 shows some proposed attempts for engineering the MNWs coercivity, by taking benefit of the following: (a–b and e–f) varying diameters, (c and g–h) modulating the diameter, and (d and i) multi-segmenting the MNWs. In all these approaches, the magnetic domain walls are being pinned at the transition sites, which generally leads to an increase in coercivity. Thus, it would be interesting to combine the aforementioned approaches for tailoring the coercivity over a much wider range than the current state-of-the-art.

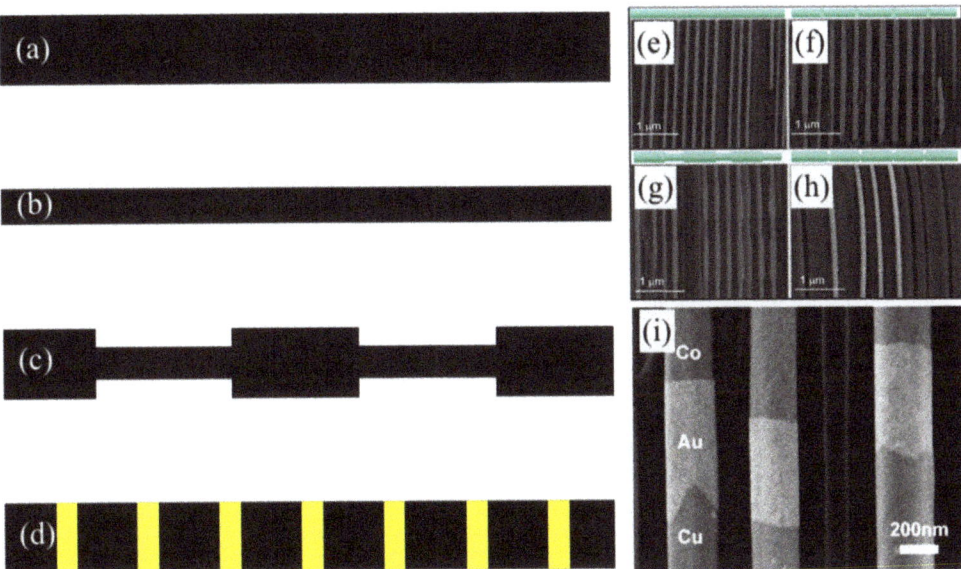

Figure 4. Illustrating the different techniques for tailoring the coercivity of the MNWs, where (**a**,**b**) changing the diameter (or aspect ratio) of single diameter MNWs; (**c**) modulated diameter MNWs; (**d**) multi-segmented MNWs; (**e**–**h**) SEM images of varying and modulating the MNWs diameter, adapted from [95]; and (**i**) a SEM image of multi-segmented MNWs, adapted from [96].

The first-order reversal curve (FORC) method is another DC measurement that has been broadly used for the qualitative, and partially quantitative, description of magnetic signatures [97–99], as shown in Figure 2b,c. In the context of the FORC method, the hysteresis loop of a magnetic nanobarcode can be considered as a picture, where each pixel is the contribution of a hysteron, such as an individual MNW, which builds the whole hysteresis loop area. Experimentally, the FORC method scans the whole area of the hysteresis loop in a two-dimensional fashion, using the following two magnetic fields: (1) reversal field, and (2) applied field [100]. Once the magnetic responses are measured, in terms of the reversal and applied fields, the second derivative of the magnetic responses are taken [101], and the results are plotted as heat-maps versus the reversal field and the applied field.

When plotting each pixel (i.e., a single MNW switching), in terms of the reversal and applied field, the reversal field is the field in which magnetization switches from +1 to −1, while the applied field is the field in which the magnetization switches reversely. In other words, half of their difference (equivalently, the width) is the coercivity, and half of their summation (equivalently, the horizontal shift) is the interaction field [42,100,102]. According to these definitions of the coercivity and interaction fields, the FORC heat-maps can be plotted in the coercivity interaction field plane, which is a 45-degree rotation of the reversal-applied fields plane. Conceptually, the FORC heat-maps indeed determine the probability of finding an MNW with a specific coercivity and interaction field pair. Consequently, since there are many MNWs with different coercivities and under different interaction fields, the coercivity and interaction fields are represented as distributions that have been used as magnetic signatures [103–105].

As mentioned earlier, the FORC heat-maps have been broadly used as qualitative descriptions of MNWs magnetic signatures. For quantitative description, the heat-maps are projected onto the coercivity and interaction fields in order to calculate the coercivity and interaction field distributions, and to be used as quantitative signatures [106,107]. This detailed analysis of the FORC method has been known as a very powerful probe for analyzing the magnetic signatures of many complex MNW-based nanobarcodes. As a result, the FORC method became a very promising method for the reliable sensing of nanobarcodes. The major advantage of the FORC method is that it provides the magnetic signatures (coercivity and interaction field) as distributions rather than single values. It is very useful because the measured signatures can be decoded [9]. This detailed sensing of magnetic nanobarcodes, provided by the FORC method, is accompanied with unpleasant downsides that dramatically hinder the usability of the FORC method for practical applications [41,108]. A few examples of these drawbacks are (1) its slow signature extraction, which makes it extremely inefficient for daily usages, and (2) complex data analysis, which causes artifacts. In this direction, several researches have been conducted, to speed up both the data collection [34,35,109] and data processing [110,111] of the FORC method, which is still a long way to go.

Among all of the approaches to speed up the FORC measurements, the projection method was proposed, to significantly accelerate the sensing of magnetic nanobarcodes signatures, particularly for biolabeling and nanobarcoding [8,27,34]. A schematic for the projection method is given in Figure 5a. The projection method employs the fundamentals of the FORC method, to extract the irreversible switching field distribution at the reversal field in lieu of coercivity and interaction field distributions. The irreversible switching field distribution as a magnetic signature was found to provide several advantages that are compatible with the expectations for novel nanobarcodes. First, using the projection method, the irreversible switching field distribution can be measured by scanning only the vicinity of the upper branch hysteresis loop—leading to a significant time reduction in comparison to scanning the whole area of the hysteresis loop, as is needed for the FORC method. It was shown that up to five data points along each reversal curve are sufficient to reliably measure the irreversible switching field distribution, which leads to a factor of 50–100X faster measurements compared to the FORC method [34]. Second, the projection method requires only one derivative to calculate the irreversible switching field distribution, while the FORC method requires two sequential derivatives followed by an integral. Due to the measurement noise, the FORC signatures are accompanied by artifacts, which are still elusive [35]. Last, but not least, the projection method indeed measures both the reversible and irreversible switching field distributions, and decomposes them. Since the irreversible switching field is residual magnetization, it is exclusively generated by the MNWs in the barcode, while the reversible switching field could be the superposition of the MNWs reversible response and the surrounding materials. Thus, the projection method provides a better signal-to-noise ratio that is suitable for further miniaturizing the nanobarcode sizes by excluding the background signals. Note, the irreversible switching field distribution is a function of the MNWs coercivity and the interaction fields between them. Therefore, in addition to the aforementioned parameters for tailoring the coercivity, the irreversible switching field distribution can be further tailored by tuning the interaction fields (e.g., by varying the interwire distance) within the MNWs, leading to a more expandable encoding capability [35,112,113].

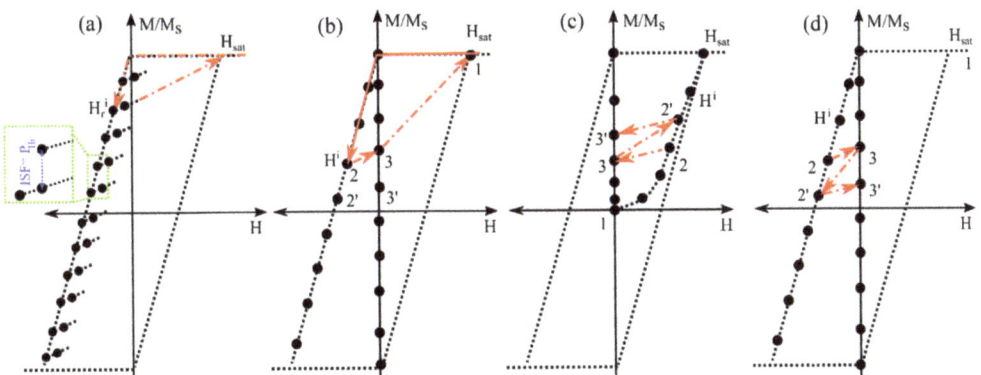

Figure 5. Schematically demonstrating the data collection protocols for (**a**) the projection method, (**b**) backward remanence method, (**c**) isothermal remanence method, and (**d**) DC demagnetization method. The projection method (**a**) provides the irreversible switching (equivalent to the residual magnetization) at the reversal field, Hr. While, the remanence methods (**b**–**d**) provide the residual magnetization at the zero applied field, H, such as points 3 and 3'. The key feature that separates the backward remanence (**b**) from the isothermal remanence (**c**) and the DC demagnetization remanence (**d**) is the saturating the whole system before applying the Hr and removing it.

It should be emphasized that the projection method measures the irreversible switching field distribution at the reversal field. The advantages of the irreversible switching field distribution for nanobarcoding have attracted attention towards measuring the irreversible switching field distribution at zero field, which is also known as the backward remanence measurement [5,6], as shown in Figure 5b. Even though the backward remanence method measures the remanence (magnetization at zero field), it is different from other remanence methods, such as isothermal remanence [114,115] (Figure 5c) or DC demagnetization remanence [116,117] (Figure 5d). Indeed, the backward remanence measurement measures the remanence in a more restricted method that was shown to be more reproducible [34], which is necessary for reliable sensing of magnetic nanobarcodes. The only difference between the backward remanence measurement and the other remanence method is that it saturates the magnetic nanobarcode at each step before applying and removing the field [6]; Figure 5b–d schematically shows the data collections for each of these remanence methods for comparison. This suppresses the stochastic effects of MNWs magnetization in an array, leading to more reliable sensing [5,6].

2.1.2. AC Measurements

AC measurements apply an alternating magnetic field and measure the response of magnetic nanobarcodes at different frequencies or a biased magnetic field. The two widely used AC measurements are magnetic particles spectroscopy [118–120], in Figure 6a–d, and magnetic resonance spectroscopy [121–123], in Figure 6e,f. Magnetic particles spectroscopy applies an alternating magnetic field, using a magnetic coil at a single frequency (sometimes two frequencies [124,125]), shown in Figure 6c, and measures the magnetic responses in real time, in Figure 6b, and the frequency domain, as shown in Figure 6d. For superparamagnetic nanoparticles, where the coercivity is zero, as in Figure 6a, the magnetization will be a function of odd multiplication of the applied frequency, which aims to sense the magnetic response of the magnetic nanoparticles and distinguish it from the applied signal [126]. Sensing the MNWs using magnetic particles spectroscopy is practically very challenging; this is because the non-zero coercivity of the MNWs causes a nonlinear dynamic response that cannot reliably be sensed and distinguished from the applied field. Furthermore, the MNWs have a non-zero coercivity, from a hundred Oe to several hundred Oe, which mandates a very large AC field for AC oscillations. Applying a large AC field at a high frequency causes heat generation, due to eddy currents. To avoid the eddy currents and to be able to use the magnetic particles spectroscopy for nanobarcoding, the MNWs must have small coercivity. However, this limits the range of encoding, leading to a limited magnetic nanobarcode that is not favorable for nanobarcoding applications.

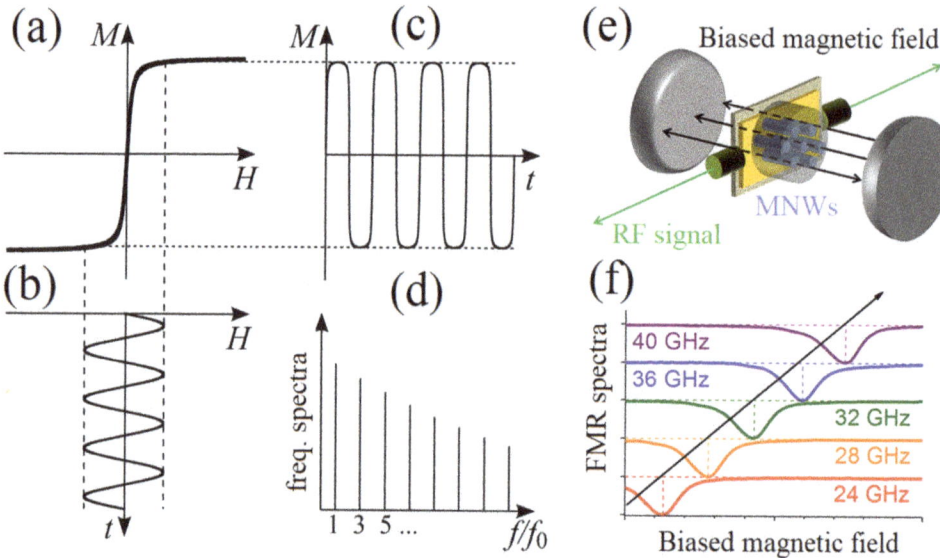

Figure 6. Schematically demonstrating (**a**–**d**) magnetic particle spectroscopy data, adapted from [127], and (**e**,**f**) ferromagnetic resonance spectroscopy data, adapted from [123]. In magnetic particle spectroscopy, superparamagnetic nanoparticles are exposed to an alternating magnetic field (**b**), which forces them to oscillate (**c**). By linearizing their response in frequency domain, multiple peaks appear at odd higher frequencies (**d**) that are being used for sensing them. In magnetic resonance spectroscopy, the MNWs are exposed to an RF signal while a biased magnetic field is applied. By sweeping the RF signal frequency or the biased magnetic field magnitude, the RF absorption of the MNWs varies due to their spins' precession, where the absorption signal is being used for sensing the MNWs.

Another AC measurement for sensing magnetic nanobarcodes is magnetic resonance spectroscopy [123,128], also known as ferromagnetic resonance measurement, shown in Figure 6e,f. Magnetic resonance spectroscopy is developed based on the traditional radiofrequency (RF) identification method, which uses the AC magnetic field of a radio-frequency signal, in the either presence or absence of a DC magnetic field, to sense the magnetic nanobarcodes. By varying the DC magnetic field, the resonance frequency of the MNWs change, and that can be used as an extra degree of freedom for secure sensing [129]. Magnetic resonance spectroscopy could be faster compared to the DC measurements for sensing. However, it tends to have a poor signal-to-noise ratio and short distance sensing, due to absorption/attenuation of the RF signals. Indeed, since magnetic resonance spectroscopy uses an RF signal for the stimulation, it inherently has the limitations of traditional RF identification.

2.2. Decoding of Magnetic Nanobarcodes

As discussed in the previous section, there is a trade-off between fast sensing and reliable sensing. A solution for avoiding this trade-off is to reliably sense multiple magnetic nanobarcodes, to speed up the decoding by reducing the number of required read-outs/measurements. In contrast to the huge progress in the encoding and sensing of magnetic nanobarcodes, the reliable decoding of them has not received much attention, even though it is crucial for the commercial transition [5,6]. Furthermore, establishing a roadmap for the reliable decoding of multiple nanobarcodes is not only beneficial for the magnetic nanobarcodes, but other nanobarcodes can also take benefits from this, to ramp up the authentication speed. Here, we again discuss the recent works to focus on the MNW-based nanobarcodes.

For reliable decoding of multiple unknown MNW-based nanobarcodes, it is necessary for the magnetic nanobarcodes to have distinct features, with minimum overlapping. This requirement discards the hysteresis loop measurement as it only provides single values, for example, saturation magnetization or coercivity. Aside from the fact that the saturation magnetization and coercivity are well known and have strong magnetic signatures for encoding, they are insufficient for reliable decoding, especially when there is more than one nanobarcode at the readout [5–7]. For example, assume the scanner reads the saturation magnetization 100 emu/cc. Since it is a single value, it does not indicate if there was only one nanobarcode with 100 emu/cc or two nanobarcodes with 50 emu/cc for each, and so on. This drawback restricts the application of the hysteresis loop measurement, regardless of it being a fast, easy, and cheap sensing method. As a result, the sensing methods, such as the projection method, that provide distributions stand out. Practically, the most favorable distributions are those that can be tailored using multiple parameters, such as saturation magnetization, coercivity, and interaction fields, to provide higher flexibilities for encoding.

The key for the reliable decoding of multiple nanobarcodes is that the readout signal of a combination must be a linear superposition of the individual components. This is usually achievable because the interwire distance between the MNWs (usually in order of 500 nm) in a nanobarcode is several orders of magnitude smaller than the distance between the nanobarcodes (it is at least the thickness of the nanobarcode, which could be in order of 1 mm). This allows the readout signal to be a linear superposition of the signatures of the composed nanobarcodes/subdivisions. Indeed, the challenge is to determine the number of nanobarcodes; Figure 7a shows a simple flowchart for this purpose. This is because increasing the number of nanobarcodes improves the fitting quality, leading to unbounded values for the number of nanobarcodes. To overcome this challenge, it was proposed to use the degree of the fitting quality (i.e., the root mean square, RMS) improvement as an indicator of the likelihood for having the expected number of nanobarcodes [130–132], as shown in Figure 7b,c. In other words, it is true that the fitting quality improves as the number of nanobarcodes increases (under-fitting); however, this improvement will not be significant as the fit number surpasses the actual number of codes (over-fitting). To overcome this, one of the proposed techniques is to consider a cutoff value for the improvement in the RMS [5,6]. Therefore, by selecting a cutoff for the RMS, to differentiate between the under-fitting and over-fitting, one can predict the number of the nanobarcodes at the readout; Figure 7 schematically illustrates such algorithms for a readout signal of two nanobarcodes.

Figure 7 schematically illustrates the procedure for decoding using a cutoff value. One first assumes that there is only one nanobarcode (N = 1) at the readout, and the measured remanence spectrum is fit to one Gaussian function to find the fitting parameters by optimizing RMS^1, where superscript one indicates N = 1. Next, N is increased to 2 and the new optimum RMS error, RMS^2, is calculated. Then, RMS^2 is compared with RMS^1 to determine how much the RMS error decreased, by increasing N from 1 to 2. If the reduction meets the cutoff, then there are at least two nanobarcodes at the readout (N ≥ 2), as shown in Figure 7b. Then, it is necessary to increase N to 3 and repeat the same procedure, to determine whether or not there are more nanobarcodes present. Note, at this step, RMS^3 and RMS^2 must be considered, and their ratio must be compared with the cutoff value, as in Figure 6c. If the reduction in RMS^3 compared to RMS^2 was not sufficient, the decoding process can be terminated, because it would appear that only two nanobarcodes were present at the readout (N = 2). This process must be continued until the ratio of RMS^N-to-RMS^{N-1} is no longer smaller than the cutoff value. The main drawback of this technique for reliable decoding is finding the correct value for the cutoff. For example, as the number of nanobarcodes at the readout increases, the magnetic signatures start overlapping, which makes the decoding difficult. It should be emphasized that this drawback is not limited only to magnetic nanobarcodes, as the reliable decoding of any nanobarcodes suffers from this problem. To resolve this problem, we proposed two alternatives. The first alternative is to use a floating cutoff value, which is a function of the predicted number of nanobarcodes. The second alternative, which could be a more effective approach, is to eliminate the need for a cutoff, which could be accomplished by using the artificial intelligent (AI) or the machine learning (ML) approaches. To accelerate the transition of MNW-based nanobarcodes to real-life applications, the reliable decoding of multiple nanobarcodes demands a huge amount of attention, with many research opportunities in computer science and signal processing domains, which are expected to flourish soon.

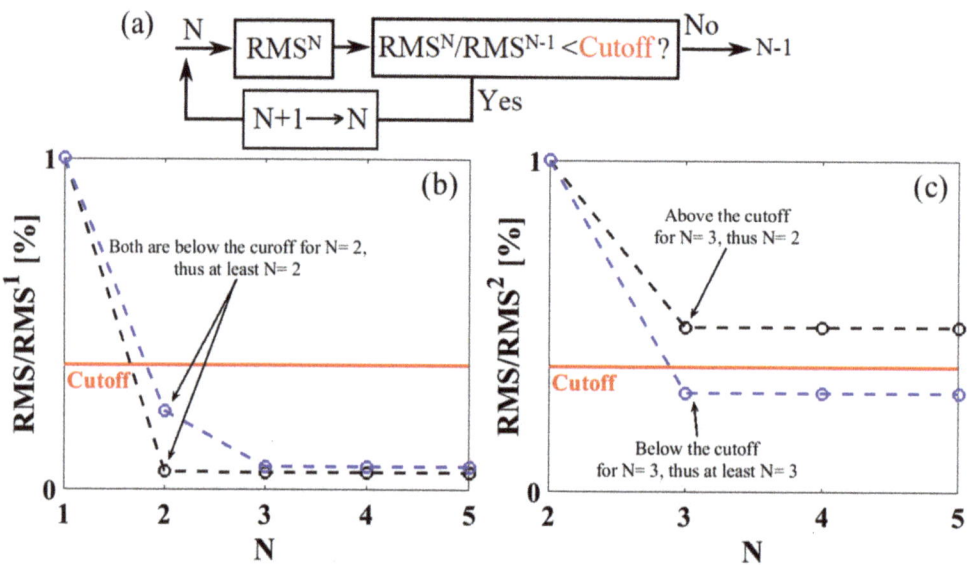

Figure 7. Depicting a decoding method based on using the fitting quality (RMS) as an indicator for determining the number of nanobarcodes at the readout, (**a**) a flowchart for decoding and (**b**,**c**) data analysis for finding the number of the nanobarcodes were produced. The algorithm assumes that one (N = 1) nanobarcode exists, thus, it fits the data with one Gaussian function and calculates the RMS1. Then, it increases N to two, and it calculates RMS2. If the RMS2/RMS1 (**b**) is larger than the cutoff value, there was only one nanobarcode at the readout. Otherwise, there are at least two nanobarcodes and the procedure must be repeated for N=3, which means RMS3/RMS2 must be evaluated (**c**).

3. Summary and Outlook

Though considerable progress and advances have been recently made toward the encoding/sensing/decoding of MNW-based nanobarcodes, there are still challenges for their use in nanobarcoding applications that need to be addressed to achieve practical translation. The extremely important merits of promising nanobarcodes are as follows: (1) expandable encoding capability, (2) fast sensing with minimal background noises, and (3) reliable decoding of multiple nanobarcodes simultaneously present at the scanner. The encoding capability and sensing merits are inherently correlated, as the former is the targeted magnetic property/signature and the latter is the magnetic measurement method for sensing/measuring the targeted property. The main considerations that must be taken into account when choosing the proper property/signature and sensing/measurement are the stability over time, expandability, and the speed and reproducibility of the signature, in addition to the cost, ease of use, and portability of the sensing instruments. These criteria make many magnetic signatures unusable, and leave a few options available, which can be categorized based on the type of sensing methods, which are DC and AC methods. Basically, the DC sensing methods are more reliable and simpler, but slower than the AC sensing methods. More importantly, the DC methods intuitively provide sensing of the hidden MNW-based nanobarcodes, with higher signal-to-noise ratios compared to the AC methods.

The need for faster sensing, based on DC methods, that provides distributions, instead of single values, to achieve reliable decoding, eliminates the hysteresis loop and first-order reversal curve (FORC) methods, which are brief and slow, respectively. Among all, the projection method and backward remanence deem more promising, as they are as fast as the hysteresis loop method, companied by detailed analyses similar to the FORC method. To further enhance the rapid authentication, it is essential to be able to reliably decode the readout signals from multiple nanobarcodes simultaneously present at the scanner. This requirement demands other fields, such as signal processing from electrical engineering and machine learning or artificial intelligence from computer science, into magnetism, to facilitate the realization of MNW-nanobarcodes translation to daily applications. Indeed, despite the huge progress in the encoding and sensing of MNW-based nanobarcodes, the reliable decoding of multiple MNW-based nanobarcodes is still in its rudimentary stage and requires much exploration.

Author Contributions: Conceptualization, M.R.Z.K. and B.J.H.S.; investigation, M.R.Z.K.; resources, B.J.H.S.; writing—original draft preparation, M.R.Z.K.; writing—review and editing, M.R.Z.K. and B.J.H.S.; supervision, B.J.H.S.; project administration, B.J.H.S.; funding acquisition, B.J.H.S. All authors have read and agreed to the published version of the manuscript.

Funding: This research was funded by National Science Foundation (NSF), grant number CMMI-1762884. The APC was funded by Molecular Diversity Preservation International (MDPI).

Institutional Review Board Statement: Not applicable.

Data Availability Statement: Not applicable.

Acknowledgments: The authors cordially acknowledge the financial support of Molecular Diversity Preservation International (MDPI) journal for generously covering the publication fee.

Conflicts of Interest: There is no conflict of interest to be declared.

References

1. Soon, J.M.; Manning, L. Developing anti-counterfeiting measures: The role of smart packaging. *Food Res. Int.* **2019**, *123*, 135–143. [CrossRef] [PubMed]
2. Lai, H.; Liao, H. A multi-criteria decision making method based on DNMA and CRITIC with linguistic D numbers for blockchain platform evaluation. *Eng. Appl. Artif. Intell.* **2021**, *101*, 104200. [CrossRef]
3. Liu, A.; Liu, T.; Mou, J.; Wang, R. A supplier evaluation model based on customer demand in blockchain tracing anti-counterfeiting platform project management. *J. Manag. Sci. Eng.* **2020**, *5*, 172–194. [CrossRef]
4. Büyüközkan, G.; Tüfekçi, G. A decision-making framework for evaluating appropriate business blockchain platforms using multiple preference formats and VIKOR. *Inf. Sci.* **2021**, *571*, 337–357. [CrossRef]
5. Zamani Kouhpanji, M.R.; Stadler, B. Magnetic Nanowires toward Authentication. *Part. Part. Syst. Charact.* **2021**, *38*, 2000227. [CrossRef]
6. Zamani Kouhpanji, M.R.; Stadler, B.J.H. Unlocking the decoding of unknown magnetic nanobarcode signatures. *Nanoscale Adv.* **2021**, *3*, 584–592. [CrossRef]
7. Zamani Kouhpanji, M.R.; Stadler, B.J.H. Magnetic nanowires for quantitative detection of biopolymers. *AIP Adv.* **2020**, *10*, 125231. [CrossRef]
8. Zamani Kouhpanji, M.R.; Stadler, B.J.H. Projection method as a probe for multiplexing/demultiplexing of magnetically enriched biological tissues. *RSC Adv.* **2020**, *10*, 13286–13292. [CrossRef]
9. Zamani Kouhpanji, M.R.; Um, J.; Stadler, B.J.H. Demultiplexing of Magnetic Nanowires with Overlapping Signatures for Tagged Biological Species. *ACS Appl. Nano Mater.* **2020**, *3*, 3080–3087. [CrossRef]
10. Miller, F.; Wintzheimer, S.; Prieschl, J.; Strauss, V.; Mandel, K. A Supraparticle-Based Five-Level-Identification Tag That Switches Information Upon Readout. *Adv. Opt. Mater.* **2021**, *9*, 2001972. [CrossRef]
11. Miller, F.; Wintzheimer, S.; Reuter, T.; Groppe, P.; Prieschl, J.; Retter, M.; Mandel, K. Luminescent Supraparticles Based on CaF$_2$–Nanoparticle Building Blocks as Code Objects with Unique IDs. *ACS Appl. Nano Mater.* **2020**, *3*, 734–741. [CrossRef]
12. Lin, J.; Lu, Y.; Li, X.; Huang, F.; Yang, C.; Liu, M.; Jiang, N.; Chen, D. Perovskite Quantum Dots Glasses Based Backlit Displays. *ACS Energy Lett.* **2021**, *6*, 519–528. [CrossRef]

13. Li, F.; Huang, S.; Liu, X.; Bai, Z.; Wang, Z.; Xie, H.; Bai, X.; Zhong, H. Highly Stable and Spectrally Tunable Gamma Phase Rb x Cs 1−x PbI 3 Gradient-Alloyed Quantum Dots in PMMA Matrix through A Sites Engineering. *Adv. Funct. Mater.* **2021**, *31*, 2008211. [CrossRef]
14. Galstyan, V. Quantum dots: Perspectives in next-generation chemical gas sensors—A review. *Anal. Chim. Acta* **2021**, *1152*, 238192. [CrossRef]
15. Kwizera, E.A.; O'Connor, R.; Vinduska, V.; Williams, M.; Butch, E.R.; Snyder, S.E.; Chen, X.; Huang, X. Molecular detection and analysis of exosomes using surface-enhanced Raman scattering gold nanorods and a miniaturized device. *Theranostics* **2018**, *8*, 2722–2738. [CrossRef]
16. van den Boorn, J.G.; Schlee, M.; Coch, C.; Hartmann, G. SiRNA delivery with exosome nanoparticles. *Nat. Biotechnol.* **2011**, *29*, 325–326. [CrossRef]
17. Yong, T.; Zhang, X.; Bie, N.; Zhang, H.; Zhang, X.; Li, F.; Hakeem, A.; Hu, J.; Gan, L.; Santos, H.A.; et al. Tumor exosome-based nanoparticles are efficient drug carriers for chemotherapy. *Nat. Commun.* **2019**, *10*, 3838. [CrossRef]
18. Khongkow, M.; Yata, T.; Boonrungsiman, S.; Ruktanonchai, U.R.; Graham, D.; Namdee, K. Surface modification of gold nanoparticles with neuron-targeted exosome for enhanced blood–brain barrier penetration. *Sci. Rep.* **2019**, *9*, 8278. [CrossRef]
19. Pregibon, D.C.; Toner, M.; Doyle, P.S. Multifunctional Encoded Particles for High-Throughput Biomolecule Analysis. *Science* **2007**, *315*, 1393–1396. [CrossRef]
20. Nolan, J.P.; Sklar, L.A. Suspension array technology: Evolution of the flat-array paradigm. *Trends Biotechnol.* **2002**, *20*, 9–12. [CrossRef]
21. Gunderson, K.L. Decoding Randomly Ordered DNA Arrays. *Genome Res.* **2004**, *14*, 870–877. [CrossRef]
22. Shikha, S.; Salafi, T.; Cheng, J.; Zhang, Y. Versatile design and synthesis of nano-barcodes. *Chem. Soc. Rev.* **2017**, *46*, 7054–7093. [CrossRef]
23. Zamani Kouhpanji, M.R.; Stadler, B.J.H. A Guideline for Effectively Synthesizing and Characterizing Magnetic Nanoparticles for Advancing Nanobiotechnology: A Review. *Sensors* **2020**, *20*, 2554. [CrossRef]
24. Lavorato, G.C.; Das, R.; Alonso Masa, J.; Phan, M.-H.; Srikanth, H. Hybrid magnetic nanoparticles as efficient nanoheaters in biomedical applications. *Nanoscale Adv.* **2021**, *3*, 867–888. [CrossRef]
25. Melnikov, G.Y.; Lepalovskij, V.N.; Svalov, A.V.; Safronov, A.P.; Kurlyandskaya, G.V. Magnetoimpedance Thin Film Sensor for Detecting of Stray Fields of Magnetic Particles in Blood Vessel. *Sensors* **2021**, *21*, 3621. [CrossRef]
26. Mollarasouli, F.; Zor, E.; Ozcelikay, G.; Ozkan, S.A. Magnetic nanoparticles in developing electrochemical sensors for pharmaceutical and biomedical applications. *Talanta* **2021**, *226*, 122108. [CrossRef]
27. Zamani Kouhpanji, M.R.; Nemati, Z.; Mahmoodi, M.M.; Um, J.; Modiano, J.; Franklin, R.; Stadler, B. Selective Detection of Cancer Cells Using Magnetic Nanowires. *ACS Appl. Mater. Interfaces* **2021**, *13*, 21060–21066. [CrossRef] [PubMed]
28. Zamani Kouhapanji, M.R.; Nemati, Z.; Modiano, J.F.; Franklin, R.R.; Stadler, B.J.H. Realizing the principles for remote and selective detection of cancer cells using magnetic nanowires. *J. Phys. Chem. B* **2021**.
29. Burks, E.C.; Gilbert, D.A.; Murray, P.D.; Flores, C.; Felter, T.E.; Charnvanichborikarn, S.; Kucheyev, S.O.; Colvin, J.D.; Yin, G.; Liu, K. 3D Nanomagnetism in Low Density Interconnected Nanowire Networks. *Nano Lett.* **2020**, *21*, 716–722. [CrossRef] [PubMed]
30. Sousa, C.T.; Leitao, D.C.; Proenca, M.P.; Ventura, J.; Pereira, A.M.; Araujo, J.P. Nanoporous alumina as templates for multifunctional applications. *Appl. Phys. Rev.* **2014**, *1*, 031102. [CrossRef]
31. Um, J.; Zamani Kouhpanji, M.R.; Liu, S.; Nemati Porshokouh, Z.; Sung, S.Y.; Kosel, J.; Stadler, B. Fabrication of Long-Range Ordered Aluminum Oxide and Fe/Au Multilayered Nanowires for 3-D Magnetic Memory. *IEEE Trans. Magn.* **2020**, *56*, 1–6. [CrossRef]
32. Yin, A.J.; Li, J.; Jian, W.; Bennett, A.J.; Xu, J.M. Fabrication of highly ordered metallic nanowire arrays by electrodeposition. *Appl. Phys. Lett.* **2001**, *79*, 1039–1041. [CrossRef]
33. Walter, E.C.; Zach, M.P.; Favier, F.; Murray, B.J.; Inazu, K.; Hemminger, J.C.; Penner, R.M. Metal Nanowire Arrays by Electrodeposition. *ChemPhysChem* **2003**, *4*, 131–138. [CrossRef]
34. Zamani Kouhpanji, M.R.; Ghoreyshi, A.; Visscher, P.B.; Stadler, B.J.H. Facile decoding of quantitative signatures from magnetic nanowire arrays. *Sci. Rep.* **2020**, *10*, 15482. [CrossRef]
35. Zamani Kouhpanji, M.R.; Stadler, B.J.H. Beyond the qualitative description of complex magnetic nanoparticle arrays using FORC measurement. *Nano Express* **2020**, *1*, 010017. [CrossRef]
36. Schotter, J.; Kamp, P.; Becker, A.; Pühler, A.; Reiss, G.; Brückl, H. Comparison of a prototype magnetoresistive biosensor to standard fluorescent DNA detection. *Biosens. Bioelectron.* **2004**, *19*, 1149–1156. [CrossRef]
37. Gaster, R.S.; Hall, D.A.; Nielsen, C.H.; Osterfeld, S.J.; Yu, H.; Mach, K.E.; Wilson, R.J.; Murmann, B.; Liao, J.C.; Gambhir, S.S.; et al. Matrix-insensitive protein assays push the limits of biosensors in medicine. *Nat. Med.* **2009**, *15*, 1327–1332. [CrossRef]
38. Gijs, M.A.M.; Lacharme, F.; Lehmann, U. Microfluidic Applications of Magnetic Particles for Biological Analysis and Catalysis. *Chem. Rev.* **2010**, *110*, 1518–1563. [CrossRef]
39. Wang, S.X.; Li, G. Advances in Giant Magnetoresistance Biosensors With Magnetic Nanoparticle Tags: Review and Outlook. *IEEE Trans. Magn.* **2008**, *44*, 1687–1702. [CrossRef]
40. Dobrotă, C.-I.; Stancu, A. Tracking the individual magnetic wires' switchings in ferromagnetic nanowire arrays using the first-order reversal curves (FORC) diagram method. *Phys. B Condens. Matter* **2015**, *457*, 280–286. [CrossRef]

41. Kouhpanji, M.R.Z.; Stadler, B. Exploring Effects of Magnetic Nanowire Arrangements and Imperfections on First-Order Reversal Curve Diagrams. *IEEE Trans. Magn.* **2021**, 1. [CrossRef]
42. Pike, R. First-order reversal-curve diagrams and reversible magnetization. *Phys. Rev. B Condens. Matter Mater. Phys.* **2003**, *68*, 1–5. [CrossRef]
43. Mayergoyz, I.D. Hysteresis models from the mathematical and control theory points of view. *J. Appl. Phys.* **1985**, *57*, 3803–3805. [CrossRef]
44. Mayergoyz, I.D. Mathematical models of hysteresis (Invited). *IEEE Trans. Magn.* **1986**, *22*, 603–608. [CrossRef]
45. Velázquez, Y.G.; Guerrero, A.L.; Martínez, J.M.; Araujo, E.; Tabasum, M.R.; Nysten, B.; Piraux, L.; Encinas, A. Relation of the average interaction field with the coercive and interaction field distributions in First order reversal curve diagrams of nanowire arrays. *Sci. Rep.* **2020**, *10*, 21396. [CrossRef]
46. Hillion, A.; Tamion, A.; Tournus, F.; Gaier, O.; Bonet, E.; Albin, C.; Dupuis, V. Advanced magnetic anisotropy determination through isothermal remanent magnetization of nanoparticles. *Phys. Rev. B Condens. Matter Mater. Phys.* **2013**, *88*, 094419. [CrossRef]
47. Zamani Kouhpanji, M.R.; Visscher, P.B.; Stadler, B.J.H. Fast and universal approach for quantitative measurements of bistable hysteretic systems. *J. Magn. Magn. Mater.* **2021**, *537*, 168170. [CrossRef]
48. Mourdikoudis, S.; Pallares, R.M.; Thanh, N.T.K. Characterization techniques for nanoparticles: Comparison and complementarity upon studying nanoparticle properties. *Nanoscale* **2018**, *10*, 12871–12934. [CrossRef]
49. Faraji, M.; Yamini, Y.; Rezaee, M. Magnetic nanoparticles: Synthesis, stabilization, functionalization, characterization, and applications. *J. Iran. Chem. Soc.* **2010**, *7*, 1–37. [CrossRef]
50. Cullity, B.D.; Graham, C.D. *Introduction to Magnetic Materials*; John Wiley & Sons, Inc.: Hoboken, NJ, USA, 2009; ISBN 9780471477419.
51. Lucatero, S.; Podlaha, E.J. Influence of Citric and Ascorbic Acids on Electrodeposited Au/FeAu Multilayer Nanowires. *J. Electrochem. Soc.* **2010**, *157*, D370. [CrossRef]
52. Khan, S.; Ahmad, N.; Safeer, A.; Iqbal, J. Compositional dependent morphology, structural and magnetic properties of Fe100−XCuX alloy nanowires via electrodeposition in AAO templates. *Appl. Phys. A* **2018**, *124*, 678. [CrossRef]
53. Nana, A.B.A.; Marimuthu, T.; Kondiah, P.P.D.; Choonara, Y.E.; Toit, L.C.D.; Pillay, V. Multifunctional magnetic nanowires: Design, fabrication, and future prospects as cancer therapeutics. *Cancers* **2019**, *11*, 1956. [CrossRef] [PubMed]
54. Palmero, E.M.; Bran, C.; del Real, R.P.; Magén, C.; Vázquez, M. Magnetic behavior of NiCu nanowire arrays: Compositional, geometry and temperature dependence. *J. Appl. Phys.* **2014**, *116*, 033908. [CrossRef]
55. Matei, E.; Enculescu, I.; Toimil-Molares, M.E.; Leca, A.; Ghica, C.; Kuncser, V. Magnetic configurations of Ni–Cu alloy nanowires obtained by the template method. *J. Nanopart. Res.* **2013**, *15*, 1863. [CrossRef]
56. Esmaeili, A.; Almasi Kashi, M.; Ramazani, A.; Montazer, A.H. Tailoring magnetic properties in arrays of pulse-electrodeposited Co nanowires: The role of Cu additive. *J. Magn. Magn. Mater.* **2016**, *397*, 64–72. [CrossRef]
57. Balke, B.; Wurmehl, S.; Fecher, G.H.; Felser, C.; Kübler, J. Rational design of new materials for spintronics: Co 2 Fe Z (Z =Al, Ga, Si, Ge). *Sci. Technol. Adv. Mater.* **2008**, *9*, 014102. [CrossRef]
58. Dupuis, V.; Khadra, G.; Hillion, A.; Tamion, A.; Tuaillon-Combes, J.; Bardotti, L.; Tournus, F. Intrinsic magnetic properties of bimetallic nanoparticles elaborated by cluster beam deposition. *Phys. Chem. Chem. Phys.* **2015**, *17*, 27996–28004. [CrossRef]
59. Wijn, H.P.J. *Magnetic Properties of Metals*; Springer: Berlin, Germany, 1991.
60. Zhang, X.; Zhang, H.; Wu, T.; Li, Z.; Zhang, Z.; Sun, H. Comparative study in fabrication and magnetic properties of FeNi alloy nanowires and nanotubes. *J. Magn. Magn. Mater.* **2013**, *331*, 162–167. [CrossRef]
61. Doludenko, I.M.; Zagorskii, D.L.; Frolov, K.V.; Perunov, I.V.; Chuev, M.A.; Kanevskii, V.M.; Erokhina, N.S.; Bedin, S.A. Nanowires Made of FeNi and FeCo Alloys: Synthesis, Structure, and Mössbauer Measurements. *Phys. Solid State* **2020**, *62*, 1639–1646. [CrossRef]
62. Huang, C.; Wang, P.; Guan, W.; Yang, S.; Gao, L.; Wang, L.; Song, X.; Murakami, R. Improved microstructure and magnetic properties of iron–cobalt nanowire via an ac electrodeposition with a multistep voltage. *Mater. Lett.* **2010**, *64*, 2465–2467. [CrossRef]
63. Ghemes, A.; Dragos-pinzaru, O.; Chiriac, H.; Lupu, N.; Grigoras, M.; Shore, D.; Stadler, B.; Tabakovic, I. Controlled Electrodeposition and Magnetic Properties of Co 35 Fe 65 Nanowires with High Saturation Magnetization. *J. Electrochem. Soc.* **2017**, *164*, 13–22. [CrossRef]
64. Vilana, J.; Gómez, E.; Vallés, E. Electrochemical control of composition and crystalline structure of CoNi nanowires and films prepared potentiostatically from a single bath. *J. Electroanal. Chem.* **2013**, *703*, 88–96. [CrossRef]
65. Elkins, J.; Mohapatra, J.; Xing, M.; Beatty, J.; Liu, J.P. Structural, morphological and magnetic properties of compositionally modulated CoNi nanowires. *J. Alloys Compd.* **2021**, *864*, 158123. [CrossRef]
66. Bai, A.; Hu, C.-C. Iron–cobalt and iron–cobalt–nickel nanowires deposited by means of cyclic voltammetry and pulse-reverse electroplating. *Electrochem. commun.* **2003**, *5*, 78–82. [CrossRef]
67. Mansouri, N.; Benbrahim-Cherief, N.; Chainet, E.; Charlot, F.; Encinas, T.; Boudinar, S.; Benfedda, B.; Hamadou, L.; Kadri, A. Electrodeposition of equiatomic FeNi and FeCo nanowires: Structural and magnetic properties. *J. Magn. Magn. Mater.* **2020**, *493*, 165746. [CrossRef]

68. Pfeifer, F.; Radeloff, C. Soft magnetic Ni-Fe and Co-Fe alloys—Some physical and metallurgical aspects. *J. Magn. Magn. Mater.* **1980**, *19*, 190–207. [CrossRef]
69. Tabakovic, I.; Venkatasamy, V. Preparation of metastable CoFeNi alloys with ultra-high magnetic saturation (Bs = 2.4–2.59 T) by reverse pulse electrodeposition. *J. Magn. Magn. Mater.* **2018**, *452*, 306–314. [CrossRef]
70. Thiem, L.V.; Tu, L.T.; Phan, M. Magnetization Reversal and Magnetic Anisotropy in Ordered CoNiP Nanowire Arrays: Effects of Wire Diameter. *Sensors* **2015**, *15*, 5687–5696. [CrossRef]
71. Xu, J.; Hong, B.; Peng, X.; Wang, X.; Ge, H.; Hu, J. Preparation and magnetic properties of gradient diameter FeCoNi alloys nanowires arrays. *Chem. Phys. Lett.* **2021**, *767*, 138368. [CrossRef]
72. Salem, M.S.; Tejo, F.; Zierold, R.; Sergelius, P.; Moreno, J.M.M.; Goerlitz, D.; Nielsch, K.; Escrig, J. Composition and diameter modulation of magnetic nanowire arrays fabricated by a novel approach. *Nanotechnology* **2018**, *29*, 065602. [CrossRef]
73. Yang, H.; Li, Y.; Zeng, M.; Cao, W.; Bailey, W.E.; Yu, R. Static and Dynamic Magnetization of Gradient FeNi Alloy Nanowire. *Sci. Rep.* **2016**, *6*, 1–9. [CrossRef]
74. Zeng, M.; Yang, H.; Liu, J.; Yu, R. Gradient magnetic binary alloy nanowire. *J. Appl. Phys.* **2014**, *115*, 17B514. [CrossRef]
75. Abdellahi, M.; Tajally, M.; Mirzaee, O. The effect of the particle size on the heating and drug release potential of the magnetic nanoparticles in a novel point of view. *J. Magn. Magn. Mater.* **2021**, *530*, 167938. [CrossRef]
76. Ghazkoob, N.; Zargar Shoushtari, M.; Kazeminezhad, I.; Lari Baghal, S.M. Structural, magnetic and optical investigation of AC pulse electrodeposited zinc ferrite nanowires with different diameters and lengths. *J. Magn. Magn. Mater.* **2021**, *537*, 168113. [CrossRef]
77. Karim, S.; Maaz, K. Magnetic behavior of arrays of nickel nanowires: Effect of microstructure and aspect ratio. *Mater. Chem. Phys.* **2011**, *130*, 1103–1108. [CrossRef]
78. Aharoni, A. Angular dependence of nucleation by curling in a prolate spheroid. *J. Appl. Phys.* **1997**, *82*, 1281–1287. [CrossRef]
79. Bran, C.; Espejo, A.P.; Palmero, E.M.; Escrig, J.; Vázquez, M. Angular dependence of coercivity with temperature in Co-based nanowires. *J. Magn. Magn. Mater.* **2015**, *396*, 327–332. [CrossRef]
80. Escrig, J.; Lavín, R.; Palma, J.L.; Denardin, J.C.; Altbir, D.; Cortés, A.; Gómez, H. Geometry dependence of coercivity in Ni nanowire arrays. *Nanotechnology* **2008**, *19*, 075713. [CrossRef]
81. Elmekawy, A.H.A.; Iashina, E.; Dubitskiy, I.; Sotnichuk, S.; Bozhev, I.; Kozlov, D.; Napolskii, K.; Menzel, D.; Mistonov, A. Magnetic properties of ordered arrays of iron nanowires: The impact of the length. *J. Magn. Magn. Mater.* **2021**, *532*, 167951. [CrossRef]
82. Sun, L.; Hao, Y.; Chien, C.-L.; Searson, P.C. Tuning the properties of magnetic nanowires. *IBM J. Res. Dev.* **2005**, *49*, 79–102. [CrossRef]
83. Vivas, L.G.; Vazquez, M.; Escrig, J.; Allende, S.; Altbir, D.; Leitao, D.C.; Araujo, J.P. Magnetic anisotropy in CoNi nanowire arrays: Analytical calculations and experiments. *Phys. Rev. B Condens. Matter Mater. Phys.* **2012**, *85*, 1–8. [CrossRef]
84. Aslam, S.; Khanna, M.; Kuanr, B.K.; Celinski, Z. One dimensional FexCo1-x nanowires; ferromagnetic resonance and magnetization dynamics. *AIP Adv.* **2017**, *7*, 056027. [CrossRef]
85. Zafar, N.; Shamaila, S.; Sharif, R.; Wali, H.; Naseem, S.; Riaz, S.; Khaleeq-Ur-Rahman, M. Effects of pH on the crystallographic structure and magnetic properties of electrodeposited cobalt nanowires. *J. Magn. Magn. Mater.* **2015**, *377*, 215–219. [CrossRef]
86. Aslam, S.; Das, A.; Khanna, M.; Kuanr, B.K. Concentration gradient Co–Fe nanowire arrays: Microstructure to magnetic characterizations. *J. Alloys Compd.* **2020**, *838*, 155566. [CrossRef]
87. Agarwal, S.; Khatri, M.S. Effect of pH and Boric Acid on Magnetic Properties of Electrodeposited Co Nanowires. *Proc. Natl. Acad. Sci. India Sect. A Phys. Sci.* **2020**, *14*, 1–6. [CrossRef]
88. Iglesias-Freire, Ó.; Bran, C.; Berganza, E.; Mínguez-Bacho, I.; Magén, C.; Vázquez, M.; Asenjo, A. Spin configuration in isolated FeCoCu nanowires modulated in diameter. *Nanotechnology* **2015**, *26*, 395702. [CrossRef]
89. Fernandez-Roldan, J.A.; De Riz, A.; Trapp, B.; Thirion, C.; Vazquez, M.; Toussaint, J.-C.; Fruchart, O.; Gusakova, D. Modeling magnetic-field-induced domain wall propagation in modulated-diameter cylindrical nanowires. *Sci. Rep.* **2019**, *9*, 5130. [CrossRef]
90. Rial, J.; Proenca, M.P. A Novel Design of a 3D Racetrack Memory Based on Functional Segments in Cylindrical Nanowire Arrays. *Nanomaterials* **2020**, *10*, 2403. [CrossRef]
91. Wang; Mukhtar; Wu; Gu; Cao Multi-Segmented Nanowires: A High Tech Bright Future. *Materials* **2019**, *12*, 3908. [CrossRef]
92. Berganza, E.; Bran, C.; Jaafar, M.; Vázquez, M.; Asenjo, A. Domain wall pinning in FeCoCu bamboo-like nanowires. *Sci. Rep.* **2016**, *6*, 29702. [CrossRef]
93. Fischer, P.; Sanz-Hernández, D.; Streubel, R.; Fernández-Pacheco, A. Launching a new dimension with 3D magnetic nanostructures. *APL Mater.* **2020**, *8*, 010701. [CrossRef]
94. Ruiz-Clavijo, A.; Ruiz-Gomez, S.; Caballero-Calero, O.; Perez, L.; Martin-Gonzalez, M. Tailoring Magnetic Anisotropy at Will in 3D Interconnected Nanowire Networks. *Phys. Status Solidi Rapid Res. Lett.* **2019**, *13*, 1900263. [CrossRef]
95. Bran, C.; Fernandez-Roldan, J.A.; del Real, R.P.; Asenjo, A.; Chubykalo-Fesenko, O.; Vazquez, M. Magnetic Configurations in Modulated Cylindrical Nanowires. *Nanomaterials* **2021**, *11*, 600. [CrossRef]
96. Jang, B.; Pellicer, E.; Guerrero, M.; Chen, X.; Choi, H.; Nelson, B.J.; Sort, J.; Pané, S. Fabrication of Segmented Au/Co/Au Nanowires: Insights in the Quality of Co/Au Junctions. *ACS Appl. Mater. Interfaces* **2014**, *6*, 14583–14589. [CrossRef]
97. Arzuza, L.C.C.; Béron, F.; Pirota, K.R. High-frequency GMI hysteresis effect analysis by first-order reversal curve (FORC) method. *J. Magn. Magn. Mater.* **2021**, *534*, 168008. [CrossRef]

98. Samanifar, S.; Kashi, M.A.; Ramazani, A. Study of reversible magnetization in FeCoNi alloy nanowires with different diameters by first order reversal curve (FORC) diagrams. *Phys. C Supercond. Appl.* **2018**, *548*, 72–74. [CrossRef]
99. Samardak, A.S.; Ognev, A.V.; Samardak, A.Y.; Stebliy, E.V.; Modin, E.B.; Chebotkevich, L.A.; Komogortsev, S.V.; Stancu, A.; Panahi-Danaei, E.; Fardi-Ilkhichy, A.; et al. Variation of magnetic anisotropy and temperature-dependent FORC probing of compositionally tuned Co-Ni alloy nanowires. *J. Alloys Compd.* **2018**, *732*, 683–693. [CrossRef]
100. Mayergoyz, I.D. The classical Preisach model of hysteresis and reversibility. *J. Appl. Phys.* **1991**, *69*, 4602–4604. [CrossRef]
101. Pike, C.R.; Roberts, A.P.; Verosub, K.L. Characterizing interactions in fine magnetic particle systems using first order reversal curves. *J. Appl. Phys.* **1999**, *85*, 6660–6667. [CrossRef]
102. Preisach, F. Uber die magnetische nachwirkung. Mitteilung aus dem Zentrallaboratorium des Wernerwerkes der Siemens Halske. *Z. Phys.* **1935**, *277*, 277–302. [CrossRef]
103. Pierrot, A.; Béron, F.; Blon, T. FORC signatures and switching-field distributions of dipolar coupled nanowire-based hysterons. *J. Appl. Phys.* **2020**, *128*, 093903. [CrossRef]
104. Shojaie Mehr, S.; Ramazani, A.; Almasi Kashi, M. Study on magnetic properties of NiFe/Cu multisegmented nanowire arrays with different Cu thicknesses via FORC analysis: Coercivity, interaction, magnetic reversibility. *J. Mater. Sci. Mater. Electron.* **2018**, *29*, 18771–18780. [CrossRef]
105. Gilbert, D.A.; Murray, P.D.; De Rojas, J.; Dumas, R.K.; Davies, J.E.; Liu, K. Reconstructing phase-resolved hysteresis loops from first-order reversal curves. *Sci. Rep.* **2021**, *11*, 4018. [CrossRef]
106. Gilbert, D.A.; Zimanyi, G.T.; Dumas, R.K.; Winklhofer, M.; Gomez, A.; Eigagi, N.; Vicent, J.L.; Liu, K. Quantitative Decoding of Interactions in Tunable Nanomagnet Arrays Using First Order Reversal Curves. *Sci. Rep.* **2014**, *4*, 1–5. [CrossRef]
107. De Biasi, E.; Curiale, J.; Zysler, R.D. Quantitative study of FORC diagrams in thermally corrected Stoner–Wohlfarth nanoparticles systems. *J. Magn. Magn. Mater.* **2016**, *419*, 580–587. [CrossRef]
108. Dobrotă, C.-I.; Stancu, A. What does a first-order reversal curve diagram really mean? A study case: Array of ferromagnetic nanowires. *J. Appl. Phys.* **2013**, *113*, 043928. [CrossRef]
109. De Biasi, E. Faster modified protocol for first order reversal curve measurements. *J. Magn. Magn. Mater.* **2017**, *439*, 259–268. [CrossRef]
110. Berndt, T.A.; Chang, L. Waiting for Forcot: Accelerating FORC Processing 100× Using a Fast-Fourier-Transform Algorithm. *Geochem. Geophys. Geosyst.* **2019**, *20*, 6223–6233. [CrossRef]
111. Groß, F.; Martínez-García, J.C.; Ilse, S.E.; Schütz, G.; Goering, E.; Rivas, M.; Gräfe, J. gFORC: A graphics processing unit accelerated first-order reversal-curve calculator. *J. Appl. Phys.* **2019**, *126*, 163901. [CrossRef]
112. Zamani Kouhpanji, M.R.; Stadler, B.J.H. Assessing the reliability and validity ranges of magnetic characterization methods. *arXiv* **2020**, arXiv:2003.06911.
113. Zamani Kouhpanji, M.R.; Visscher, P.B.; Stadler, B.J.H. Underlying magnetization responses of magnetic nanoparticles in assemblies. *arXiv* **2020**, arXiv:2002.07742.
114. Stockhausen, H. Some new aspects for the modelling of isothermal remanent magnetization acquisition curves by cumulative log Gaussian functions. *Geophys. Res. Lett.* **1998**, *25*, 2217–2220. [CrossRef]
115. De Toro, J.A.; Vasilakaki, M.; Lee, S.S.; Andersson, M.S.; Normile, P.S.; Yaacoub, N.; Murray, P.; Sánchez, E.H.; Muñiz, P.; Peddis, D.; et al. Remanence Plots as a Probe of Spin Disorder in Magnetic Nanoparticles. *Chem. Mater.* **2017**, *29*, 8258–8268. [CrossRef]
116. Geshev, J.; Mikhov, M. Remanence curves for a disordered system of three- and four-axial fine particles. Henkel-type plots. *J. Magn. Magn. Mater.* **1992**, *104–107*, 1569–1570. [CrossRef]
117. Harres, A.; Cichelero, R.; Pereira, L.G.; Schmidt, J.E.; Geshev, J. Remanence plots technique extended to exchange bias systems. *J. Appl. Phys.* **2013**, *114*, 043902. [CrossRef]
118. Reeves, D.B.; Weaver, J.B. Magnetic nanoparticle sensing: Decoupling the magnetization from the excitation field. *J. Phys. D Appl. Phys.* **2014**, *47*, 045002. [CrossRef]
119. Wawrzik, T.; Ludwig, F.; Schilling, M.; Häfeli, U.; Schütt, W.; Zborowski, M. Multivariate Magnetic Particle Spectroscopy for Magnetic Nanoparticle Characterization. *AIP Conf. Proc.* **2010**, *1311*, 267.
120. Paysen, H.; Wells, J.; Kosch, O.; Steinhoff, U.; Trahms, L.; Schaeffter, T.; Wiekhorst, F. Towards quantitative magnetic particle imaging: A comparison with magnetic particle spectroscopy. *AIP Adv.* **2018**, *8*, 056712. [CrossRef]
121. Encinas-Oropesa, A.; Demand, M.; Piraux, L.; Ebels, U.; Huynen, I. Effect of dipolar interactions on the ferromagnetic resonance properties in arrays of magnetic nanowires. *J. Appl. Phys.* **2001**, *89*, 6704–6706. [CrossRef]
122. Raposo, V.; Zazo, M.; Flores, A.G.; Garcia, J.; Vega, V.; Iñiguez, J.; Prida, V.M. Ferromagnetic resonance in low interacting permalloy nanowire arrays. *J. Appl. Phys.* **2016**, *119*, 143903. [CrossRef]
123. Um, J.; Zhang, Y.; Zhou, W.; Zamani Kouhpanji, M.R.; Radu, C.; Franklin, R.R.; Stadler, B.J.H. Magnetic Nanowire Biolabels Using Ferromagnetic Resonance Identification. *ACS Appl. Nano Mater.* **2021**, *4*, 3557–3564. [CrossRef]
124. Nikitin, P.I.; Vetoshko, P.M.; Ksenevich, T.I. New type of biosensor based on magnetic nanoparticle detection. *J. Magn. Magn. Mater.* **2007**, *311*, 445–449. [CrossRef]
125. Krause, H.-J.; Wolters, N.; Zhang, Y.; Offenhäusser, A.; Miethe, P.; Meyer, M.H.F.; Hartmann, M.; Keusgen, M. Magnetic particle detection by frequency mixing for immunoassay applications. *J. Magn. Magn. Mater.* **2007**, *311*, 436–444. [CrossRef]

126. Paysen, H.; Loewa, N.; Weber, K.; Kosch, O.; Wells, J.; Schaeffter, T.; Wiekhorst, F. Imaging and quantification of magnetic nanoparticles: Comparison of magnetic resonance imaging and magnetic particle imaging. *J. Magn. Magn. Mater.* **2019**, *475*, 382–388. [CrossRef]
127. Biederer, S.; Knopp, T.; Sattel, T.F.; Lüdtke-Buzug, K.; Gleich, B.; Weizenecker, J.; Borgert, J.; Buzug, T.M. Magnetization response spectroscopy of superparamagnetic nanoparticles for magnetic particle imaging. *J. Phys. D Appl. Phys.* **2009**, *42*, 205007. [CrossRef]
128. Hernando, A.; Lopez-Dominguez, V.; Ricciardi, E.; Osiak, K.; Marin, P. Tuned scattering of Electromagnetic Waves by a Finite Length Ferromagnetic Microwire. *IEEE Trans. Antennas Propag.* **2016**, *64*, 1112–1115. [CrossRef]
129. Parsa, N.; Toonen, R.C. Ferromagnetic Nanowires for Nonreciprocal Millimeter-Wave Applications: Investigations of Artificial Ferrites for Realizing High-Frequency Communication Components. *IEEE Nanotechnol. Mag.* **2018**, *12*, 28–35. [CrossRef]
130. Heslop, D.; Dillon, M. Unmixing magnetic remanence curves without a priori knowledge. *Geophys. J. Int.* **2007**, *170*, 556–566. [CrossRef]
131. Dempster, A.P.; Laird, N.M.; Rubin, D.B. Maximum Likelihood from Incomplete Data Via the EM Algorithm. *J. R. Stat. Soc. Ser. B* **1977**, *39*, 1–22. [CrossRef]
132. Robertson, D.J.; France, D.E. Discrimination of remanence-carrying minerals in mixtures, using isothermal remanent magnetisation acquisition curves. *Phys. Earth Planet. Inter.* **1994**, *82*, 223–234. [CrossRef]

Article

Effect of the Substrate Crystallinity on Morphological and Magnetic Properties of $Fe_{70}Pd_{30}$ Nanoparticles Obtained by the Solid-State Dewetting

Gabriele Barrera [1,*], Federica Celegato [1], Matteo Cialone [2], Marco Coïsson [1], Paola Rizzi [3] and Paola Tiberto [1]

[1] Advanced Materials Metrology and Life Sciences, Istituto Nazionale di Ricerca Metrologica (INRiM), Strada delle Cacce 91, I-10135 Torino, Italy; f.celegato@inrim.it (F.C.); m.coisson@inrim.it (M.C.); p.tiberto@inrim.it (P.T.)
[2] CNR SPIN Genova, c.so F. M. Perrone 24, I-16152 Genova, Italy; matteo.cialone@spin.cnr.it
[3] Chemistry Department and NIS, Università di Torino, Via Pietro Giuria 7, I-10125 Torino, Italy; paola.rizzi@unito.it
* Correspondence: g.barrera@inrim.it

Abstract: Advances in nanofabrication techniques are undoubtedly needed to obtain nanostructured magnetic materials with physical and chemical properties matching the pressing and relentless technological demands of sensors. Solid-state dewetting is known to be a low-cost and "top-down" nanofabrication technique able to induce a controlled morphological transformation of a continuous thin film into an ordered nanoparticle array. Here, magnetic $Fe_{70}Pd_{30}$ thin film with 30 nm thickness is deposited by the co-sputtering technique on a monocrystalline (MgO) or amorphous (Si_3N_4) substrate and, subsequently, annealed to promote the dewetting process. The different substrate properties are able to tune the activation thermal energy of the dewetting process, which can be tuned by depositing on substrates with different microstructures. In this way, it is possible to tailor the final morphology of FePd nanoparticles as observed by advanced microscopy techniques (SEM and AFM). The average size and height of the nanoparticles are in the ranges 150–300 nm and 150–200 nm, respectively. Moreover, the induced spatial confinement of magnetic materials in almost-spherical nanoparticles strongly affects the magnetic properties as observed by in-plane and out-of-plane hysteresis loops. Magnetization reversal in dewetted FePd nanoparticles is mainly characterized by a rotational mechanism leading to a slower approach to saturation and smaller value of the magnetic susceptibility than the as-deposited thin film.

Keywords: FePd thin film; solid-state dewetting; substrate crystallinity; magnetic nanoparticles

1. Introduction

Nanotechnology and sensors are research areas with a high degree of multi- and interdisciplinarity and, therefore, they are often and continuously combined [1–3]. Therefore, the development of advanced sensors based on nanostructured materials is increasingly expanding, improving the detection sensitivity of several chemical, physical and biological quantities in several external conditions [4–10]. In this framework, nanostructures-based bio-sensors are already used to sense various signals from a wide range of biological environments with many technological advantages in terms of reducing cost, easing in technologies, and improving measurement efficiencies [8,11–13].

Recent advances and tuning of innovative nanofabrication techniques have allowed to finely control the size and shape of the nanostructured materials, tailoring their physical and chemical properties to the practical demands [14–16]. Moreover, the more suitable nanofabrication technique should take into account also the economic issues, whose aim is the cost reduction of the sensor with its diffusion in the global market [14].

In this context, solid-state dewetting is a physical, "top-down" and low-cost nanofabrication technique [17,18] able to induce a morphological transformation of a thin film into an ordered nanoparticles array, usable as a key component in catalysts, photonic and magnetic applications as well as in several sensors [19–22]. The thermally activated morphological transformation is driven by a reduction in the surface energy of the thin film and, strictly depends on the interface energy between the thin film and the underlying substrate [18]. Therefore, the final shape and size of the nanoparticles (NPs) lying on the substrate are strongly influenced by well-known factors, including the thickness and composition of the thin film, temperature, time, and atmosphere of annealing treatment, as well as the physical and chemical properties of the substrate surface [23,24].

The solid-state dewetting process applies to a wide class of materials and alloys [19,22,24–27]. Interestingly, this process occurring in noble metals (Au and Ag) supports the development of the plasmonic sensors based on surface-enhanced Raman scattering (SERS) effect by the fabrication of suitable nanostructured substrates with high performance and low cost [21,28,29]. On the other hand, the solid-state dewetting applied to single- and multi-layer magnetic thin film strengthens the understanding of the correlation between the technologically relevant magnetic properties, such as anisotropy, susceptibility, coercivity, remanence and the morphological features of the obtained NPs [24,25,30].

Using the solid-state dewetting process to obtain bimetallic NPs, one element of which is magnetic, is an interesting scientific goal. The magnetic bimetallic NPs are promising materials in the technological and research communities because they show both the superimposition of the properties of the single elements and new interesting combined effects [31–34]. Among others, the bimetallic FePd NPs, thanks to their well-known multifunctional properties [35,36], have been already used as glucose sensors [37], biomedical agents [38,39], high-efficiency (bio-)catalysts [40], SERS-active substrate [41,42], and in magnetorheological fluid [36].

The aim of this work is to propose the solid-state dewetting process as an efficient nanofabrication technique able to obtain an array of bimetallic $Fe_{70}Pd_{30}$ nanoparticles starting from a continuous thin film deposited by the sputtering technique on a substrate. This research study is mainly devoted to understanding how the microstructure of the underlying substrate affects this morphological transformation and can be used as a tool to meet technological needs. In particular, monocrystalline MgO and amorphous Si_3N_4 substrates are used to control the kinetics of the dewetting process and to tailor the final morphology of magnetic FePd nanoparticles. The structural and morphological properties of the as-deposited FePd thin film and subsequently the size, shape, density and distribution of the obtained FePd NPs attached to the underlying substrate are investigated by X-ray diffractometer, scanning electron microscopy (SEM) and atomic force microscopy (AFM). Moreover, the evolution of the in-plane and out-of-plane magnetic properties, directly related to the spatial confinement induced by the dewetting process, are carefully investigated.

2. Materials and Methods

The co-sputtering deposition technique on monocrystalline MgO or amorphous Si_3N_4 substrate was used to grow FePd alloy in a thin-film form with 30 nm thickness. The two substrates were simultaneously loaded into the sputtering chamber. The power density of the two targets was fixed based on their relative deposition rates, at 250 W for Fe element and 15 W for the Pd element, in order to obtain the desired $Fe_{70}Pd_{30}$ stoichiometry. The base pressure of the sputtering was set at 7×10^{-7} mbar and the Ar gas pressure at 1.2×10^{-2}. The deposition time, to obtain the desired 30 nm thickness, was calculated from the deposition rate of FePd (1.36 Å/s) which was previously experimentally evaluated by an atomic force microscopy (AFM) measurement. The two as-deposited samples are named as follows: FePd/MgO and FePd/SiN.

The as-deposited thin films were submitted to annealing in a furnace (carbolite) (heating rate 51 °C/min) under vacuum atmosphere (2×10^{-6} mbar) to promote the solid-state-dewetting process. The selected annealing temperatures (T_A) are in the range

750–860 °C, and the annealing time (t_A) is set to 55 min. Despite the high temperatures reached during the annealing process, the high vacuum atmosphere in the furnace chamber severely hinders the oxidation of the FePd surface.

The crystal structure of the as-deposited FePd thin film was investigated by means of the grazing incidence X-ray diffraction (GIXRD) technique with Cu-K α radiation (PANalytical X'Pert Pro MPD). The XRD spectra were collected at room temperature. The Scherrer formula was used to estimate the average grains size [43]. The morphological characterization of the continuous as-deposited thin films and dewetted samples was performed by scanning electron microscopy (SEM—FEI Inspect-F). The corresponding images were analyzed by an open source software ImageJ [44]. The energy dispersive X-ray spectrometer (EDS) equipped in the SEM was used to check the stoichiometry of the as-deposited FePd alloy.

The surface roughness of the as-deposited samples and the height maps of the dewetted ones were measured by AFM (Bruker Multimode V Nanoscope 8) operating in intermittent-contact mode at atmospheric pressure and room temperature. In-plane and out-of-plane room temperature magnetic hysteresis loops were measured by means of a high-sensitivity alternating gradient field magnetometer (AGFM, Princeton Measurements Corporation) operating in the field range $-18 \leq H \leq 18$ kOe. The magnetic signal of the sample holder and substrate was effectively subtracted.

3. Results

3.1. As-Deposited $Fe_{70}Pd_{30}$ Thin Film

3.1.1. Structural and Morphological Characterization

The Fe:Pd ratio in the alloy, experimentally evaluated by the EDS technique, is $Fe_{70}Pd_{30}$ for the thin film deposited both on the monocrystalline MgO and amorphous Si_3N_4 substrates.

The surface morphology of the as-deposited FePd/MgO and FePd/SiN samples is depicted by top-view SEM and AFM images (see Figure 1). In particular, the SEM micrographs (Figure 1a,b) show a uniform and flat surface without visible defects and macro-structures in both samples. AFM images (Figure 1c,d) confirm the high surface flatness, revealing a roughness with a round shape with a root mean square (R_q) values of 1.4 nm and 1.1 nm for FePd/MgO and FePd/SiN samples, respectively. These measured R_q values are too low to induce a contrast/brightness variation in the SEM images, which appear to be mainly monochrome. The absence of visible defects or cracks indicates a low accumulation of strain at the substrate/film interface during the FePd deposition for both substrates [45,46].

The crystal structure of the as-deposited FePd/MgO and FePd/SiN samples is determined by the analysis of the spectra obtained by GIXRD and shown in Figure 2a. A body-centered cubic (BCC) structure is found in both cases. This evidence indicates the formation of a supersaturated solid solution of α-(Fe,Pd) as predicted by the FePd phase diagram and previously reported in the literature [47,48]. Considering the relative intensity of the diffraction peaks, no preferential orientation is observed. The used deposition parameters (see Section 2), such as a relatively high deposition rate and the room temperature of the substrates, lead to a polycrystalline structure of the growing thin film independently from the microstructure of the substrate. An enlargement of the spectra in the range 40–47° is shown in Figure 2b to compare the (110) peaks for the FePd film deposited on crystalline MgO (black line) and amorphous Si_3N_4 (red line) substrates. The peak is broader for the film deposited on the MgO substrate with respect to the corresponding one deposited on the Si_3N_4. In particular, the full width at half maximum (FWHM) value of the (110) peak is $(2.56 \pm 0.05)°$ and $(1.63 \pm 0.05)°$ for FePd/MgO and FePd/SiN sample, respectively. In a rough approximation, using the Scherrer formula [43], the average dimension ($<D>$) of the crystallites of the as-deposited FePd thin film can be estimated at about $<D>$ = 33 and 52 nm for FePd/MgO and FePd/SiN sample, respectively. It is worth noting that the $<D>$ values by Scherrer's formula are generally underestimated [43] but, nevertheless, it can be said with confidence that the average size of the crystallites deposited on the amorphous Si_3N_4 substrate is greater if compared to the one of the FePd film deposited on the crystalline MgO substrate. This behavior provides evidence of the different structural features of the substrates which affect the crystalline grains dimension in the as-deposited FePd films. Therefore, the slight mismatch in the lattice parameter between the single crystalline MgO substrate (0.421 nm) [49] and the FePd thin film (0.376 nm) [50] provides substrate-induced stress/strain to the as-deposited film, resulting in the formation of grains with a smaller size. Conversely, the Si_3N_4 substrate with its amorphous texture does not provide any substrate-induced stress/strain to the film that grows effectively relaxed, leading to grains with a larger volume.

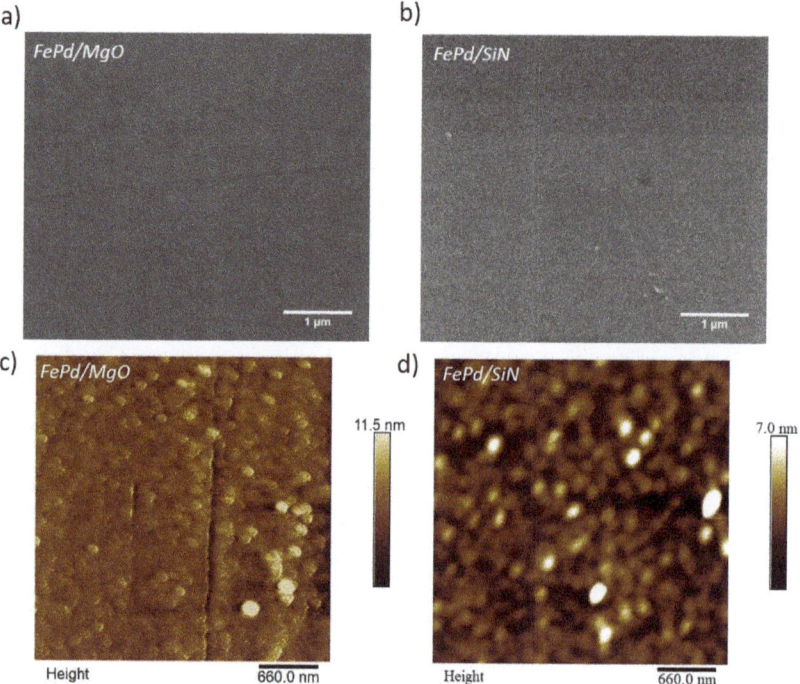

Figure 1. SEM (panels **a,b**) and AFM (panels **c,d**) images of as-deposited FePd thin films on MgO or Si_3N_4 substrate.

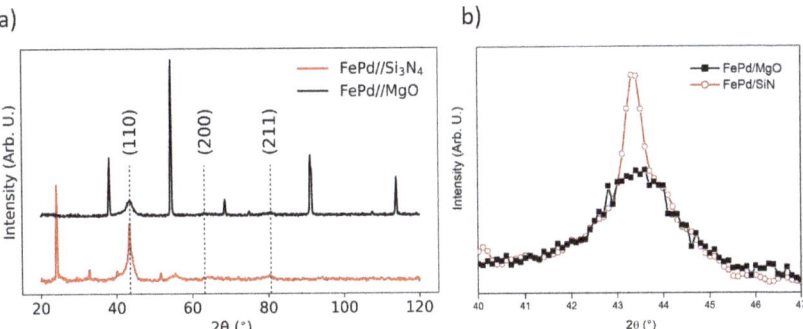

Figure 2. (a) XRD spectra of the as-deposited FePd samples on MgO (black curve, full symbols) and Si$_3$N$_4$ (red curve, open symbols) substrate, all the unlabeled peaks belong to the substrates; (b) enlargement around the (110) peak.

3.1.2. Magnetic Properties

Room-temperature in-plane hysteresis loops of the as-deposited FePd/MgO and FePd/SiN samples are shown in Figure 3. In both samples, the magnetization reversal occurs mainly for a single irreversible jump of the magnetization, indicating that the domain walls movement is the main mechanism governing the magnetization process. The magnetic susceptibility at the coercive field (χ_{Hc}) is $\approx 2.7 \times 10^{-3}$ Oe^{-1} and $\approx 4.2 \times 10^{-2}$ Oe^{-1} for FePd/MgO and FePd/SiN, respectively.

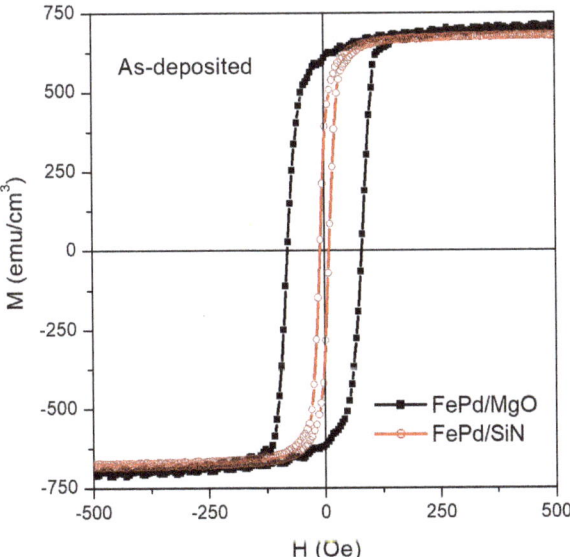

Figure 3. Room-temperature in-plane hysteresis loops of the as-deposited FePd samples on MgO (black curve) and Si$_3$N$_4$ (red curve) substrate.

The coercive field (H_c) and the normalized remanence (M_r/M_{10kOe}) of the as-deposited FePd thin film are influenced by the different structural properties of the underlying substrate. Both parameters reach higher values in the FePd/MgO sample with respect to the ones in the FePd/SiN sample: H_c = 80 and 10 Oe; M_r/M_{10kOe} = 0.84 and 0.47 for FePd/MgO and FePd/SiN, respectively. The difference in the H_c values can be correlated to the average grain size of the as-deposited thin film [49]: the smaller grains in the FePd/MgO sample induce a large number of grain boundaries, crystal imperfections, and defects that hinder the motion of the domain walls and, consequently, increase the coercivity. On the other hand, the larger grains in the FePd/SiN make the magnetization reversal easier, leading to a lower coercive field value. The hindrance of the domain walls movement in the FePd/MgO sample, concurring to the increase in its H_c value, also arises from the residual micro-stress [51] in the as-deposited FePd thin film induced by the slight mismatch of the lattice parameter and the one of the underlying substrate. Conversely, the amorphous texture of the Si_3N_4 substrate reduces the residual micro-stress, making the domain walls motion easier.

3.2. Dewetted $Fe_{70}Pd_{30}$ Thin Film

3.2.1. Structural and Morphological Characterization

The thermally-assisted breakup of the highly flat continuous $Fe_{70}Pd_{30}$ layer by the solid-state-dewetting process is shown in Figure 4a,b and Figure 4c–e for the FePd/MgO and FePd/SiN samples, respectively. The high temperature values drive the morphological transformation by minimizing the free energy at the interface between the substrate and the thin film [18].

Clearly, the structural, compositional, and superficial features of the underlying substrate determine the kinetics of the solid-state-dewetting, the temperatures at which it starts and the different morphological properties of the final magnetic nanoparticles. The crystalline MgO substrate favors, already at T_A = 750 °C, the formation of well-separated magnetic nanoparticles (Figure 4a), although their irregular shape indicates that their interconnections have just separated. By increasing the annealing temperature up to T_A = 820 °C, a more spherical-like shape of the NPs (Figure 4b) is developed by means of a process in which the free energy is further reduced and the system approaches equilibrium [52]. The exposure of the underlying MgO substrate is about 82% for both samples.

Conversely, annealing at the same temperature (T_A = 750 and 820 °C), the thermal energy provided to the as-deposited $Fe_{70}Pd_{30}$ layer on amorphous Si_3N_4 is not high enough to form separated magnetic nanoparticles. Indeed, the morphology of the $FePd_{750°C}$/SiN sample shows nucleated small empty space in the FePd film with irregular shape (i.e., holes that expose the underlying substrate) (Figure 4c) likely located at the grain boundaries of the as-deposited polycrystalline thin film [53,54]. Such formation of empty spaces in the continuous layer is the main feature that indicates the starting point of the solid-state dewetting process; in this case, the annealing parameters induce the exposure of the underlying Si_3N_4 substrate of about 12.8%. The increase in the annealing temperature up to T_A = 820 °C drives the spontaneous growth of the size of the holes (Figure 4d); consequently, in the $FePd_{820°C}$/SiN sample, the exposed area of the underlying substrate is increased up to 16.2%. Only a further increase in the thermal energy (T_A = 860 °C) provided to the as-deposited $Fe_{70}Pd_{30}$ thin film leads to the complete growth of the empty spaces with their consequent interconnection and the formation of well-separated FePd particles (Figure 4e). The exposure of the underlying Si_3N_4 substrate is, in this case, of about 83.4%.

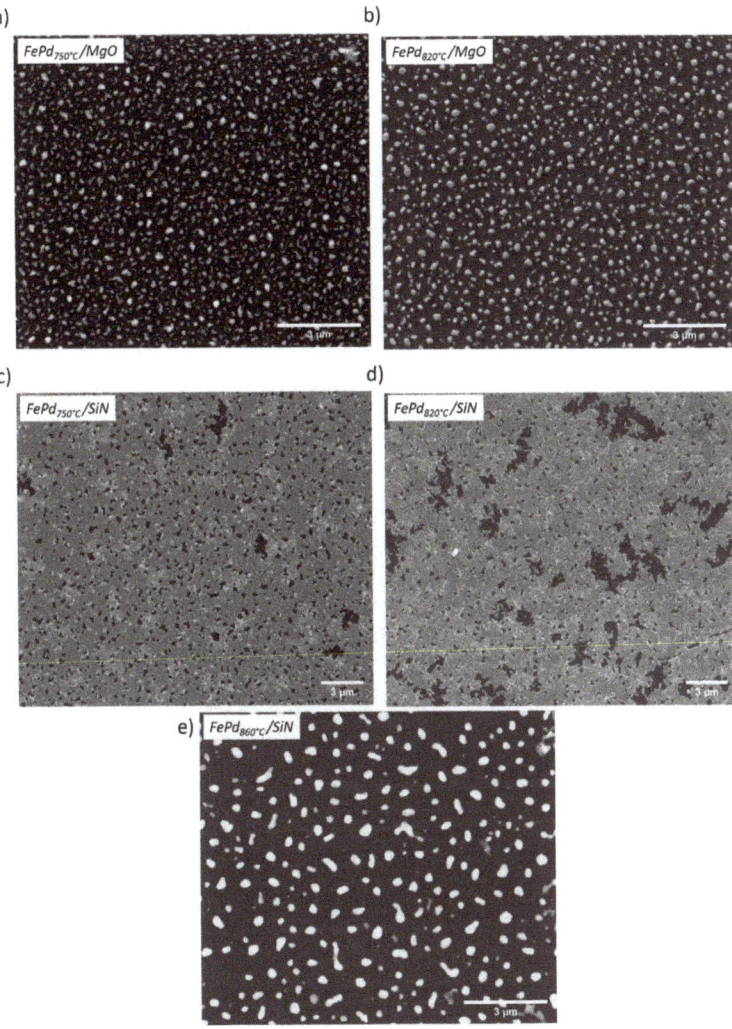

Figure 4. Morphology of $Fe_{70}Pd_{30}$ thin film deposited on MgO (panels **a,b**) and Si_3N_4 (panels **c–e**) substrate and annealed at selected temperature (see labels) for $t_A = 55$ min.

In order to compare the morphology obtained in $FePd_{820°C}/MgO$ and $FePd_{860°C}/SiN$ samples, a statistical analysis of the SEM images is performed. The distributions of the NPs diameter ($<D>$) and the center-to-center distance ($<d_{cc}>$) among the first neighborhood of nanoparticles are shown in Figure 5. The diameter distributions (Figure 5a) are fitted with a Gaussian curve with a mean value of $<D> = 164$ and 296 nm and a standard deviation of $\sigma_D = 128$ and 187 nm for the $FePd_{820°C}/MgO$ and $FePd_{860°C}/SiN$ samples, respectively. Similarly, the distributions of the center-to-center distance (Figure 5b) are fitted with a Gaussian curve with a mean value of $<d_{cc}> = 213$ and 478 nm and a standard deviation of $\sigma_{cc} = 129$ and 276 nm for the $FePd_{820°C}/MgO$ and $FePd_{860°C}/SiN$ samples, respectively. Consequently, the surface density (ρ) of NPs results in being much higher in the $FePd_{820°C}/MgO$ ($\rho \approx 60$ NPs/μm^2) than in the $FePd_{860°C}/SiN$ samples ($\rho \approx 15$ NPs/μm^2).

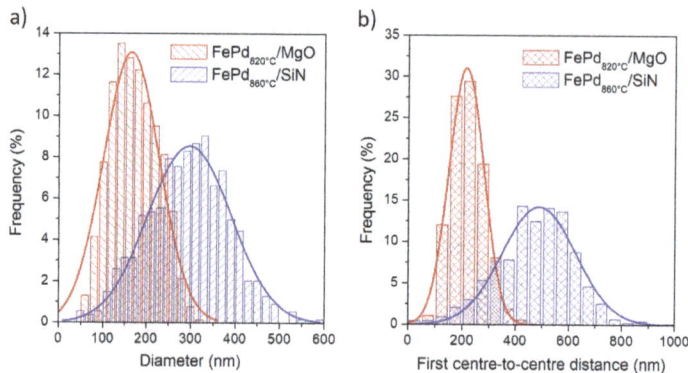

Figure 5. (a) Nanoparticles diameter distribution and (b) distribution of the center-to-center distance among first neighborhood of nanoparticles for the $FePd_{820°C}/MgO$ and $FePd_{860°C}/SiN$ samples. The Gaussian fits are plotted as full lines.

Therefore, the crystalline MgO substrate is observed to induce smaller ($<D>$ reduced by a factor of ≈62%), closer ($<d_{cc}>$ reduced by a factor of ≈55%), and more evenly distributed FePd magnetic NPs than the amorphous Si_3N_4 substrate.

The height map of $Fe_{70}Pd_{30}$ nanoparticles lying on the MgO or Si_3N_4 substrate is obtained by the AFM technique; representative AFM images for the $FePd_{820°C}/MgO$ and $FePd_{860°C}/SiN$ samples are shown in Figure 6a,b. Statistical analysis of several AFM images allows to obtain the height distributions of the nanoparticles, which are shown in Figure 6c. Both distributions are roughly approximated by a Gaussian curve with a mean value of $<h>$ = 156 and 206 nm and standard deviation of σ_h = 65 and 91 nm for the $FePd_{820°C}/MgO$ and $FePd_{860°C}/SiN$ samples, respectively.

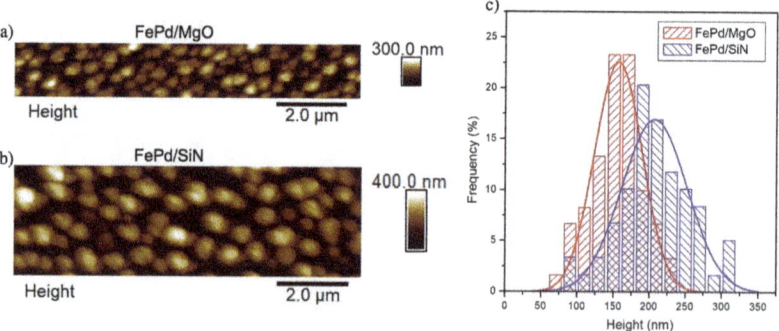

Figure 6. AFM images of dewetted samples: (a) $FePd_{820°C}/MgO$ and (b) $FePd_{860°C}/SiN$ samples; (c) nanoparticles height distribution with Gaussian fit (full line).

The estimation of the nanoparticles average aspect ratio ($<h>/<D>$) by combining the diameter and height information results in 0.95 ± 1.14 and 0.70 ± 0.75 for the $FePd_{820°C}/MgO$ and $FePd_{860°C}/SiN$ samples, respectively, revealing an almost spherical shape for the NPs obtained on the MgO substrate whereas a more oblate shape is revealed for the ones on the Si_3N_4 substrate.

These morphological evidences demonstrate that the substrate crystallinity influences the energy at the substrate/thin-film interface [23] as well as the thin-film structural properties (see Section 3.1). Consequently, the thermally activated atomic diffusion during the solid-state dewetting process can be controlled, leading to <h>/<D> and ρ of the dewetted FePd particles to meet the technological demands.

Moreover, the thermal energy provided to the thin films during the solid-state dewetting could lead to surface segregation effects of the bimetallic alloy [55], i.e., the enrichment of the surface in one of the components in comparison to the nominal concentration. In this context, the segregation of Au in Fe- and Ni-rich dewetted alloy is the subject of several studies [56–58]. In the case of FePd alloy, segregation of the Fe atoms on the surface was found, due to their higher oxide-forming tendency, compared to Pd atoms [59]. Therefore, a surface segregation effect in the dewetted samples studied in this work cannot be excluded a priori; however, further characterizations with more effective and in-depth techniques, such as X-ray photoemission spectroscopy (XPS) [60], atom probe [61] or atomic-resolution HAADF transmission microscopy [62] should be performed to have a more comprehensive picture.

3.2.2. Magnetic Properties

Room-temperature in-plane hysteresis loops of the annealed samples are shown in Figure 7. All curves are normalized to the magnetic moment value measured at H = 10 kOe. The $M(H)$ curves (Figure 7a,b) show a progressive and significant reduction in the magnetic susceptibility at the coercive field (χ_{Hc}) as a function of T_A (especially in the FePd/SiN samples; see Table 1) leading to a wider magnetic field interval in which the magnetic saturation is reached. This magnetic behavior indicates the appearance of a rotational mechanism of magnetization, which replaces the single irreversible reversal mechanism associated with the domain wall motion in the as-deposited thin film (see Figure 3 and discussion above) and, consequently, leads to a slower approach to saturation magnetization.

The measured magnetic properties are in strong correlation with the evolution of the $Fe_{70}Pd_{30}$ film morphology induced by the solid-state-dewetting process and observed in Figure 4. The remarkable increase of the H_c value in the dewetted $FePd_{750°C}$/SiN and $FePd_{820°C}$/SiN samples (see Table 1), compared to the corresponding as-deposited thin film, is ascribed to the nucleation of the holes that act as pinning sites for the domain walls during the magnetization process of the whole sample. In particular, smaller holes (with a size smaller than the domain wall thickness) can directly pin the walls, whereas larger holes hinder the wall motion through the subsidiary domains around them, reducing the overall magnetostatic energy [51]. Instead, the rounder shape of the FePd well-separated nanoparticles, observed at the end of the solid-state-dewetting process in the $FePd_{860°C}$/SiN sample, induces a decrease in the effective magnetic anisotropy with a consequent reduction in the coercive field [25,63]. In this case, the magnetization process occurs independently in each nanoparticle mainly by rotational mechanisms: the magnetization vector rotates toward the applied field direction against the restoring force of the effective anisotropy, which includes a combination of the shape and the crystal anisotropy. As a consequence, a lower χ_{Hc} value is observed. Similar rotation mechanisms with comparable χ_{Hc} value are observed in the spherical shape magnetic NPs of the $FePd_{750°C}$/MgO and $FePd_{820°C}$/MgO samples (see Figure 4a,b). As expected, the improvement in the NPs roundness, induced increasing T_A from 750 °C to 820 °C for FePd thin film on the MgO substrate, leads to a slight decrease in the χ_{Hc} and H_C values (see Table 1). By comparing the coercive field values of the dewetted $FePd_{820°C}$/MgO and $FePd_{860°C}$/SiN samples, a multidomain configuration of the nanoparticles in both cases can be hypothesized since the decrease in the NPs average size <D> leads to an increase in the coercivity [51].

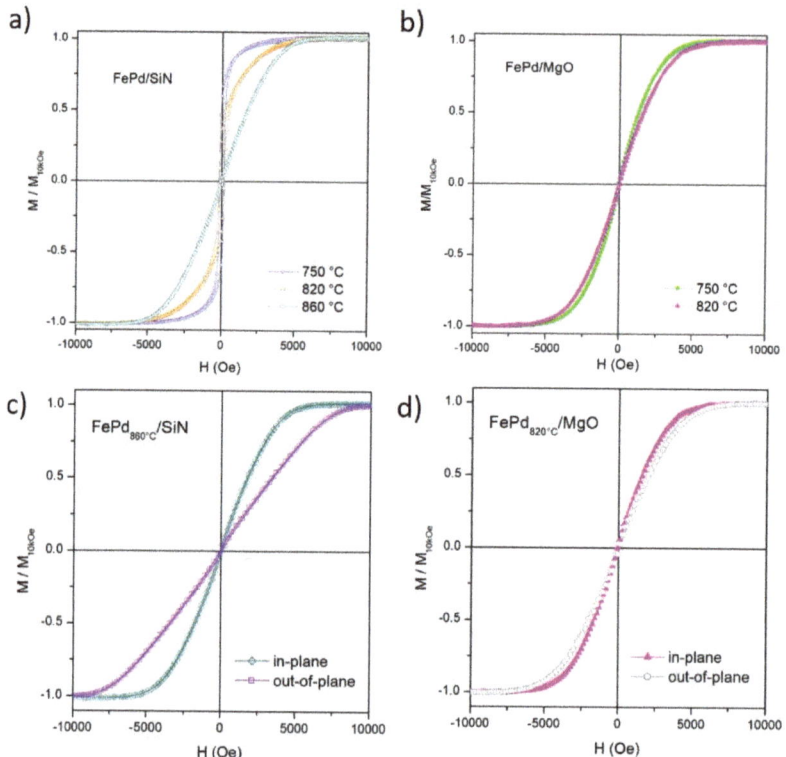

Figure 7. Room-temperature hysteresis loops: in-plane configuration for (**a**) FePd/SiN and (**b**) FePd/MgO samples as a function of the T_A; in-plane and out-of-plane magnetic hysteresis loops comparison for (**c**) FePd$_{860°C}$/SiN and (**d**) FePd$_{820°C}$/MgO.

Table 1. Annealing temperature (T_A), magnetic susceptibility at the coercive field (χ_{Hc}), coercive field (H_c) for FePd/SiN and FePd/MgO samples.

	FePd/SiN		FePd/MgO	
T_A	χ_{Hc}	H_c	χ_{Hc}	H_c
750 °C	9.1×10^{-3}	147	5.3×10^{-4}	92
820 °C	4.2×10^{-3}	177	4.1×10^{-4}	88
860 °C	3.5×10^{-4}	73	-	-

Room-temperature out-of-plane magnetic hysteresis loops of the FePd$_{860°C}$/SiN and FePd$_{820°C}$/MgO samples are measured and matched with the corresponding ones along the in-plane direction; see Figure 7c,d. In the FePd$_{860°C}$/SiN sample, the in-plane curve shows higher χ_{Hc} and lower saturation field, indicating that this direction is an anisotropy easy-axis of magnetization. Conversely, the in-plane and the out-of-plane hysteresis loops of the FePd$_{820°C}$/MgO sample appear almost perfectly superimposed. Such a features indicates a prevalent isotropic magnetic behavior with a random distribution in the space of the easy magnetization axis (only a very slight preference for the in-plane direction is still measurable). The area enclosed between the first branch of the in-plane hysteresis loop and the first branch of the out-of-plane one in the applied magnetic field interval 0–10 kOe is used to evaluate the effective anisotropy energy (E_{eff}) that the applied magnetic field spends to move the magnetization away from the easy-axis toward saturation along the hard axis [24,63].

As expected, the E_{eff} value of the FePd$8_{860°C}$/SiN sample ($\approx 1.1 \times 10^6$ erg/cm^3) is largely higher than that for FePd$_{820°C}$/SiN sample (2.6×10^5 erg/cm^3). These anisotropic results are in very good agreement with the morphological evidence reported and discussed in the previous section. The evident magnetic anisotropic behavior with the easy axis along the in-plane direction in FePd$_{860°C}$/SiN sample is excellently linked to the calculated oblate shape ($<h>/<D>$ = 0.70 ± 0.75) of their FePd nanoparticles, whose major axis is in the plane of the film. Instead, the almost perfect spherical shape ($<h>/<D>$ = 0.95 ± 0.14) of the nanoparticles in the FePd$_{820°C}$/MgO sample impact the magnetic isotropic behavior.

4. Conclusions

The proposed low-cost and "top-down" solid-state dewetting process is successfully used to nanostructure the as-deposited FePd thin film into an ordered nanoparticle array. The overall results indicate that the structural, compositional, and superficial features of the underlying substrate combined with the annealing parameters determine the kinetics of the solid-state-dewetting and, consequently, the final morphology, spatial arrangements and magnetic properties of the FePd nanoparticles.

The as-deposited Fe$_{70}$Pd$_{30}$ thin film with 30 nm thickness was successfully grown by the co-sputtering technique on both monocrystalline MgO and amorphous Si$_3$N$_4$ substrates. Morphological and structural characterizations reveal a flat, homogeneous, and continuous FePd layer, indicating a low strain accumulation at the substrate/film interface independently from the underlying substrate. On the other hand, the substrate affects the average size of the crystalline grains of the FePd thin film, resulting in bigger ones for the FePd/SiN, compared to the FePd/MgO. The magnetization reversal process is dominated in both as-deposited samples by the magnetic domain walls motion; the smaller grains size in the FePd/MgO sample results in a higher value of the coercive field.

The crystalline MgO substrate favors, by submitting the as-deposited FePd thin film to the dewetting process (annealing at T_A= 820 °C for t_A = 55 min), the formation of well-separated magnetic nanoparticles with a spherical-like shape ($<D>$ = 164 nm and $<h>$ = 156 nm) and high surface density ($\rho \approx$ 60 NPs/µm^2). Conversely, exploiting the same annealing parameters, the layer deposited on the amorphous Si$_3$N$_4$ substrate leads to a hindering of the activation of the solid-state dewetting process leading only to the primary nucleation of the holes. An increase in the annealing temperature up to T_A = 860 °C for t_A = 55 min is required to complete propagation of the holes and to obtain the formation of well-separated FePd particles, which result in being higher, almost double in size, oblate in shape ($<D>$ = 296 nm and $<h>$= 206 nm) and with a considerable lower surface density ($\rho \approx$ 15 NPs/µm^2).

The spatial confinement of the magnetic materials induced by the solid-state dewetting remarkably affects the in-plane and out-of-plane magnetic properties. The magnetization reversal process in the magnetic FePd NPs array is mainly dominated by rotational mechanisms, leading to a slower approach to magnetic saturation with a significant reduction of the magnetic susceptibility at the coercive field, compared with the as-deposited thin film. The magnetization process occurs independently in each nanoparticle overcoming the shape and the crystal anisotropy. The coercive field is observed to increase as long as the holes and interconnection that act as pinning sites for the domain walls are still present in the sample. With the ending of the dewetting process, the spherical-like shape of the FePd nanoparticles leads to a reduction in the effective anisotropy with a consequent reduction in the coercive field. Such an increase in the coercivity as a function of the reduction in nanoparticle size (by comparing the FePd$_{820°C}$/MgO and FePd$_{860°C}$/SiN samples) is compatible with a multidomain configuration of the magnetization in each individual NP. Magnetic anisotropic features taken by the in-plane and out-of-plane hysteresis loops excellently support the average aspect ratio ($<h>/<D>$) values obtained by the morphological analysis: a slight magnetic anisotropic behavior is observed for the oblate shape of the nanoparticles in the FePd$_{860°C}$/SiN sample, whereas an almost perfect anisotropic behavior is in excellent agreement with the almost perfect spherical shape of the nanoparticles in the FePd$_{820°C}$/MgO sample.

In conclusion, the present structural, morphological, and magnetic characterizations prove that the substrate plays a primary role in the tuning of the dewetting process; therefore, a comprehensive structural analysis of the substrates and their interface with the magnetic layer will be one of the next crucial steps that allow to finely control the final morphology of magnetic FePd nanoparticles to meet the technological demands.

Author Contributions: Conceptualization, P.T., M.C. (Marco Coïsson) and P.R.; thin film sputtering deposition, F.C.; solid-state dewetting process, G.B. and F.C.; XRD characterization and structural data analysis, M.C. (Matteo Cialone) and P.R.; SEM and AFM investigation, F.C.; SEM and AFM images analysis, G.B.; DC magnetic measurements and data analysis, G.B., M.C. (Marco Coïsson) and P.T.; writing—original draft preparation, G.B.; supervision, P.T., P.R. and M.C. (Marco Coïsson). All authors have read and agreed to the published version of the manuscript.

Funding: This research received no external funding.

Institutional Review Board Statement: Not applicable.

Informed Consent Statement: Not applicable.

Data Availability Statement: The data presented in this study are available on request from the authors.

Conflicts of Interest: The authors declare no conflict of interest.

References

1. Arregui, F.J. *Sensors Based on Nanostructured Materials*; Springer Nature: Cham, Switzerland, 2009; p. 326.
2. Gloag, L.; Mehdipour, M.; Chen, D.; Tilley, R.D.; Gooding, J.J. Advances in the Application of Magnetic Nanoparticles for Sensing. *Adv. Mater.* **2019**, *31*, 1904385. [CrossRef]
3. Abdel-Karim, R.; Reda, Y.; Abdel-Fattah, A. Review—Nanostructured Materials-Based Nanosensors. *J. Electrochem. Soc.* **2020**, *167*, 037554. [CrossRef]
4. Huang, X.J.; Choi, Y.K. Chemical sensors based on nanostructured materials. *Sens. Actuators Chem.* **2007**, *122*, 659–671. [CrossRef]
5. Hahn, Y.B.; Ahmad, R.; Tripathy, N. Chemical and biological sensors based on metal oxide nanostructures. *Chem. Commun.* **2012**, *48*, 10369–10385. [CrossRef]
6. Ponzoni, A.; Comini, E.; Concina, I.; Ferroni, M.; Falasconi, M.; Gobbi, E.; Sberveglieri, V.; Sberveglieri, G. Nanostructured metal oxide gas sensors, a survey of applications carried out at SENSOR lab, brescia (Italy) in the security and food quality fields. *Sensors* **2012**, *12*, 17023–17045. [CrossRef]
7. Elosua, C.; Arregui, F.J.; Del Villar, I.; Ruiz-Zamarreño, C.; Corres, J.M.; Bariain, C.; Goicoechea, J.; Hernaez, M.; Rivero, P.J.; Socorro, A.B.; et al. Micro and nanostructured materials for the development of optical fibre sensors. *Sensors* **2017**, *17*, 2312. [CrossRef]
8. Nasiri, N.; Clarke, C. Nanostructured gas sensors for medical and health applications: Low to high dimensional materials. *Biosensors* **2019**, *9*, 43. [CrossRef] [PubMed]
9. Antonacci, A.; Arduini, F.; Moscone, D.; Palleschi, G.; Scognamiglio, V. Nanostructured (Bio)sensors for smart agriculture. *Trac—Trends Anal. Chem.* **2018**, *98*, 95–103. [CrossRef]
10. Huynh, T.P. Chemical and biological sensing with nanocomposites prepared from nanostructured copper sulfides. *Nano Futures* **2020**, *4*, 032001. [CrossRef]
11. Barua, S.; Dutta, H.S.; Gogoi, S.; Devi, R.; Khan, R. Nanostructured MoS$_2$-Based Advanced Biosensors: A Review. *ACS Appl. Nano Mater.* **2018**, *1*, 2–25. [CrossRef]
12. Di Pietrantonio, F.; Cannatà, D.; Benetti, M. Biosensor technologies based on nanomaterials. *Funct. Nanostruct. Interfaces Environ. Biomed. Appl.* **2019**, 181–242. [CrossRef]
13. Lee, H.B.; Meeseepong, M.; Trung, T.Q.; Kim, B.Y.; Lee, N.E. A wearable lab-on-a-patch platform with stretchable nanostructured biosensor for non-invasive immunodetection of biomarker in sweat. *Biosens. Bioelectron.* **2020**, *156*, 112133. [CrossRef] [PubMed]
14. Santos, A.; Deen, M.J.; Marsal, L.F. Low-cost fabrication technologies for nanostructures: State-of-the-art and potential. *Nanotechnology* **2015**, *26*, 042001. [CrossRef]
15. Stepanova, M. *Nanofabrication: Techniques and Principles*; Springer: Berlin/Heidelberg, Germany, 2012.
16. Prasad, S.; Kumar, V.; Kirubanandam, S.; Barhoum, A. Engineered nanomaterials: Nanofabrication and surface functionalization. In *Emerging Applications of Nanoparticles and Architectural Nanostructures: Current Prospects and Future Trends*; Elsevier: Amsterdam, The Netherlands, 2018; pp. 305–340. [CrossRef]
17. Kim, D.; Giermann, A.L.; Thompson, C.V. Solid-state dewetting of patterned thin films. *Appl. Phys. Lett.* **2009**, *95*, 2–4. [CrossRef]
18. Thompson, C.V. Solid-State Dewetting of Thin Films. *Annu. Rev. Mater. Res.* **2012**, *42*, 399–434. [CrossRef]

19. Andalouci, A.; Roussigné, Y.; Farhat, S.; Chérif, S.M. Magnetic and magneto-optical properties of assembly of nanodots obtained from solid-state dewetting of ultrathin cobalt layer. *J. Phys. Condens. Matter* **2019**, *31*, 495805. [CrossRef] [PubMed]
20. Garfinkel, D.A.; Pakeltis, G.; Tang, N.; Ivanov, I.N.; Fowlkes, J.D.; Gilbert, D.A.; Rack, P.D. Optical and Magnetic Properties of Ag-Ni Bimetallic Nanoparticles Assembled via Pulsed Laser-Induced Dewetting. *ACS Omega* **2020**, *5*, 19285–19292. [CrossRef] [PubMed]
21. Pengphorm, P.; Nuchuay, P.; Boonrod, N.; Boonsit, S.; Srisamran, P.; Thongrom, S.; Pewkum, P.; Kalasuwan, P.; van Dommelen, P.; Daengngam, C. Fabrication of 3D surface-enhanced Raman scattering (SERS) substrate via solid-state dewetting of sputtered gold on fumed silica surface. *J. Phys. Conf. Ser.* **2021**, *1719*, 012082. [CrossRef]
22. Oh, H.; Pyatenko, A.; Lee, M. A hybrid dewetting approach to generate highly sensitive plasmonic silver nanoparticles with a narrow size distribution. *Appl. Surf. Sci.* **2021**, *542*, 148613. [CrossRef]
23. Nsimama, P.D.; Herz, A.; Wang, D.; Schaaf, P. Influence of the substrate on the morphological evolution of gold thin films during solid-state dewetting. *Appl. Surf. Sci.* **2016**, *388*, 475–482. [CrossRef]
24. Barrera, G.; Celegato, F.; Coïsson, M.; Cialone, M.; Rizzi, P.; Tiberto, P. Formation of free-standing magnetic particles by solid-state dewetting of $Fe_{80}Pd_{20}$ thin films. *J. Alloys Compd.* **2018**, *742*. [CrossRef]
25. Esterina, R.; Liu, X.M.; Adeyeye, A.O.; Ross, C.A.; Choi, W.K. Solid-state dewetting of magnetic binary multilayer thin films. *J. Appl. Phys.* **2015**, *118*, 144902. [CrossRef]
26. Bhalla, N.; Jain, A.; Lee, Y.; Shen, A.Q.; Lee, D. Dewetting Metal Nanofilms—Effect of Substrate on Refractive Index Sensitivity of Nanoplasmonic Gold. *Nanomaterials* **2019**, *9*, 1530. [CrossRef] [PubMed]
27. Song, X.; Liu, F.; Qiu, C.; Coy, E.; Liu, H.; Aperador, W.; Zaleski, K.; Li, J.J.; Song, W.; Lu, Z.; et al. Nanosurfacing Ti alloy by weak alkalinity-activated solid-state dewetting (AAD) and its biointerfacial enhancement effect. *Mater. Horiz.* **2021**, *8*, 912–924. [CrossRef]
28. Quan, J.; Zhang, J.; Qi, X.; Li, J.; Wang, N.; Zhu, Y. A study on the correlation between the dewetting temperature of Ag film and SERS intensity. *Sci. Rep.* **2017**, *7*, 1–12. [CrossRef]
29. Andrikaki, S.; Govatsi, K.; Yannopoulos, S.N.; Voyiatzis, G.A.; Andrikopoulos, K.S. Thermal dewetting tunes surface enhanced resonance Raman scattering (SERRS) performance. *RSC Adv.* **2018**, *8*, 29062–29070. [CrossRef]
30. Wang, L.W.; Cheng, C.F.; Liao, J.W.; Wang, C.Y.; Wang, D.S.; Huang, K.F.; Lin, T.Y.; Ho, R.M.; Chen, L.J.; Lai, C.H. Thermal dewetting with a chemically heterogeneous nano-template for self-assembled $L1_0$ FePt nanoparticle arrays. *Nanoscale* **2016**, *8*, 3926–3935. [CrossRef] [PubMed]
31. Preger, C.; Bulbucan, C.; Meuller, B.O.; Ludvigsson, L.; Kostanyan, A.; Muntwiler, M.; Deppert, K.; Westerström, R.; Messing, M.E. Controlled Oxidation and Self-Passivation of Bimetallic Magnetic FeCr and FeMn Aerosol Nanoparticles. *J. Phys. Chem. C* **2019**, *123*, 16083–16090. [CrossRef]
32. Sharma, G.; Kumar, A.; Sharma, S.; Naushad, M.; Prakash Dwivedi, R.; ALOthman, Z.A.; Mola, G.T. Novel development of nanoparticles to bimetallic nanoparticles and their composites: A review. *J. King Saud Univ. Sci.* **2019**, *31*, 257–269. [CrossRef]
33. Ragothaman, M.; Mekonnen, B.T.; Palanisamy, T. Synthesis of magnetic Fe-Cr bimetallic nanoparticles from industrial effluents for smart material applications. *Mater. Chem. Phys.* **2020**, *253*, 123405. [CrossRef]
34. Elanchezhiyan, S.S.; Muthu Prabhu, S.; Kim, Y.; Park, C.M. Lanthanum-substituted bimetallic magnetic materials assembled carboxylate-rich graphene oxide nanohybrids as highly efficient adsorbent for perfluorooctanoic acid adsorption from aqueous solutions. *Appl. Surf. Sci.* **2020**, *509*, 144716. [CrossRef]
35. Tiberto, P.; Celegato, F.; Barrera, G.; Coisson, M.; Vinai, F.; Rizzi, P. Magnetization reversal and microstructure in polycrystalline $Fe_{50}Pd_{50}$ dot arrays by self-assembling of polystyrene nanospheres. *Sci. Technol. Adv. Mater.* **2016**, *17*, 462–472. [CrossRef] [PubMed]
36. Shao, Z.; An, L.; Li, Z.; Huang, Y.; Hu, Y.; Ren, S. Eutectic crystallized FePd nanoparticles for liquid metal magnet. *Chem. Commun.* **2020**, *56*, 6555–6558. [CrossRef] [PubMed]
37. Yang, C.; Feng, W.; Li, Y.; Tian, X.; Zhou, Z.; Lu, L.; Nie, Y. A promising method for diabetes early diagnosis via sensitive detection of urine glucose by Fe Pd/rGO. *Dye. Pigment.* **2019**, *164*, 20–26. [CrossRef]
38. Yang, Y.; Lyu, M.; Li, J.H.; Zhu, D.M.; Yuan, Y.F.; Liu, W. Ultra-small bimetallic iron-palladium (FePd) nanoparticle loaded macrophages for targeted tumor photothermal therapy in NIR-II biowindows and magnetic resonance imaging. *RSC Adv.* **2019**, *9*, 33378–33387. [CrossRef]
39. Yamamoto, S.; Takao, S.; Muraishi, S.; Xu, C.; Taya, M. Synthesis of $Fe_{70}Pd_{30}$ nanoparticles and their surface modification by zwitterionic linker. *Mater. Chem. Phys.* **2019**, *234*, 237–244. [CrossRef]
40. Kwon, J.; Mao, X.; Lee, H.A.; Oh, S.; Tufa, L.T.; Choi, J.Y.; Kim, J.E.; Kim, C.Y.; Kim, J.G.; Hwang, D.Y.; et al. Iron–Palladium magnetic nanoparticles for decolorizing rhodamine B and scavenging reactive oxygen species. *J. Colloid Interface Sci.* **2021**, *588*, 646–656. [CrossRef]
41. Zheng, Y.; Mourdikoudis, S.; Zhang, Z. Plasmonic Metallic Heteromeric Nanostructures. *Small* **2020**, *16*, 2002588. [CrossRef]
42. Cialone, M.; Celegato, F.; Scaglione, F.; Barrera, G.; Raj, D.; Coïsson, M.; Tiberto, P.; Rizzi, P. Nanoporous FePd alloy as multifunctional ferromagnetic SERS-active substrate. *Appl. Surf. Sci.* **2021**, *543*, 148759. [CrossRef]
43. Smilgies, D.M. Scherrer grain-size analysis adapted to grazing-incidence scattering with area detectors. *J. Appl. Crystallogr.* **2009**, *42*, 1030–1034. [CrossRef]
44. Schneider, C.; Rasband, W.; Eliceiri, K. NIH Image to ImageJ: 25 Years of image analysis. *Nat. Methods* **2012**, *9*, 671–675. [CrossRef]

45. Thornton, J.A. Influence of substrate temperature and deposition rate on structure of thick sputtered Cu coatings. *J. Vac. Sci. Technol.* **1975**, *12*, 830–835. [CrossRef]
46. Ibru, T.; Kalaitzidou, K.; Kevin Baldwin, J.; Antoniou, A. Stress-induced surface instabilities and defects in thin films sputter deposited on compliant substrates. *Soft Matter* **2017**, *13*, 4035–4046. [CrossRef] [PubMed]
47. Cialone, M.; Celegato, F.; Coïsson, M.; Barrera, G.; Fiore, G.; Shvab, R.; Klement, U.; Rizzi, P.; Tiberto, P. Tailoring magnetic properties of multicomponent layered structure via current annealing in FePd thin films. *Sci. Rep.* **2017**, *7*, 16691. [CrossRef] [PubMed]
48. Cialone, M.; Fernandez-Barcia, M.; Celegato, F.; Coisson, M.; Barrera, G.; Uhlemann, M.; Gebert, A.; Sort, J.; Pellicer, E.; Rizzi, P.; et al. A comparative study of the influence of the deposition technique (electrodeposition versus sputtering) on the properties of nanostructured Fe70Pd30 films. *Sci. Technol. Adv. Mater.* **2020**, *21*, 424–434. [CrossRef] [PubMed]
49. Weisheit, M.; Schultz, L.; Fähler, S. Textured growth of highly coercive $L1_0$ ordered FePt thin films on single crystalline and amorphous substrates. *J. Appl. Phys.* **2004**, *95*, 7489–7491. [CrossRef]
50. Seki, K.; Kura, H.; Sato, T.; Taniyama, T. Size dependence of martensite transformation temperature in ferromagnetic shape memory alloy FePd. *J. Appl. Phys.* **2008**, *103*, 063910. [CrossRef]
51. Cullity, B.D.; Graham, C.D. *Introduction to Magnetic Materials*; Wiley-IEEE Press: Hoboken, NJ, USA, 2009.
52. Porter, A.A.; Easterling, K.E.; Sherif, M.Y. *Phase Transformations in Metals and Alloys*, 3rd ed.; Routledge: London, UK, 2009; p. 536.
53. Srolovitz, D. Capillary instabilities in thin films. I. Energetics. *J. Appl. Phys.* **1986**, *60*, 247–254. [CrossRef]
54. Gazit, N.; Klinger, L.; Rabkin, E. Chemically-induced solid-state dewetting of thin Au films. *Acta Mater.* **2017**, *129*, 300–311. [CrossRef]
55. Wynblatt, P. Interfacial segregation effects in wetting phenomena. *Annu. Rev. Mater. Res.* **2008**, *38*, 173–196. [CrossRef]
56. Amram, D.; Klinger, L.; Rabkin, E. Phase transformations in Au(Fe) nano-and microparticles obtained by solid state dewetting of thin Au-Fe bilayer films. *Acta Mater.* **2013**, *61*, 5130–5143. [CrossRef]
57. Herre, P.; Will, J.; Dierner, M.; Wang, D.; Yokosawa, T.; Zech, T.; Wu, M.; Przybilla, T.; Romeis, S.; Unruh, T.; et al. Rapid fabrication and interface structure of highly faceted epitaxial Ni-Au solid solution nanoparticles on sapphire. *Acta Mater.* **2021**, *220*, 117318. [CrossRef]
58. Barda, H.; Rabkin, E. Improving the thermal stability of nickel thin films on sapphire by a minor alloying addition of gold. *Appl. Surf. Sci.* **2019**, *484*, 1070–1079. [CrossRef]
59. Li, Y.; Yang, B.; Xia, M.; Yang, F.; Bao, X. Oxidation-induced segregation of FeO on the Pd-Fe alloy surface. *Appl. Surf. Sci.* **2020**, *525*, 146484. [CrossRef]
60. Vahalia, U.; Dowben, P.A.; Miller, A. Surface segregation in binary alloys. *J. Vac. Sci. Technol. A* **1986**, *4*, 1675–1679. [CrossRef]
61. Sakurai, T.; Hashizume, T.; Kobyashi, A.; Sakai, A.; Hyodo, S. Surface segregation of Ni-Cu binary alloys studied by an atom-probe. *Phys. Rev. B* **1986**, *34*, 8379. [CrossRef] [PubMed]
62. Zheng, J.; Zhu, Y.; Zeng, X.; Chen, B. Segregation of solute atoms in Mg–Ce binary alloy: Atomic-scale novel structures observed by HAADF-STEM. *Philos. Mag.* **2017**, *97*, 1498–1508. [CrossRef]
63. Oh, Y.J.; Ross, C.A.; Jung, Y.S.; Wang, Y.; Thompson, C.V. Cobalt nanoparticle arrays made by templated solid-state dewetting. *Small* **2009**, *5*, 860–865. [CrossRef]

Article

The Size Dependence of Microwave Permeability of Hollow Iron Particles

Anastasia V. Artemova *, Sergey S. Maklakov, Alexey V. Osipov, Dmitriy A. Petrov, Artem O. Shiryaev, Konstantin N. Rozanov and Andrey N. Lagarkov

Institute for Theoretical and Applied Electromagnetics RAS, Moscow 125412, Russia; squirrel498@gmail.com (S.S.M.); avosipov@mail.ru (A.V.O.); dpetrov-itae@yandex.ru (D.A.P.); artemshiryaev@mail.ru (A.O.S.); k.rozanov@yandex.ru (K.N.R.); maklakov@itae.ru (A.N.L.)
* Correspondence: avometras@gmail.com

Abstract: Hollow ferromagnetic powders of iron were obtained by means of ultrasonic spray pyrolysis. A variation in the conditions of the synthesis allows for the adjustment of the mean size of the hollow iron particles. Iron powders were obtained by this technique, starting from the aqueous solution of iron nitrate of two different concentrations: 10 and 20 wt.%. This was followed by a reduction in hydrogen. An increase in the concentration of the solution increased the mean particle size from 0.6 to 1.0 microns and widened particle size distribution, but still produced hollow particles. Larger particles appeared problematic for the reduction, although admixture of iron oxides did not decrease the microwave permeability of the material. The paraffin wax-based composites filled with obtained powders demonstrated broadband magnetic loss with a complex structure for lesser particles, and single-peak absorption for particles of 1 micron. Potential applications are 5G technology, electromagnetic compatibility designs, and magnetic field sensing.

Keywords: ultrasonic spray pyrolysis; hollow particles; ferromagnetic powder; microwave permeability; Curie temperature

1. Introduction

Ferromagnetic powders of submicron size are in demand due to potential applications in magnetic data storage systems, catalysis, magnetic field sensors, biomedical treatment, and microwave absorption materials [1–4]. Diverse chemical and physical techniques for the preparation of ferromagnetic powders exist. The most common techniques are chemical vapor deposition [5], sol-gel technology [6], mechanochemistry [7], and the ultrasonic spray pyrolysis (USP) technique.

The method of ultrasonic spray pyrolysis is designed to produce powders of metals or metal oxides from aqueous solutions. This method allows for adjustment of the properties of the final product: the shape, size, and composition of particles [8,9]. The technology consists of ultrasonic atomization of a precursor solution that produces an aerosol, then drying the aerosol droplets under high temperature, followed by collecting particles on a filter. The droplet size is controlled by the frequency of ultrasonic treatment, and by the concentration of the precursor solution [8–10]. Heating of aerosol causes diffusion of the dissolved substance to the surface of the droplet and evaporation of the solvent. In other terms, the droplets undergo drying, thus producing spherical solid particles. Depending on the temperature and duration of the droplet being in the hot zone, different particle size, shape, and morphology is formed. Interestingly, it was reported [11], that using not a pure compound as a solvent but a mixture of two compounds (for example, water and ethanol), made it possible to change the shape of obtained particles from spheres to flakes. It is proposed that not only surface tension of the solvent affects the mechanism of particle formation, but so do differences in the evaporation rate of the components of the solvent.

The spray pyrolysis enables one to obtain powders of metal oxides [11–14] and metals [10,15–17] with a mean diameter from nanometers to micrometers [18]. It is possible to obtain solid or hollow metal oxide particles [10]. Metal oxides can be reduced to metals via a one-stage process, or by applying post-deposition reduction during the second stage [15] that enhances control on phase and average size formation. The one-stage process can combine USP and hydrogen flow reduction [8,9], or reducing agents can be added to the precursor solution [16]. The concentration and type of reduction agents influence not only the size and morphology of particles, but also the reduction degree of particles. Additionally, thin films [19,20] and core-shell structures can be produced by the addition of alcohol [21] or silica sol [22] to the solution in the one-step USP process.

Properties of nano- and micrometer particles may differ particularly from their bulk counterparts [23]. Shape, size, and structure as well as chemical composition have critical value for static and dynamic magnetic properties [24,25]. Considering the effect of the chemical composition of the precursor ($Fe(NO_3)_3$, FeCl, etc.) on structure, as well as size- and structure-dependence of magnetic properties [16,21] of the final product, the spray pyrolysis technique offers a versatile possibility to control magnetic properties and electromagnetic parameters of functional materials. Developing new methods to effectively produce magnetic particles with complex structures and desired magnetic properties is still a challenge, especially in the vastly expanding field of magnetic sensing for biology and health applications.

Here, the USP technique followed by hydrogen reduction was applied to obtain iron particles with hollow structures. Variations in the solution of the precursor, $Fe(NO_3)_3$, produced powders with mean particle sizes of 0.6 and 1 µm. The difference in the particle size changed magnetic loss peak quality factor as well as imaginary permittivity of the composites based on paraffin wax matrix.

2. Materials and Methods

2.1. Ultrasonic Spray Pyrolysis and Reduction with Hydrogen

The iron powders were obtained using a two-stage synthesis similar to what was previously reported [17]. In the first stage, two aqueous solutions of iron (III) nitrate nonahydrate, $Fe(NO_3)_3 \cdot 9H_2O$ with concentrations of 10 and 20 wt.% (0.5 and 1 M, respectively) were prepared. An aerosol was obtained from these solutions using an ultrasonic dispenser (Timberk, China). Then, the aerosol was transferred through a tube furnace Nabertherm RT 50/250/13 (Nabertherm GmbH, Lilienthal, Germany) with airflow (16 L/min). The stainless steel tube 1000 mm (Ø 50 mm) in length was used as reactor, and the length of the heat zone in the furnace was about 30 cm. As a result of brief heat treatment at 1000 °C and chemical decomposition of $Fe(NO_3)_3$, a Fe_2O_3 oxide was formed and collected on a filter. In the second stage, iron oxide powders were reduced to metal in a tube furnace, Carbolite HZS 12/600 (Carbolite Gero, Neuhausen, Germany), at 450 °C in a flow of hydrogen. Metal powders were passivated in nitrogen for 12 h at room temperature. Hereafter, samples which were obtained starting from 10 wt.% solution and 20 wt.% solution are denoted as $Fe^{10\%}$ and $Fe^{20\%}$, respectively.

2.2. Characterization

The chemical phase composition was studied through the X-ray diffraction (XRD) method using a Difray 401 diffractometer (JSC Scientific instruments, Saint Petersburg, Russia) with Cr K_α radiation (λ = 0.229 nm). Bragg–Brentano geometry was applied for the measurements from 14 to 140 degrees, 2θ. The detector was curved and position-sensitive. The XRD pattern was used for the semiquantitative phase analysis and evaluation of the average size of the crystallites.

Scanning electron microscope (SEM) JEOL JCM-7000 (JEOL, Tokyo, Japan) equipped with the device for energy-dispersive X-ray (EDX) analysis was applied for the study of morphology and to estimate particle size distribution of the powders. To study the size distribution of particles, diameters of approximately 1000 particles were measured from the SEM images. The EDX analysis was measured from an area of 100 × 100 μm, uniformly covered with particles; measurements were average when studying five different areas.

The investigation of the thermal stability of metal powders from 30 to 1000 °C with a heating rate of 10 K/min in Ar and air flows was carried out using synchronous thermal analysis (STA) instrument Netzsch STA 449 F3 Jupiter (Netzsch, Selb, Germany). STA stands for simultaneous measurement of thermogravimetric (TG) and differential scanning calorimetry (DSC) curves. A special setup, designed by Netzsch to measure the Curie temperature, was used. For this purpose, external magnets were fastened onto the furnace of the STA device. At the temperature of the Curie transition, the ferromagnet lost magnetic interaction with these magnets. At this moment, the weight sensor detected an abrupt change of mass, and the thermal sensor detected a weak peak that was caused by the phase transition. The Curie temperature was defined by step-like mass change, measured in Ar flow.

The magnetic hysteresis at room temperature was measured using the vibrating sample magnetometer (VSM) in a magnetic field of ±15 kOe. The complex microwave permeability ($\mu' + i \times \mu''$) and permittivity ($\varepsilon' + i \times \varepsilon''$) were measured from the wax-based composites with filler fractions of 33, 50, and 66 wt.% in a standard 7/3 coaxial airline with an HP 8720 vector network analyzer (Keysight, Santa Rosa, CA, USA) by the Nicolson–Ross–Weir method [26,27] in the frequency range of 0.1–30 GHz. Higher-order modes in coaxial waveguide were accounted, following previously reported procedures [28].

3. Results and Discussion

3.1. X-ray Difraction Analysis

The XRD phase analysis demonstrated that the obtained powders consisted of α-Fe and Fe_3O_4 phases (Figure 1, Table 1). The positions of reflections of α-Fe in obtained samples (110) 68.78° 2θ and (200) 106.03° 2θ coincided well with the tabular values (ICDD No. 60696), as did the lattice constant *a*, which is 2.866 Å, both here and in the ICDD card No. 60696. For Fe_3O_4, the tabular value for constant *a* varies from 8.33 [29] to 8.432 Å [30]. Here, the lattice constant was within this range. The crystallite size of powders which was calculated using the Scherrer formula [31] was approximately 30 nm for all phases.

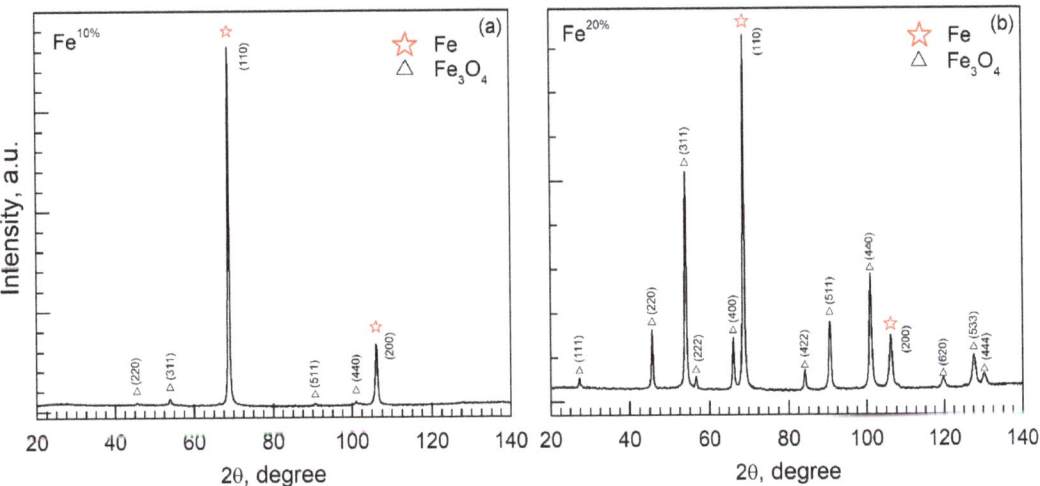

Figure 1. X-ray diffraction measured from the (**a**) $Fe^{10\%}$ and (**b**) $Fe^{20\%}$ samples.

Table 1. XRD data, where a—lattice constant, CS—crystallite size, ω^{Fe} and ω^{oxide}—mass fraction of iron and Fe_3O_4 oxide in each sample, calculated by means of the RIR method, Ms—saturation magnetization, Hc—coercivity.

Sample	Phases	a, Å	CS, nm	ω^{Fe}, wt.%	ω^{oxide}, wt.%	Ms, emu/g	Hc, Oe
$Fe^{10\%}$	α-Fe (#60696)	2866	37	95	5	140	160
	Fe3O4 (#11111)	8371	23				
$Fe^{20\%}$	α-Fe (#60696)	2866	29	30	70	103	225
	Fe3O4 (#11111)	8414	29				

The reference intensity ratio (RIR) method [32] with corundum as a reference standard for semiquantitative phase analysis was applied. For estimation, the following reference value for α-Fe, Fe_3O_4 is 10.77 (ICDD No. 60696), and 4.81 (ICDD No. 11111) were used, respectively. The estimation of relative metal to oxide (ω^{Fe}, ω^{oxide}) content in obtained samples is provided in Table 1. Whereas both powders ($Fe^{10\%}$ and $Fe^{20\%}$) were reduced under identical conditions (H_2 flow, 450 °C), a considerable lack of metal phase in $Fe^{20\%}$ was observed. Shatrova et al. [33] examined the H_2 reduction of Co powder prepared through the spray pyrolysis technique at different temperatures, and demonstrated that the purity of the metal depended on the reduction temperature. In addition, particles of different sizes (and with different wall thickness, with respect to hollow structures) may react differently to passivation with the nitrogen stage. Here, identical reduction and passivation conditions were chosen intentionally for direct comparison. Further studies involving higher reduction temperatures are required.

3.2. SEM and EDX Analyses

The prepared samples possessed a spherical shape (Figure 2), which is typical for spray pyrolysis [18]. The obtained particles demonstrated porous and inhomogeneous surfaces. "Broken" particles confirmed the hollow structure.

An increase in the precursor solution concentration increased the mean particle size (see Figure 2a vs. Figure 2c), as well as width of particle size distribution (Figure 3). The average sizes of particles of $Fe^{10\%}$ and $Fe^{20\%}$ were estimated at 0.6 and 1 μm, respectively. Full width at half maximum is increase from 0.35 to 0.62, respectively. The thicknesses of the walls of the particle were approximated at 200 ($Fe^{10\%}$) and 500 nm (see Figure 2b,d), using SEM microphotographs. The size of particles and thickness of the walls both were larger than the crystallite size. SEM images also illustrated a grain structure, with grains around 100 nm. However, since these images do not allow for a clear distinction between the real grain size, surface morphology, and surface roughness, the X-ray data is more accurate.

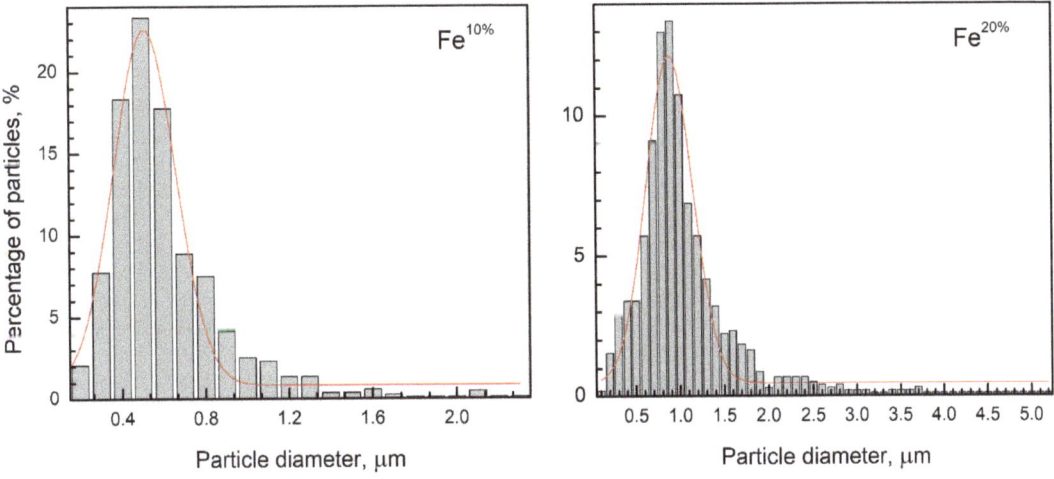

Figure 2. SEM-microphotographs of obtained iron powders (**a,b**) Fe[10%] and (**c,d**) Fe[20%].

Figure 3. Particle size distribution, measured from the SEM images, of the Fe[10%] and Fe[20%] powders, where red line is Gaussian function.

Local EDX analysis illustrated the presence of iron and oxygen. The oxygen to iron ratio was estimated at 10/90 ($Fe^{10\%}$) and 30/70 at. % ($Fe^{20\%}$).

3.3. STA Analysis

Thermal analysis in airflow was performed for $Fe^{10\%}$ and $Fe^{20\%}$ (Figure 4). The mass gain occurred in at least two steps for both samples. These steps were accompanied by exothermic maxima in DSC curves. Step-like oxidation is typical for iron powders (see Figure 3 from [34], for example). According to [34,35], iron oxidation occurs in these steps: $Fe \rightarrow FeO \rightarrow Fe_3O_4 \rightarrow Fe_2O_3$. However, direct observation of the intermediate phases is not found in the literature, partially because these oxidation steps intermix with each other, and because no technique to stop oxidation at an intermediate stage seems to be known. The total mass gain is important, and is a characteristic of the metal's purity. For comparison, oxidation of the commercial carbonyl iron (CI, ≥ 97.0 wt. % Fe) [36] produces a weight increase of 38%. Additional measurements during this study illustrated a weight increase of 38.8% at 1000 °C for CI, which was in a good agreement with the ref. [36] data.

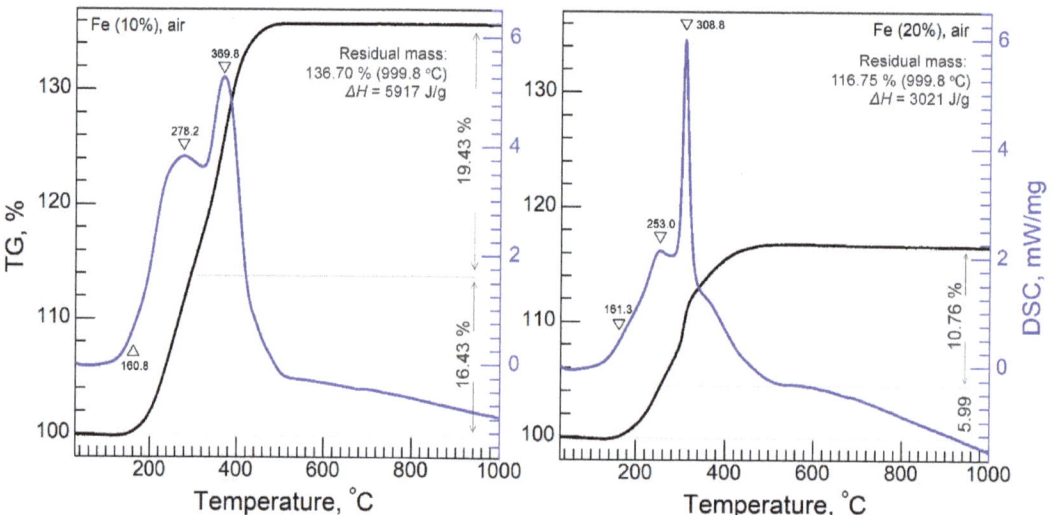

Figure 4. STA analysis of $Fe^{10\%}$ and $Fe^{20\%}$ powders in air.

Considering problem of intermediate oxidation stages, simple numeric estimation of mass gain during oxidation of Fe_3O_4 to Fe_2O_3, produces 3.2–3.3 wt.%, which was confirmed in [37]. The measured weight increase during formation of the highest-oxidation state of iron was far from 3%—it was actually 10–20%. This implies that the oxidation mechanism of the obtained samples is far more complex than that proposed in [34,35]. The temperature where maximum weight was achieved was 500 °C. The formation of Fe_2O_3 above 500 °C was confirmed with the XRD analysis.

The total mass gains for $Fe^{10\%}$ and $Fe^{20\%}$ were 36.7 and 16.8%, respectively. Comparing total mass gain between samples $Fe^{10\%}$ and $Fe^{20\%}$, it can be concluded that the $Fe^{10\%}$ sample was of higher purity than $Fe^{20\%}$.

The total released energy for $Fe^{10\%}$ ($\Delta H = 5917$ J/g) was twice as much as that of $Fe^{20\%}$ ($\Delta H = 3021$ J/g). Although there is data suggesting that an increase in the particle size of iron from 1–3 to 4–10 µm [35] increases enthalpy of oxidation from 4.8 to 5.4 J/g, just the opposite effect was observed in this study. This was probably due to different purity of samples $Fe^{10\%}$ and $Fe^{20\%}$, and because of the much lower particle size than that of powders studied in [35]. The onset temperature (~160 °C) of the exothermic process was equal for both powders.

Measurements in an inert atmosphere for Fe$^{10\%}$ and Fe$^{20\%}$ were carried out (Figure A1a). Insignificant mass loss of 1% was measured for both samples. This was due to emission of nitrogen and carbon admixtures, and coincided with the loss of 2% reported in [38]. Several weak endothermic minima observed in the DSC curve were probably related to melting or sintering of particles.

The Curie temperature (T_C) of prepared powders was defined by a step-like mass increase on the TG curve [39], accompanied with an exothermic DSC peak (Figure A1b). The T_C for Fe$^{10\%}$ powder corresponded to bulk iron that is about 770 °C (1043 K) [40]. The T_C for Fe$^{20\%}$ was lower by 30% of this value (564 °C) (see Figure A1 in Appendix A). The Curie temperature for Fe$_3$O$_4$ is known to be 440 °C (713 K) [41]. Since the Curie temperature is dependent on the Fe purity [39], the Fe$^{20\%}$ sample may be concluded to possess lower purity than Fe$^{10\%}$.

3.4. Magnetic Properties

Magnetic hysteresis loops of the composites containing 66 wt.% of obtained powders are displayed in Figure 5.

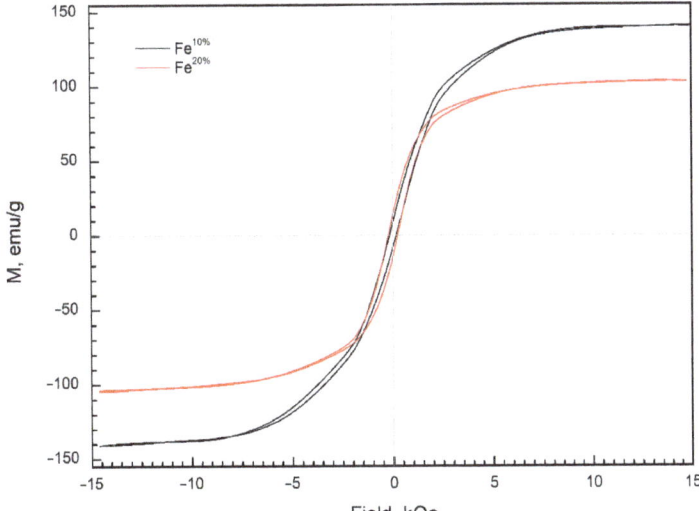

Figure 5. Hysteresis loops, measured from composites based on paraffin wax and filled with Fe$^{10\%}$ and Fe$^{20\%}$ powders.

Considering the concentration of the powders in composites, the saturation magnetizations (M_S) of the Fe$^{10\%}$ and Fe$^{20\%}$ powders were 140 and 103 emu/g, respectively (Table 1). The saturation magnetization of the carbonyl iron (CI) ranges from 175 to 188 emu/g [36,42], depending on the particle size [43]. The M_S of Fe$^{10\%}$ and Fe$^{20\%}$ were lower than the aforementioned range for CI, but higher than that reported for Fe$_3$O$_4$ (40 and 92 emu/g) [24,44].

The H_C for the Fe$^{10\%}$ and Fe$^{20\%}$ particles were 160 and 225 Oe, respectively (Table 1). The high value of the coercive force H_C and moderate values of M_S were probably related to the presence of the oxide phase [17].

Chemical purity of iron in both Fe$^{10\%}$ and Fe$^{20\%}$ samples can be estimated only simply. Comparing weight gain during oxidation, relative oxygen content in EDX, and XRD intensities and saturation magnetization between both samples, it appears clear that the Fe$^{10\%}$ contained a larger metal fraction than the Fe$^{20\%}$ sample.

3.5. Microwave Measurement of Composites

The measured frequency dependencies of complex microwave characteristics of composite with 66 wt.% powder concentration are illustrated in Figure 6. An increase in the concentration (from 33 to 66 wt.%) of the fraction leads to a linear increase in the amplitude of the imaginary part of permeability.

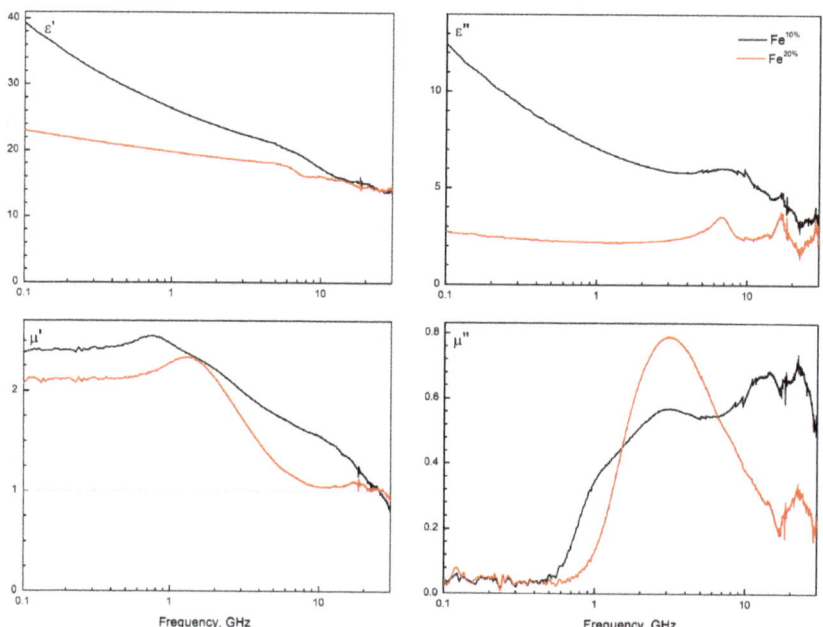

Figure 6. The measured frequency dependences of microwave permeability ($\mu' + i \times \mu''$) and permittivity ($\varepsilon' + i \times \varepsilon''$) of the composites comprising powders $Fe^{10\%}$ or $Fe^{20\%}$, and paraffin wax matrix. The curves are normalized by weight ratio 3:1 between metallic powder ($Fe^{10\%}$, $Fe^{20\%}$) and wax matrix.

Both powders demonstrated magnetic loss of high amplitude. Changes in size and structure of particles significantly changed the structure of the magnetic loss peak. The imaginary part of permeability of the composite filled with $Fe^{10\%}$ powder demonstrated a broad peak with a complex structure. Concurrently, composite filled with $Fe^{20\%}$ powder demonstrated a narrower peak with a single loss maximum. Both powders produced composites with low magnetic loss below 0.5 GHz, which is unusual for spherical iron (see carbonyl iron studies, [1,45]). This makes the samples promising materials for application in many areas, one being antennas.

Composite filled with $Fe^{10\%}$ demonstrated significantly higher imaginary permittivity, which was due to agglomeration of particles in the composite structure and therefore increased percolation conductivity. High saturation magnetization and low coercivity, as well as high complex permeability, illustrated that iron particles were micro-sized (not nano-sized). Also, imaginary permittivity was substantially lower for the sample with increased oxide concentration. This probably meant that contact conductivity was significantly suppressed with an increase in oxide phase concentration. In simpler terms, the oxide phase was predominantly on the surface of larger iron particles. Importantly, in a series of composites (33, 50, and 66 wt.% of a filler), the structure of the frequency dispersion and the position of the magnetic loss peak in imaginary permeability depended neither on filler concentration nor on the conductivity of certain samples.

In dynamic magnetism, a quantitative parameter that describes the overall magnetic performance is Acher's parameter K_A. This parameter was calculated following [46]:

$$K_A = \frac{1}{kp(\gamma 4\pi M_s)^2} \times \frac{2}{\pi} \times \int_{f_1}^{f_2} \mu''(f) f df \qquad (1)$$

where p is the volume fraction of inclusions and k is a randomization factor, $\gamma \approx 3$ GHz/kOe, and f_1 to f_2 is the frequency range of measurement.

The physical meaning of this parameter is how many of all the magnetic moments in the sample are involved in a precession that is forced by the incident microwave magnetic field. The maximum value is 1 and can only be observed in thin films. Spherical particles can possess much lower values.

In this study, the K_A parameter was estimated at 0.09 for composites based on $Fe^{10\%}$ powder and 0.05 for $Fe^{20\%}$. This suggested that the $Fe^{10\%}$-based composites were more effective in absorbing electromagnetic energy microwaves than the $Fe^{20\%}$-based composites. The K_A value, close to 0.1, was high [47] enough to signify those magnetic moments which were aligned along the shell of particles. This indirectly confirmed the hollow structure of the particles.

Several studies are devoted to resonances in hollow spherical particles; these three studies, [23,24,48], are worth mentioning. In conclusion, not only hollowness of magnetic particles excites resonances besides ferromagnetic, but the very distribution of particle size sophisticates the structure of magnetic loss. Considering this and the calculated $_IK_a$ value, the $Fe^{10\%}$ sample appeared to possess higher hollowness than $Fe^{10\%}$, which increased the potential electromagnetic performance of the former.

3.6. Hollow or Not

The problem with the analysis of the hollowness of the obtained particles is that no direct and reliable method to do so exists. In this study, there were three signals that illustrated the hollow structure of the particles. Firstly, shatters of the hollow shells that can be seen in Figure 1b,d, as well as particles with the structure of a deflated balloon, illustrated that there are particles with a hollow structure in both samples.

The second signal is the low density of the samples. The density of wax-based composites was lower than that can be expected from composites filled with dense iron. Moreover, using the measured weights of the filler, matrix, and composite, and applying density of the wax (2.2 g/cm^3 [2]), and effective density of the composite (calculated independently from a known volume of sample), the mean thickness of the walls of hollow particles of both powders was estimated. This can be done by considering the composite as a three-component mixture: the wax, iron, and voids filled with air. The density of inclusions was defined for each filler, then the proportion of voids was calculated. The thickness of the wall was estimated assuming that all of the particles possess the mean size (0.6 μm for $Fe^{10\%}$ and 1.0 μm for $Fe^{20\%}$). Following this procedure, the thickness of the walls of hollow particles was estimated at about 0.125 μm for both the $Fe^{10\%}$ and $Fe^{20\%}$ powders. This value was approximately in accordance with the electron microscopy data. The third 'signal' is that the frequency dispersion of complex permeability demonstrated features that may be interpreted as attributes of thin magnetic layers.

The formation of dense or hollow particles depends on the rate ratio of drying and precursor decomposition steps. Therefore, the temperature regime and the gas flow rate should be optimized to obtain the predominant amount of one of these two types. To further enhance this synthetic technique, additional studies are required.

4. Conclusions

Ferromagnetic hollow powders with the mean particle sizes of 0.6 and 1 μm were synthesized from iron nitrate solution with concentrations of 10 and 20 wt.%, respectively, by ultrasonic spray pyrolysis and reduction in hydrogen. According to XRD analysis, the impurity content (residual iron oxide) in the $Fe^{20\%}$ sample was higher than in the $Fe^{10\%}$ sample. This means that magnetic interaction between iron particles was negligible.

The weight increases during the oxidation of $Fe^{10\%}$ and $Fe^{20\%}$ in the air were 36.7 and 16.8 %, respectively. The experimentally determined Curie temperature was 770 °C for $Fe^{10\%}$, which corresponded to the tabular value for iron, and 564 °C for $Fe^{20\%}$. The difference between the values of released energy and weight increase during oxidation, as well as the Curie temperature of prepared iron powders, were caused by the presence of iron oxide in the $Fe^{20\%}$, and by the differences in the sizes of the particles.

The saturation magnetization of the $Fe^{10\%}$ and $Fe^{20\%}$ samples were 140 and 103 emu/g, respectively. The coercivity of the powder particles was moderately high (160 and 225 Oe) and was typical for powders prepared by spray pyrolysis. Both powders demonstrated low magnetic losses below 0.5 GHz. The $Fe^{10\%}$-based samples with the paraffin wax matrix demonstrated a broad spectrum of frequency dispersion of imaginary permeability with a complex pattern. At the same time, $Fe^{20\%}$-based composites demonstrate a single-peak curve. The prepared material can be applied in 5G technologies and in solutions for electromagnetic compatibility. Considering small particle size, high microwave permeability, high saturation magnetization, and potentially high biocompatibility, the obtained powders may be applied in biosensors, based on magnetic field sensing.

Author Contributions: Conceptualization, S.S.M. and A.V.A.; methodology, S.S.M. and A.V.A.; software, A.V.A.; validation, A.V.A., S.S.M. and K.N.R.; formal analysis, A.V.A., S.S.M., A.V.O. and D.A.P.; investigation, A.V.A., A.O.S., A.V.O. and S.S.M.; resources, A.V.A., S.S.M. and A.V.O.; data curation, A.V.A., A.V.O., A.O.S. and S.S.M.; writing—original draft preparation, A.V.A.; writing—review and editing, A.V.A., S.S.M. and K.N.R.; visualization, A.V.A.; supervision, S.S.M., K.N.R. and A.N.L.; project administration, S.S.M. and A.V.A.; funding acquisition, S.S.M. and A.N.L. All authors have read and agreed to the published version of the manuscript.

Funding: This study was financially supported by the Russian Science Foundation (RSF) under project No. 21-19-00138 (https://rscf.ru/en/project/21-19-00138/, accessed on 1 January 2022).

Institutional Review Board Statement: Not applicable.

Informed Consent Statement: Not applicable.

Data Availability Statement: Data is contained within the article.

Conflicts of Interest: The authors declare no conflict of interest.

Appendix A

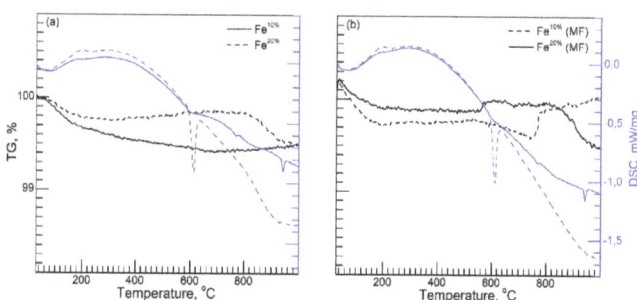

Figure A1. TG and DSC graphs of $Fe^{10\%}$ and $Fe^{20\%}$ powders in Ar (**a**) without magnetic field (MF) and (**b**) with external MF.

References

1. Maklakov, S.S.; Lagarkov, A.N.; Maklakov, S.A.; Adamovich, Y.A.; Petrov, D.A.; Rozanov, K.N.; Ryzhikov, I.A.; Zarubina, A.Y.; Pokholok, K.V.; Filimonov, D.S. Corrosion-resistive magnetic powder Fe@SiO$_2$ for microwave applications. *J. Alloys Compd.* **2017**, *706*, 267–273. [CrossRef]

2. Dolmatov, A.V.; Maklakov, S.S.; Zezyulina, P.A.; Osipov, A.V.; Petrov, D.A.; Naboko, A.S.; Polozov, V.I.; Maklakov, S.A.; Starostenko, S.N.; Lagarkov, A.N. Deposition of a SiO$_2$ Shell of Variable Thickness and Chemical Composition to Carbonyl Iron: Synthesis and Microwave Measurements. *Sensors* **2021**, *21*, 4624. [CrossRef] [PubMed]
3. Matysiak, E.; Donten, M.; Kowalczyk, A.; Bystrzejewski, M.; Grudzinski, I.P.; Nowicka, A.M. A novel type of electrochemical sensor based on ferromagnetic carbon-encapsulated iron nanoparticles for direct determination of hemoglobin in blood samples. *Biosens. Bioelectron.* **2015**, *64*, 554–559. [CrossRef] [PubMed]
4. Koren, K.; Mistlberger, G.; Borisov, S.M.; Klimant, I. Ferromagnetic and permanent magnetic spheres as platform for single- and dual-analyte optical sensors. *Procedia Eng.* **2010**, *5*, 997–1000. [CrossRef]
5. Huang, W.; Gatel, C.; Li, Z.-A.; Richter, G. Synthesis of magnetic Fe and Co nano-whiskers and platelets via physical vapor deposition. *Mater. Des.* **2021**, *208*, 109914. [CrossRef]
6. Söderberg, O.; Ge, Y.; Haimi, E.; Heczko, O.; Oja, M.; Laine, J.; Suhonen, T.; Aaltonen, A.; Kalliokari, K.; Borak, B.; et al. Morphology of ferromagnetic sol—gel submicron silica powders doped with iron and nickel particles. *Mater. Lett.* **2007**, *61*, 3171–3173. [CrossRef]
7. Ulbrich, K.F.; Campos, C.E.M. Nanocrystalline Ni$_3$S$_2$ prepared by mechanochemistry and its behavior at high temperatures and high pressure. *J. Magn. Magn. Mater.* **2020**, *493*, 165706. [CrossRef]
8. Gurmen, S.; Ebin, B.; Stopić, S.; Friedrich, B. Nanocrystalline spherical iron–nickel (Fe–Ni) alloy particles prepared by ultrasonic spray pyrolysis and hydrogen reduction (USP-HR). *J. Alloys Compd.* **2009**, *480*, 529–533. [CrossRef]
9. Gurmen, S.; Guven, A.; Ebin, B.; Stopić, S.; Friedrich, B. Synthesis of nano-crystalline spherical cobalt–iron (Co–Fe) alloy particles by ultrasonic spray pyrolysis and hydrogen reduction. *J. Alloys Compd.* **2009**, *481*, 600–604. [CrossRef]
10. Kastrinaki, G.; Lorentzou, S.; Karagiannakis, G.; Rattenbury, M.; Woodhead, J.; Konstandopoulos, A.G. Parametric synthesis study of iron based nanoparticles via aerosol spray pyrolysis route. *J. Aerosol Sci.* **2018**, *115*, 96–107. [CrossRef]
11. Krasnikova, I.V.; Mishakov, I.V.; Bauman, Y.I.; Karnaukhov, T.M.; Vedyagin, A.A. Preparation of NiO-CuO-MgO fine powders by ultrasonic spray pyrolysis for carbon nanofibers synthesis. *Chem. Phys. Lett.* **2017**, *684*, 36–38. [CrossRef]
12. Yudin, A.; Shatrova, N.; Khaydarov, B.; Kuznetsov, D.; Dzidziguri, E.; Issi, J.-P. Synthesis of hollow nanostructured nickel oxide microspheres by ultrasonic spray atomization. *J. Aerosol Sci.* **2016**, *98*, 30–40. [CrossRef]
13. Tischendorf, R.; Simmler, M.; Weinberger, C.; Bieber, M.; Reddemann, M.; Fröde, F.; Lindner, J.; Pitsch, H.; Kneer, R.; Tiemann, M.; et al. Examination of the evolution of iron oxide nanoparticles in flame spray pyrolysis by tailored in situ particle sampling techniques. *J. Aerosol Sci.* **2021**, *154*, 105722. [CrossRef]
14. Oh, S.W.; Bang, H.J.; Bae, Y.C.; Sun, Y.-K. Effect of calcination temperature on morphology, crystallinity and electrochemical properties of nano-crystalline metal oxides (Co$_3$O$_4$, CuO, and NiO) prepared via ultrasonic spray pyrolysis. *J. Power Sources* **2007**, *173*, 502–509. [CrossRef]
15. Shatrova, N.; Yudin, A.; Levina, V.; Kuznetsov, D.; Novakova, A.; Dzidziguri, E.; Perov, N.; Issi, J.-P. Characteristics of Co$_3$O$_4$ and cobalt nanostructured microspheres: Morphology, structure, reduction process, and magnetic properties. *Mater. Res. Bull.* **2018**, *99*, 189–195. [CrossRef]
16. Septiani, E.L.; Kikkawa, J.; Cao, K.L.A.; Hirano, T.; Okuda, N.; Matsumoto, H.; Enokido, Y.; Ogi, T. Direct synthesis of submicron FeNi particles via spray pyrolysis using various reduction agents. *Adv. Powder Technol.* **2021**, *32*, 4263–4272. [CrossRef]
17. Kosevich, A.; Petrusevich, E.; Maklakov, S.; Naboko, A.; Kolesnikov, E.; Petrov, D.; Zezyulina, P.; Pokholok, K.; Filimonov, D.; Han, M. Low Weight Hollow Microspheres of Iron with Thin Dielectric Coating: Synthesis and Microwave Permeability. *Coatings* **2020**, *10*, 995. [CrossRef]
18. Nandiyanto, A.B.D.; Okuyama, K. Progress in developing spray-drying methods for the production of controlled morphology particles: From the nanometer to submicrometer size ranges. *Adv. Powder Technol.* **2011**, *22*, 1–19. [CrossRef]
19. Akl, A.A. Microstructure and electrical properties of iron oxide thin films deposited by spray pyrolysis. *Appl. Surf. Sci.* **2004**, *221*, 319–329. [CrossRef]
20. Dghoughi, L.; Elidrissi, B.; Bernède, C.; Addou, M.; Lamrani, M.A.; Regragui, M.; Erguig, H. Physico-chemical, optical and electrochemical properties of iron oxide thin films prepared by spray pyrolysis. *Appl. Surf. Sci.* **2006**, *253*, 1823–1829. [CrossRef]
21. Yu, F.; Wang, J.; Sheng, Z.; Su, L. Synthesis of carbon-encapsulated magnetic nanoparticles by spray pyrolysis of iron carbonyl and ethanol. *Carbon* **2005**, *43*, 3018–3021. [CrossRef]
22. Han, S.; Yoo, K.S.; Kim, D.; Kim, J.; Othman, M.R. Metal-silica spherical particles development by spray pyrolysis: Effect of metal species on surface area and toluene adsorption. *J. Anal. Appl. Pyrolysis* **2021**, *156*, 105049. [CrossRef]
23. McKeever, C.; Ogrin, F.Y.; Aziz, M.M. Microwave magnetization dynamics in ferromagnetic spherical nanoshells. *Phys. Rev. B* **2019**, *100*, 054425. [CrossRef]
24. Li, Z.W.; Yang, Z.H. Microwave absorption properties and mechanism for hollow Fe$_3$O$_4$ nanosphere composites. *J. Magn. Magn. Mater.* **2015**, *387*, 131–138. [CrossRef]
25. Chen, G.; Xu, D.; Chen, P.; Guo, X.; Yu, Q.; Qiu, H. Constructing and optimizing hollow bird-nest-patterned C@Fe$_3$O$_4$ composites as high-performance microwave absorbers. *J. Magn. Magn. Mater.* **2021**, *532*, 167990. [CrossRef]
26. Nicolson, A.M.; Ross, G.F. Measurement of the Intrinsic Properties of Materials by Time-Domain Techniques. *IEEE Trans. Instrum. Meas.* **1970**, *19*, 377–382. [CrossRef]
27. Weir, W.B. Automatic measurement of complex dielectric constant and permeability at microwave frequencies. *Proc. IEEE* **1974**, *62*, 33–36. [CrossRef]

28. Petrov, D.A.; Rozanov, K.N.; Koledintseva, M.Y. Influence of Higher-order Modes in Coaxial Waveguide on Measurements of Material Parameters. In Proceedings of the 2018 IEEE Symposium on Electromagnetic Compatibility, Signal Integrity and Power Integrity (EMC, SI & PI), Long Beach, CA, USA, 30 July–3 August 2018; pp. 66–70.
29. Lindsley, D.H. Chapter 1. The CRYSTAL CHEMISTRY and STRUCTURE of OXIDE MINERALS as EXEMPLIFIED by the Fe-Ti OXIDES. In *Oxide Minerals*; Rumble, D., Ed.; De Gruyter: Berlin, Germany, 1976; 60p.
30. Jeon, S.-H.; Son, Y.-H.; Choi, W.-I.; Song, G.; Hur, D. Simulating Porous Magnetite Layer Deposited on Alloy 690TT Steam Generator Tubes. *Materials* **2018**, *11*, 62. [CrossRef]
31. Patterson, A.L. The Scherrer Formula for X-Ray Particle Size Determination. *Phys. Rev.* **1939**, *56*, 978–982. [CrossRef]
32. Hubbard, C.R.; Snyder, R.L. RIR-Measurement and Use in Quantitative XRD. *Powder Diffr.* **1988**, *3*, 74–77. [CrossRef]
33. Shatrova, N.; Yudin, A.; Levina, V.; Dzidziguri, E.; Kuznetsov, D.; Perov, N.; Issi, J.-P. Elaboration, characterization and magnetic properties of cobalt nanoparticles synthesized by ultrasonic spray pyrolysis followed by hydrogen reduction. *Mater. Res. Bull.* **2017**, *86*, 80–87. [CrossRef]
34. Wen, D.; Song, P.; Zhang, K.; Qian, J. Thermal oxidation of iron nanoparticles and its implication for chemical-looping combustion. *J. Chem. Technol. Biotechnol.* **2011**, *86*, 375–380. [CrossRef]
35. Huang, D.H.; Tran, T.N.; Yang, B. Investigation on the reaction of iron powder mixture as a portable heat source for thermoelectric power generators. *J. Therm. Anal. Calorim.* **2014**, *116*, 1047–1053. [CrossRef]
36. Mrlík, M.; Ilčíková, M.; Cvek, M.; Pavlínek, V.; Zahoranová, A.; Kroneková, Z.; Kasak, P. Carbonyl iron coated with a sulfobetaine moiety as a biocompatible system and the magnetorheological performance of its silicone oil suspensions. *RSC Adv.* **2016**, *6*, 32823–32830. [CrossRef]
37. Monazam, E.R.; Breault, R.W.; Siriwardane, R. Kinetics of Magnetite (Fe_3O_4) Oxidation to Hematite (Fe_2O_3) in Air for Chemical Looping Combustion. *Ind. Eng. Chem. Res.* **2014**, *53*, 13320–13328. [CrossRef]
38. Chung, K.H.; Wu, C.S.; Malawer, E.G. Thermomagnetometry and thermogravimetric analysis of carbonyl iron powder. *Thermochim. Acta* **1989**, *154*, 195–204. [CrossRef]
39. Somunkiran, I.; Buytoz, S.; Dagdelen, F. Determination of curie temperatures and thermal oxidation behavior of Fe-Cr matrix composites produced by hot pressing. *J. Alloys Compd.* **2019**, *777*, 302–308. [CrossRef]
40. Velasco, S.; Román, F.L. Determining the Curie Temperature of Iron and Nickel. *Phys. Teach.* **2007**, *45*, 387–389. [CrossRef]
41. Manohar, A.; Krishnamoorthi, C. Low Curie-transition temperature and superparamagnetic nature of Fe_3O_4 nanoparticles prepared by colloidal nanocrystal synthesis. *Mater. Chem. Phys.* **2017**, *192*, 235–243. [CrossRef]
42. Kim, S.Y.; Kwon, S.H.; Liu, Y.D.; Lee, J.-S.; You, C.-Y.; Choi, H.J. Core–shell-structured cross-linked poly(glycidyl methacrylate)-coated carbonyl iron microspheres and their magnetorheology. *J. Mater. Sci.* **2014**, *49*, 1345–1352. [CrossRef]
43. Bombard, A.J.F.; Joekes, I.; Alcântara, M.R.; Knobel, M. Magnetic Susceptibility and Saturation Magnetization of some Carbonyl Iron Powders used in Magnetorheological Fluids. *Mater. Sci. Forum* **2003**, *416–418*, 753. [CrossRef]
44. Tuo, Y.; Liu, G.; Dong, B.; Zhou, J.; Wang, A.; Wang, J.; Jin, R.; Lv, H.; Dou, Z.; Huang, W. Microbial synthesis of Pd/Fe_3O_4, Au/Fe_3O_4 and $PdAu/Fe_3O_4$ nanocomposites for catalytic reduction of nitroaromatic compounds. *Sci. Rep.* **2015**, *5*, 13515. [CrossRef] [PubMed]
45. Semenenko, V.N.; Chistyaev, V.A.; Politiko, A.A.; Kibets, S.G.; Kisel, V.N.; Gallagher, C.P.; McKeever, C.; Hibbins, A.P.; Ogrin, F.Y.; Sambles, J.R. Complex Permittivity and Permeability of Composite Materials Based on Carbonyl Iron Powder Over an Ultrawide Frequency Band. *Phys. Rev. Appl.* **2021**, *16*, 014062. [CrossRef]
46. Iakubov, I.T.; Lagarkov, A.N.; Maklakov, S.A.; Osipov, A.V.; Rozanov, K.N.; Ryzhikov, I.A.; Starostenko, S.N. Microwave permeability of composites filled with thin Fe films. *J. Magn. Magn. Mater.* **2006**, *300*, e74–e77. [CrossRef]
47. Rozanov, K.N.; Koledintseva, M.Y. Application of generalized Snoek's law over a finite frequency range: A case study. *J. Appl. Phys.* **2016**, *119*, 073901. [CrossRef]
48. McKeever, C.; Ogrin, F.Y.; Aziz, M.M. Influence of surface anisotropy on exchange resonance modes in spherical shells. *J. Phys. D Appl. Phys.* **2018**, *51*, 305003. [CrossRef]

Article

Effect of Temperature on Microwave Permeability of an Air-Stable Composite Filled with Gadolinium Powder

Sergey N. Starostenko [1,*], Dmitriy A. Petrov [1], Konstantin N. Rozanov [1], Artem O. Shiryaev [1] and Svetlana F. Lomaeva [2]

[1] Institute for Theoretical and Applied Electromagnetics, Izhorskaya 13/19, 125412 Moscow, Russia; dpetrov-itae@yandex.ru (D.A.P.); k_rozanov@mail.ru (K.N.R.); artemshiryaev@mail.ru (A.O.S.)
[2] Physical-Technical Institute, Udmurt Federal Research Center UB RAS, T. Baramzina Str., 34, 426067 Izhevsk, Russia; lomayevasf@mail.ru
* Correspondence: snstar@mail.ru

Abstract: A composite containing about 30% volume of micrometer-size powder of gadolinium in paraffin wax is synthesized mechanochemically. The composite permittivity and permeability are measured within the frequency range from 0.01 to 15 GHz and the temperature range from ~0 °C to 35 °C. The permittivity is constant within the measured ranges. Curie temperature of the composite is close to 15.5 °C, the phase transition is shown to take place within a temperature range about ±10 °C. The effect of temperature deviation from Curie point on reflection and transmission of a composite layer filled with Gd powder is studied experimentally and via simulation. Constitutive parameters of the composite are measured in cooled coaxial lines applying reflection-transmission and open-circuit-short-circuit techniques, and the measured low-frequency permeability is in agreement with the values retrieved from the published magnetization curves. The effect of temperature on permeability spectrum of the composite is described in terms of cluster magnetization model based on the Wiener mixing formula. The model is applied to design a microwave screen with variable attenuation; the reflectivity attenuation of 4.5 mm-thick screen increases from about −2 dB to −20 dB at 3.5 GHz if the temperature decreases from 25 °C to 5 °C.

Keywords: microwave permeability; Curie temperature; cluster magnetization; Hopkinson effect; mixing model; tunable screen

1. Introduction

Adaptive composites with tunable constitutive parameters look promising for many applications including microwave screens, sensors, waveguide elements, etc. [1,2]; however, few of them are used in practice. One of the problems is that the tuning needs electric bias about 10–30 kV/cm for ferroelectrics [3] or magnetic bias about 2–4 kOe for composites with permeable microwires [3]. Here the temperature tuning of microwave permeability μ is under consideration. It may seem that such tuning is possible at a temperature of several hundred degrees centigrade and has too high a transit time compared to bias tuning. The temperature may be decreased by the proper selection of a magnetic substance; the time for heating or cooling for several degrees is comparable to the transit time for application of several kOe bias.

Gadolinium (Gd) is a ferromagnet with a Curie temperature of about 20.2 °C, and its saturation magnetization is 15% higher than that of iron, therefore powders of Gd and its alloys look like promising fillers [4,5] for microwave applications, especially as adaptive composites with parameters governed by temperature.

The magnetization of bulk gadolinium has been thoroughly studied, as Gd is the first material used for magnetic refrigeration; nevertheless, the published magnetization curves [6–8] differ significantly and there are practically no data on gadolinium permeability. To retrieve at least a reference value of permeability μ to compare with the measured data a set of published magnetization curves is digitized and the static permeability is estimated as:

$$\mu_{static} \underset{H \to 0}{\approx} 1 + M/[(1-N) \times H] \tag{1}$$

Here M is the sample magnetization for the strength of magnetic bias H, N is the demagnetization factor (the shape factor) defined by sample shape, and orientation is relative to bias direction [6,7]. The shape factor range is $0 \leq N \leq 1$, for spheres $N = 1/3$ [9,10].

The magnetization data at low bias are scarce, but the hysteresis loop looks narrow, so the estimations with Equation (1) should be valid up to a bias of about 18,000 G [11]. The problem is that the higher the bias, the higher the observed Curie temperature (see curve d in Figure 1). Another problem is that monocrystalline gadolinium is anisotropic; the magnetization curve may depend on the sample shape and orientation [6,12]. The retrieved susceptibility $\chi = \mu - 1$ curve (Figure 1, line a) is calculated as a mean value for two orientations of easy magnetization axis. The susceptibility estimates with Equation (1) (Figure 1) show that the permeability of polycrystalline Gd is much lower than that of iron-group metals. Below $-100\ °C$ $\mu \approx 4 \div 7$, at $t \approx 0\ °C$ the permeability is about $\mu \approx 1.5 \div 4$ and falls rapidly with temperature increase (Figure 1).

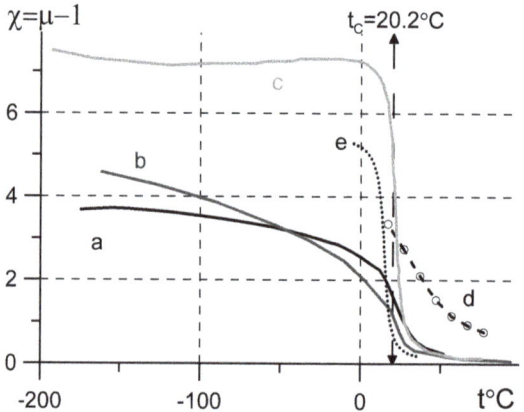

Figure 1. Temperature dependence of Gd susceptibility retrieved from the published magnetization data: curve a is retrieved for bias $H = 0.4$ T as mean value for two orientations of the monocrystalline sample [6]; curve b is retrieved for $H = 0.5$ T [7]; curve c is retrieved for $H = 0.02$ T [8]; curve d is retrieved for $H = 1.8$ T [11]. The dotted curve e shows the susceptibility of the Gd particle retrieved from susceptibility measurements of the composite under study (see Section 3).

The published data on high-frequency permeability of gadolinium were obtained in a wide temperature range with two techniques: the low-frequency impedance measurements of a bulk rod of gadolinium [13–15] and the ferromagnetic resonance (FMR) measurements of tiny Gd spheres at 9 GHz [16,17] or monocrystalls at 32 GHz [18]. The impedance measurements [13–15] performed in the frequency range below 100 kHz are very sensitive to the purity and structure of the metal surface, therefore the obtained results depend on sample treatment and differ significantly for various samples. Moreover, the calculations [13,14] neglect a slight conductivity change of Gd in the vicinity of Curie temperature [19] and result in permeability values ($\mu \approx 1400 + i300$ at 10 kHz and $-63\ °C$ [13]) about 1000 times higher than the estimates from magnetization curves (Figure 1).

The samples for FMR measurements are usually paraffin-bound composites filled below 1% vol of Gd [16,17]. The measurements result in the relative dependence of permeability on external bias; recalculation of the field-domain data into the permeability dependence on frequency using LLG formula [20] leads to rough estimations only.

The problem of permeability dependence on temperature for composites that are filled with particles of various shapes and sizes is that the Curie temperature of the particles is not a constant equal to 20.2 ÷ 21 °C [21], but there is a dispersion of the measured temperatures depending on many factors that are difficult to define (inhomogeneity of atomic structure [22], the random orientation of easy axis and domains in a polydomain sample [17], etc.). The most essential factors are accounted for by Equation (2), derived from analysis of the Brillouin function [23]:

$$T_{C\_meas}(N) \underset{H_{ext} \neq 0}{=} T_C \left(1 - \frac{J+1}{3J} \frac{\mu_B(M_0 N - H_{ext})}{kT_C} \right) \quad (2)$$

Here T_C is Curie temperature (°K), H_{ext} is the strength of external bias, J is the spin magnetic moment, μ_B is the effective Bohr magneton number, N is demagnetization factor (shape factor) defined by inclusion shape, $k = 1.380649 \times 10^{-23}$ J/K is the Boltzmann constant, M_0 is the saturation magnetization.

For zero bias ($H_{ext} = 0$) Equation (2) shows that the measured Curie temperature linearly decreases with the increase of demagnetization factor N and cannot exceed 293.2 °K (20.2 °C). The minimal Curie temperature is obtained for flakes or films perpendicular to the external field (with $N \approx 1$); the maximal temperature is obtained for fibres or films parallel to the external field (with $N \approx 0$). The maximal difference of the measured Curie temperature from 20.2 °C due to particle shape for $J = 7/2$, $\mu_B = 7.55$ and $M_{0Gd} = 2.42 \cdot \times 10^6$ A/m is about 5°.

The values of T_C, M_0, μ_{Bohr} and even J for Gd in Equation (1) are not known as accurately as Zverev [23] has taken for the above estimate, e.g., the effective Bohr magneton number in for Gd varies according to published data from 7.55 to 7.98 [19,24]. The saturation magnetization M_0 is a distributed value because of spin-waves, defects of crystalline structure, random orientation easy axis and domain magnetization [19,21,22]. As result the reported Curie temperature falls out of the 5 °K range defined in [23] (e.g., the measured Curie temperature for Gd films is about 44 °C [25]) and the dependence of low-frequency permeability of Gd on temperature is neither a step-like drop, nor a linear decrease according to Equation (1) to $\chi_0 = 0$ at $T = T_C$, but looks like a flattened distribution curve with the maximal density close to T_C values. The similarity is confirmed by the dependence of coil inductance as a function of temperature, where the coil core is a dispersion of Gd-filings in oil [26].

Neither the static permeability of Gd, nor the permeability dependence on frequency have been studied. Therefore, the research aims are to synthesize practicable amounts of air stable composite with Gd powder (the synthesis a challenge itself as the fine powder is pyroforic in air); to study the dependence of constitutive parameters of the composite on frequency and temperature (the effect of temperature is here an additional factor to refine mixing rules) and to estimate performance of the composite as a microwave screen with temperature-tunable attenuation.

2. Synthesis of Gd Powder and Preparation of Composite Samples

Commercial Gd of 99.6% purity is available in shape of 3–4 kg ingots. Synthesis of a composite filled with micrometer-size powder of Gd is a chemical challenge itself, as Gd powders are pyrophoric in air. The developed technique, in contrast to the vacuum vaporisation technique [16,17], is aimed to produce amounts of micrometer-size gadolinium powder sufficient for preparation of samples for free-space microwave measurements (up to about 50 g of powder per cycle). Similar to the published procedure [27], the technique is based on the mechanochemical treatment of bulk gadolinium.

The ingot is mechanically lathed into shavings of about 0.5 mm thickness. The shavings are grinded for an hour in a Fritsch P7 planetary mill with 12 mm steel balls in a steel mortar, and KCl is added into the mortar as an abrasive agent. To prevent ignition of Gd particles, the grinding is performed in argon media. The grinding product is washed with an alcohol-water mixture to dissolve KCl, then dried with acetone and stored in petroleum-ether to prevent oxidation. The synthesized powder slowly oxidizes in air, but readily self-ignites on kicking or rubbing.

The powder microstructure, morphology and impurity distribution is analyzed with a scanning electronic microscope (SEM) VEGA 3 LMN (TESCAN) equipped with an energy-dispersive X-ray spectroscopy (EDS) system. SEM images are shown in Figure 2; EDS shows traces of surface oxidation and carbidisation of obtained powder. The images show that the powder consists of particles of various shapes and sizes, the small particles have an approximate uniform size of about 0.1 mcm along three axes, while the larger particles are close to flakes of about 2 mcm diameter. The flakes are partially agglomerated into clusters up to 20 mcm in size.

Figure 2. SEM images of Gd powder.

X-ray phase analysis of the obtained powder is performed with a MiniFlex (Rigaku) X-ray diffractometer using Co Kα radiation. An X-ray diffraction powder pattern with main Gd and GdH$_2$ peaks indexed is shown in Figure 3, and the measured and reference data on peak position, intensity and corresponding Miller indices (hkl) are shown in Table 1. The analysis shows that the obtained powder is metal Gd with about 10% vol. additions of gadolinium hydride GdH$_2$. The increase of grinding time leads to the partial transformation of metal into GdH$_2$.

The paraffin-bound composite with 30% vol ($p_{incl} = 0.3$) of the above powder is prepared by mixing it in an ultrasonic bath heated to 100 °C. The volume fraction (the filling factor) p_{incl} of Gd in a composite is calculated accounting for a GdH$_2$ impurity of 10%. The composite is stable in air (there is neither a weight increase nor constitutive parameter change in the washer-shaped samples stored for more than a year at room conditions).

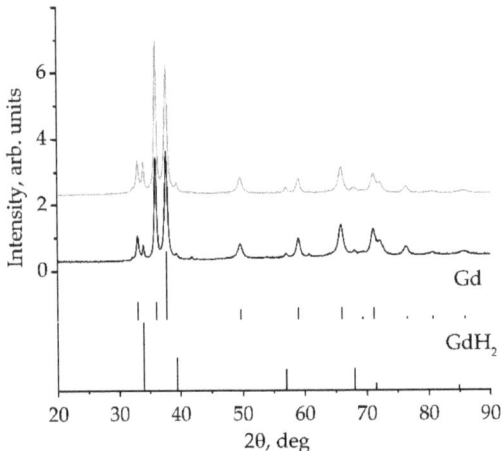

Figure 3. X-ray diffraction powder patterns of Gd powder (using Co Kα radiation) with main peaks indexed. X-axis presents the irradiation angle, Y-axis presents the pattern intensity. The grey top curve corresponds to 2 h grinding, the black bottom one corresponds to 1 h grinding. The lines below the curves mark the tabulated peak positions for Gd and GdH$_2$, respectively.

Table 1. The tabulated peak positions, intensities and Miller indexes for Gd and GdH$_2$.

1 h			2 h			Gd			GdH$_2$		
2θ$_{exp}$	d$_{exp}$	I$_{exp}$	2θ$_{exp}$	d$_{exp}$	I$_{exp}$	d	I	(hkl)	d	I	(hkl)
33.06	3.142	21	33.05	3.144	23	3.145	25	(100)			
34.00	3.059	20	34.00	3.059	14				3.062	100	(111)
36.04	2.891	100	36.00	2.894	94	2.889	25	(002)			
37.76	2.764	84	37.75	2.764	100	2.762	100	(101)			
39.42	2.652	6	39.35	2.656	5				2.652	47	(200)
49.68	2.129	10	49.65	2.130	13	2.127	14	(102)			
57.04	1.873	4	57.05	1.873	3				1.875	30	(220)
59.04	1.815	9	59.00	1.816	17	1.816	16	(110)			
65.94	1.643	17	65.90	1.644	28	1.642	16	(103)			
68.08	1.597	4	68.10	1.597	5				1.599	31	(311)
69.44	1.570	2	69.52	1.569	4	1.572	2	(200)			
71.16	1.537	13	71.15	1.537	24	1.537	16	(112)			

3. Measurement Techniques

The complex microwave permittivity ε and permeability μ of samples under study are determined with a vector network analyzer (Keysight N5224B) in standard 7 × 3 mm coaxial cells. At the frequency range 0.3–15 GHz, the measurements are performed in a coaxial airline applying the reflection-transmission technique. The thickness of washer-shaped samples is about 1.5–3 mm.

At the range 10–300 MHz, the measurements are performed in a cell with a detachable short applying short-circuit technique for permeability measurements and open-circuit techniques for permittivity measurements. The sample thickness is about 10–15 mm, and the washers are additionally pressed inside the cell to reduce the air gaps formed while the sample is slid into the cell.

The above supplementary techniques are applied to improve the accuracy of low-frequency data and to decrease the effect of air gaps on the measured permittivity. The static ε and μ are determined as median values in the frequency range of 10–200 MHz where the loss is negligible.

The transmission cell is equipped with an external ring-shaped cooling jacket; the cooling is performed by a flow of nitrogen vaporized from a Dewar vessel. The obtained maximal cooling is about $-190\,°C$; the heating above the room temperature is performed with a flow of hot air.

The cell with a detachable short is cooled by immersing it in a glass with coolant (ice-alcohol mixture), and the S11 measurements are performed while the cell is naturally warmed up to the room temperature, then the coolant is replaced with warm (about 35 °C) water and the measurements are performed with natural cooling down to room temperature.

The temperature is measured with type K thermocouples soldered to the cells' external surface close to the measured sample. The measurement accuracy is about $\pm 0.5°$ for short-circuit-open-circuit measurements at 0 °C; the uncertainty decreases with the increase to the room temperature. The temperature uncertainty for reflection-transmission measurements is about $\pm 2°$ in the temperature range from -10 to $+30\,°C$ and about $\pm 5°$ at the temperatures below $-20\,°C$.

4. Measurement Results and Fitting of the Obtained Data

To shorten the formulae and to simplify calculations, the data treatment is performed in terms of magnetic susceptibility $\chi = \mu - 1$; the temperature data are presented in Celsius scale.

The measured dependence of complex susceptibility on frequency and temperature for the paraffin-based composite filled with 30% vol of Gd powder (p_{incl} = 0.3) is presented in Figure 4.

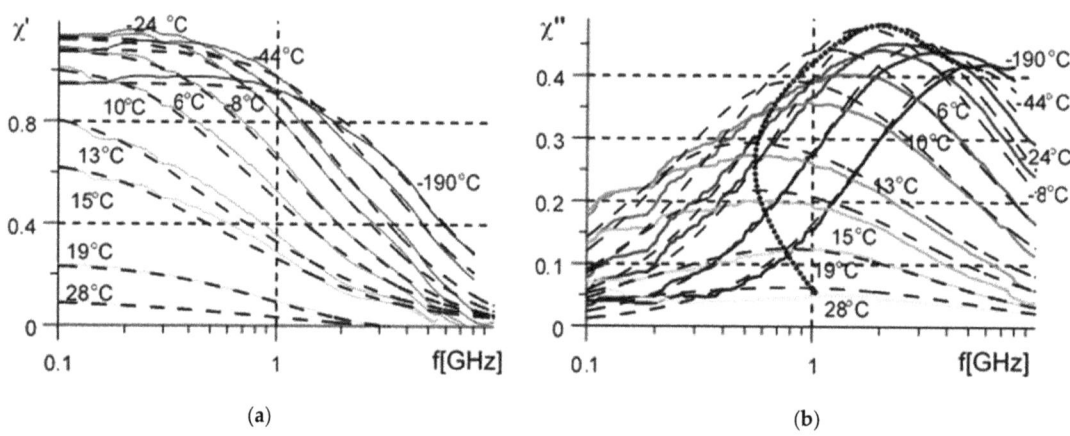

Figure 4. Dependence (**a**) of the real χ' and (**b**) imaginary χ'' susceptibility parts on frequency measured at temperatures $-190, -44, -24, -8, 0, +6, 10, 13, 15, 19, 28\,°C$, the higher the temperature, the lighter the line. The thin dashed lines present the approximations with Equation (3). The temperatures close to Curie temperature (10–20 °C) are marked close to the corresponding susceptibility curves, and the dotted line indicates χ'' peak position.

The composite permittivity does not vary with frequency and temperature within the measurement accuracy and is equal to $\varepsilon \approx 15$. A rough estimation of inclusion shape factor with the Maxwell Garnett mixing model with account for $\varepsilon_{binder} \approx 2.2$ results in $N_{incl} \approx \varepsilon_{binder} \times p_{incl}/(1 - p_{incl})/(\varepsilon - \varepsilon_{binder}) \approx 0.07$; the appropriate model can be defined studying the samples with a wide range of filling factors only.

The dependence of complex permeability on frequency is fitted with Havriliak-Negami formula [28] generalized with account for the resonance term f/F_{rez} Equation (3). The advantage of Equation (3) is that it describes a variety of lineshapes with five fitting parameters only.

$$\chi_0(f) = \chi_0 \left(1 - \left(\frac{f}{F_{rez}}\right)^{2\delta} + i\left(\frac{f}{F_{rel}}\right)^{\delta}\right)^{-\alpha} = \chi_0 \left(1 - \left(\frac{f}{F}\right)^{2\delta} + i\left(\Gamma\frac{f}{F}\right)^{\delta}\right)^{-\alpha} \quad (3)$$

Here parameters α and δ characterize spectrum width and symmetry correspondingly (the parameter limitations are: $\alpha > 0$, $\delta > 0$, $\alpha \times \delta \leq 2$), F_{rez} is resonance frequency, F_{rel} is relaxation frequency and χ_0 is static susceptibility. The relaxation and resonance frequencies are related by the damping factor: $F_{rel} = F_{rez}/\Gamma$. For $\alpha = \delta = 1$ and $F_{rel} \gg F_{rez}$ Equation (3) describes the Lorentzian dispersion, and for $F_{rel} \ll F_{rez}$, Equation (3) describes Debye dispersion. Parameters α and δ are included into the fitting together with the Lorentzian resonance parameter F_{rez} to account at least partially for distribution of cluster shapes and therefore to increase the approximation accuracy for the measured data (Figure 4).

The shape distribution may be described accurately with a Bergman-Milton geometrical spectral function [10], but its determination is a separate task falling out of the research scope, while the accuracy of fitting with Equation (3) is enough to estimate the effect of temperature on reflection and transmission spectra of a microwave screen.

The measured dependence of static susceptibility on temperature is shown with circles in Figure 5. The real part of microwave susceptibility (Figure 4a) extrapolated to zero frequency is shown with crosses. The difference may be attributed to the higher measurement error with electrically thin samples compared to that of short-circuit measurements with much thicker samples.

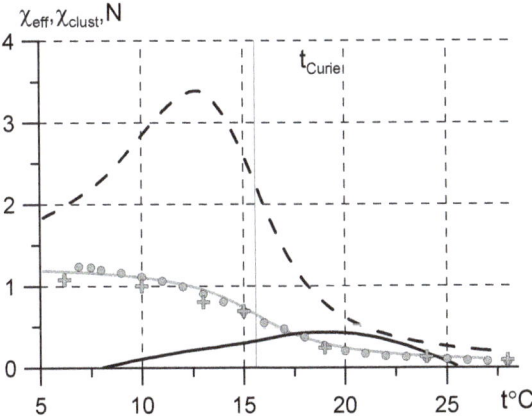

Figure 5. Temperature dependence of the composite static susceptibility (grey line), of cluster shape factor (solid black line), and of intrinsic susceptibility of cluster (dashed black line). The dots and crosses present the measured static susceptibility of the composite and the real part of microwave susceptibility (Figure 4a) extrapolated to zero frequency correspondingly; the lines present simulation data.

The dependence of static susceptibility on temperature is fitted (the grey line in Figure 5) with a Cauchy cumulative distribution function (CDF, Equation (4)); fitting with other distributions with two free parameters result in much higher inconvergence with the measured data.

$$\chi_{eff}(\tau) = \mu(\tau) - 1 = w \times CDF[(-\tau_{mean}, \sigma), -\tau] = \frac{w}{2} + \frac{w}{\pi}\arctan\left[\frac{\tau - \tau_{mean}}{\sigma}\right] \quad (4)$$

Here σ is the standard deviation related to the width of phase transition zone, τ is the distribution argument functionally related to the measurement temperature t, but $t \neq \tau$; $w = \chi_{0°C}$ is the scaling factor equal in case under study to susceptibility measured at $t \approx 0\ °C$.

An extrapolation with Maxwell Garnett formula of the composite ($p_{incl} = 0.3$) susceptibility onto susceptibility of the gadolinium inclusion $\chi_{Gd} = \chi_{composite}/[p - N_{incl}\chi_{composite}(1-p)]$ with account for the inclusion shape-factor determined from permittivity measurements ($N_{incl} \approx 0.07$) results in the dotted curve in Figure 1. The extrapolation is performed with a simplest mixing model unverified for the composite under study, the model does not account for effects related to structure changes in the composite with increase of filling factor (percolation, inversion of matrix structure, etc. [10]) and for the possible difference in magnetization of bulk metal and that of micrometer size grinded particles. Nevertheless, the estimated susceptibility appears to be close to the data retrieved from the published magnetization curves (Figure 1).

5. Cluster Magnetization Model

The fitting of the measured susceptibility with a continuous distribution function is selected in accordance with a cluster magnetization model valid in the vicinity of the Curie temperature [29,30]. The model assumes that magnetic clusters, i.e., assemblies of a multitude of interacting spin centres which are magnetically decoupled from their environment, exist even in a bulk magnet. The number, shape, and intrinsic magnetization of clusters depend on bias and temperature of the magnet relative to its Curie temperature. In developing the cluster model in terms of mixing models, we have to relate the argument of Cauchy distribution $-\infty < \tau < +\infty$ (Equation (4)) to the measurement temperature $0 < t < \infty$ or the volume fraction of magnetized clusters $0 < p < 1$, to the shape factor $0 < N < 1$ and to intrinsic susceptibility $0 < \chi < \infty$ of the clusters.

In case of a composite, the effective volume fraction of permeable clusters depends on temperature $p(t)$ and may be equal to the volume fraction of metal inclusions p_{incl} at temperatures much lower than Curie temperature only. The cluster shape and demagnetizing factor differ from that of metal inclusion $N_{cl} \neq N_{incl}$ as close to Curie temperature according to the cluster model the inclusion is magnetized partially (the clusters are magnetically decoupled from an impermeable metal matrix).

In order to relate the distribution argument τ to the measurement temperature t or to the effective volume fraction p, it is necessary to renormalize the function argument and to rewrite the distribution function (Equation (4)) correspondingly:

$$\chi_{eff}(t) = \frac{w}{2} + \frac{w}{\pi}\arctan\left[-\frac{1}{\sigma}Log\left[\frac{t+t_0}{t_C+t_0}\right]\right] \quad (5)$$

Here σ is the standard deviation, t is the measurement temperature, $t_0 = 273°$; the distribution mean is reached at the Curie point $t = t_C$. In the composite under study $t_C = 15.5\ °C$ (see Figure 5), the value is about $5°$ lower than the Curie temperature reported for bulk gadolinium [21,23].

The effective volume fraction of clusters p and the measurement temperature t are interchangeable parameters of the distribution function argument τ in Equation (4). The argument τ is the same for both p and t, $\tau = 0$ corresponds to $p = 0.5$ or $t = t_C$; τ, p and t parameters are related:

$$\tau = Log\left[\frac{p}{1-p}\right] = -Log\left[\frac{t+t_0}{t_C+t_0}\right] \quad (6)$$

50

Therefore Equation (4) may be written as well in terms of volume fraction common for mixing models:

$$\chi(p) = \frac{w}{2} + \frac{w}{\pi}\arctan\left[\frac{1}{\sigma}Log\left[\frac{p}{1-p}\right]\right] \quad (7)$$

Here σ is the same standard deviation as in Equation (4), parameter p is the effective volume fraction of permeable clusters and the maximum distribution density is reached for $p = 0.5$.

Note that the relation (Equation (6)) between the argument τ of Cauchy distribution (Equation (4)) and temperature t (Equation (5)) or effective volume fraction of magnetized clusters p (Equation (7)) is not the only one; e.g., it is possible to cut off the distribution "tails" which are far from t_C and to reduce the cumulative distribution so that the change of χ_{eff} will occupy the whole $5\,°C < t < 25\,°C$ range or the whole filling range $0 < p < 1$. The term "effective volume fraction" is used here to emphasize that p depends on the selected functional relation between the temperature t and the argument τ of the distribution function (Equation (4)). It is a separate task to determine the real volume fraction of magnetized clusters in metal; the effective volume fraction p is introduced here only to relate the cluster magnetization model to mixing models and to illustrate the transition from p to temperature t (Equations (8) and (9)).

Summing all magnetic moments in the sample it is possible to calculate [9] its effective susceptibility as:

$$\chi_{eff}(p_1) = \int_0^{p_1} \frac{\chi(p)\partial p}{1 + N_{cl}(p)\chi(p)} = \frac{p \times \chi(p)}{1 + N_{cl}(p)\chi(p)} \quad (8)$$

The integral part of Equation (8) is as a matter of fact the Wiener mixing model with an accounting for inclusion shape and the infinite number of mixture components [10]. The integral equation may be solved analytically (the right side of Equation (8)) applying the integral mean theorems. The values $\chi(p)$ and $N_{cl}(p)$ are here the mean values of intrinsic susceptibility and shape factor for permeable clusters in range of cluster volume fractions $0 < p < p_1$.

Equation (8) is readily written with an account for Equation (6) in terms of measurement temperature t instead of the effective volume fraction p:

$$\chi_{eff}(t_1) = \frac{\chi(t)}{[1 + N_{cl}(t)\chi(t)]} \times \frac{t_0 + t_C}{2t_0 + t + t_C} \quad (9)$$

Similarly to Equation (8) the $\chi(t)$ and $N_{cl}(t)$ are here the mean values of intrinsic susceptibility and shape factor for magnetized clusters in temperature range $t_0 < t < t_1$, χ_{eff} is the sample susceptibility measured at $t = t_1$ or $p = p_1$, t_C is Curie temperature

Assuming that the magnetic spectra (Figure 4) are symmetric ($\alpha \approx 1$) it is possible to simplify Equation (3) and to write it with account of Equation (9) in terms of static susceptibility as function of temperature:

$$\chi_{eff}(f,t) = \frac{\chi_{eff}(0,t)}{1 - \left(\frac{f}{F(1+N_{cl}(t)\chi_{cl}(0,t))^{1/2\delta}}\right)^{2\delta} + i\left(\frac{\Gamma}{(1+N_{cl}(t)\chi_{cl}(0,t))^{1/2\delta}}\frac{f}{F(1+N_{cl}(t)\chi_{cl}(0,t))^{1/2\delta}}\right)^{\delta}} \quad (10)$$

Taking into account that the effective filling factor p is inversely proportional to the measurement temperature t (Equation (6)), the measured resonance frequency and damping factor of gadolinium particle are related to the intrinsic resonance frequency and damping factor of a magnetic cluster as

$$F_{eff}(t) = F_{cl}[1 + N_{cl}(t)\chi_{cl}(0,t)]^{-1/2\delta} \text{ and } \Gamma_{eff}(t) = \Gamma_{cl} \times (1 + N_{cl}(t)\chi_{cl}(0,t))^{1/2\delta}$$

The fitting of the measured dependence of complex susceptibility on frequency and temperature (the dashed lines in Figure 4) shows that the damping factor $\Gamma_{eff} \approx 4 \div 5$ is too high to determine the resonance frequency F_{eff} accurately (the spectrum is close to Cole-Cole relaxation and F_{eff} exceeds the upper limit of the measurement frequency band), whereas the frequency of the loss peak (the relaxation frequency $F_{rel} = F_{rez}/\Gamma$) is determined readily (the dotted line in Figure 4), therefore it is possible to write the peak frequency in terms of cluster susceptibility and shape factor

$$F_{rel\_eff}(t) = \frac{F_{cl}}{\Gamma_{cl}[1 + N_{cl}(t)\chi_{cl}(0,t)]^{1/\delta}} = \frac{F_{rel\_cl}}{[1 + N_{cl}(t)\chi_{cl}(0,t)]^{1/\delta}} \quad (11)$$

Equation (11) shows that the increase of the product $N_{cl}(t) \times \chi_{cl}(t)$ results in the low-frequency shift of the loss peak compared to the peak frequency of magnetized clusters and in the increase of the damping factor. Note that the shift is the opposite of the one observed in composites: the higher the product $N_{incl}(p) \times \chi_{incl}(p)$, the higher the peak frequency of a composite compared to the inclusion peak frequency [10].

The joint solution of Equations (5), (9) and (11) makes it possible to retrieve the dependence of static intrinsic susceptibility $\chi_{cl}(0,t)$ and shape factor $N_{cl}(t)$ of a mean cluster on temperature. To obtain the continuous lines the discrete data relating the loss peak frequency and temperature presented in Figure 4 are interpolated polynomially. The results are presented in Figure 5 with solid and dashed black lines, respectively.

Theoretically, these very results could be obtained from the microwave measurements (Equation (10)) only; the problem is in the higher temperature error associated with the reflection-transmission measurements compared to the low-frequency short-circuit measurements and in the lower accuracy for susceptibility measurements of electrically thin samples (compare the data in Figures 4 and 5).

The dependence of a cluster shape factor on temperature ($0 \leq N_{cl}(t) \leq 0.5$, Figure 5) shows that the clusters are fiber-like and that close to the Curie temperature their orientation becomes unrelated to the magnetizing field because of the temperature oscillations of the crystal lattice. The volume fraction, intrinsic susceptibility and demagnetization factor of clusters decrease rapidly with temperature growth above the Curie point.

The dependence of cluster susceptibility on temperature reveals that the Hopkinson effect [31] (susceptibility peak close to Curie temperature presented by the dashed line in Figure 5) is unobserved in the direct susceptibility measurements (the solid grey line in Figure 5) because of the formation of variety of magnetic clusters (Equation (8)). The shift of the Hopkinson peak and of the shape-factor maximum from the Curie point as well as the shape-factor value slightly below zero at the ends of the temperature range (Figure 5) may be explained by the inaccuracy of microwave spectrum measurements and by the rough approximation of the spectra with the Cole-Cole model (approximate account for the peak frequency with Equation (11) instead of a more accurate account for the resonance frequency, damping and asymmetry factors with Equation (3)).

Preliminary measurements of susceptibility at temperatures below $0\,°C$ show (Figure 4) that the effects unrelated to clusterisation (the freezing of domain walls, the increase of anisotropy field, the transformation of easy axis magnetic anisotropy to the easy plane one [19,32], etc.) dominate there. The effects lead to a decrease of susceptibility with temperature decreases below ~$0\,°C$. This is the reason that the scaling factor w in Equations (4)–(7) is assumed to be equal to the susceptibility measured at $t \approx 0\,°C$. The above effects fall out of the cluster magnetization model (Equations (4)–(11)); their applicability for a tunable screen looks doubtful, therefore the measurements below $0\,°C$ are not analysed here.

6. Performance Estimation for a Tunable Microwave Screen

The performance of a microwave screen depends on the Fresnell equations (Equations (14) and (15)) on constitutive parameters $\varepsilon \approx 15$ and μ of the composite, on the screen thickness d, frequency f and load impedance z_{load}. The permeability of the composite $\mu = \chi_{eff} + 1$ is calculated as the function of temperature t and frequency f

using Equation (10) with the parameters of magnetic spectrum (Equation (3)) measured at $t = 0\,°C$ (Figure 4); the temperature dependence of resonance frequency F and damping factor Γ is calculated (Equation (10)) using the cluster susceptibility $\chi_{cl}(t)$ and shape-factor $N_{cl}(t)$ determined from the quasistatic measurements (Figure 5). The width and symmetry parameters are practically constant ($\alpha \approx 1$, $\delta \approx 0.7$) within the range $5° < t < 20°$.

The absorber performance $R(d,f,t)$ is calculated for the composite screen on a metal substrate ($z_{load} = 0$), and the transmission performance $T(d,f,t)$ is calculated for the twice as thick screen in a free space ($z_{load} = 1$):

$$R(d,f,t) = \frac{\sqrt{\mu(f,t)/\varepsilon} \times \text{Tanh}\left[\frac{2\pi i f}{c} d\sqrt{\varepsilon\mu(f,t)}\right] - 1}{\sqrt{\mu(f,t)/\varepsilon} \times \text{Tanh}\left[\frac{2\pi i f}{c} d\sqrt{\varepsilon\mu(f,t)}\right] + 1} \quad (12)$$

$$T(d,f,t) = \frac{Exp\left[\frac{2\pi i f}{c} d\left(1 - \sqrt{\varepsilon\mu(f,t)}\right)\right] \times \left[1 - \left(\frac{\sqrt{\mu(f,t)/\varepsilon}-1}{\sqrt{\mu(f,t)/\varepsilon}+1}\right)^2\right]}{1 - \left[Exp\left[-\frac{2\pi i f}{c} d\sqrt{\varepsilon\mu(f,t)}\right] \times \frac{\sqrt{\mu(f,t)/\varepsilon}-1}{\sqrt{\mu(f,t)/\varepsilon}+1}\right]^2} \quad (13)$$

c in the above equations is the light velocity $c = 3 \times 10^8$ m/s.

The dependence of calculated reflection coefficient (in logarithmic scale) on frequency and screen thickness is presented in Figure 6 as contour maps, the darker the filling, the lower the reflectivity. The maps are calculated for the temperature below and above the Curie temperature (0 °C and 20 °C, respectively).

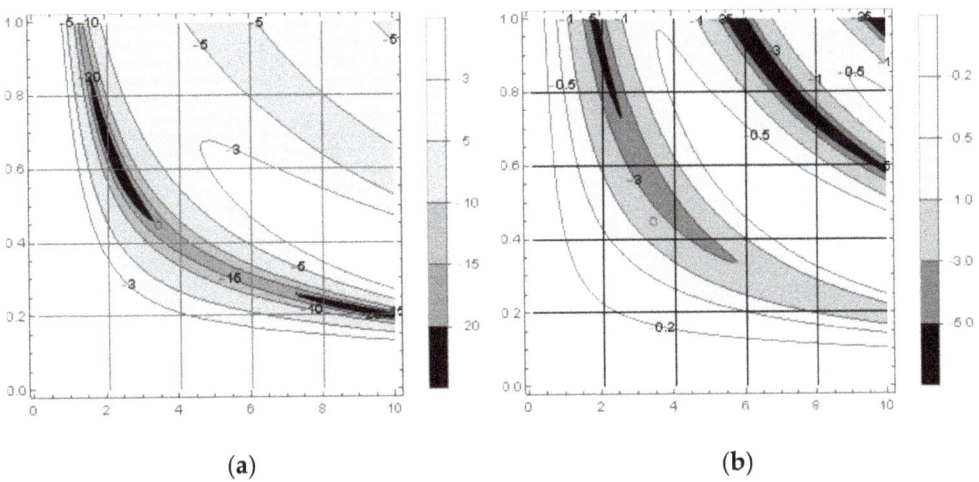

(a)   (b)

Figure 6. Contour maps for reflection coefficient (dB) of a shorted layer as a function of frequency and layer thickness (x and y axis correspondingly) calculated for 0 °C (a) and 20 °C (b).

The reflectivity control range of the composite under study is close to that of the composites with permeable microwires [4], while the transparency tuneability is lower (Figure 7a). The reason is that the temperature increase causes the simultaneous decrease of both real and imaginary permeability parts. The layer loss $Exp\left[\frac{2\pi i f}{c} d \times \text{Im}\left[\sqrt{\varepsilon\mu(f,t)}\right]\right]$ decreases with temperature increase, but the layer specific impedance $\text{Re}\left[\sqrt{\mu/\varepsilon}\right]$ decreases as well, thus increasing the reflection coefficient from the layer in a free space and decreasing the total effect on the transmission coefficient. Therefore the transparency variation range due to permeability control is in principle lower than the range due to permittivity control [4].

According to Figure 6, the minimal reflectivity of about −20 dB is achieved at 0 °C for an approximately 4.5 mm-thick layer at the frequency of about 3.5 GHz (the position is marked with the circle in the contour maps); at 20 °C the layer reflectivity is about −2 dB.

The calculated at 3.5 GHz frequency tuneability is shown in Figure 7 (the left graph) as the dependence of reflection (for 4.5 mm-thick layer) and transmission (for 9 mm-thick layer) coefficients on temperature. The right graph in Figure 7 shows the reflectivity curves of 4.5 mm-thick sample measured in the shorted coaxial cell in the frequency range of 2–7 GHz at 5 °C and 25 °C.

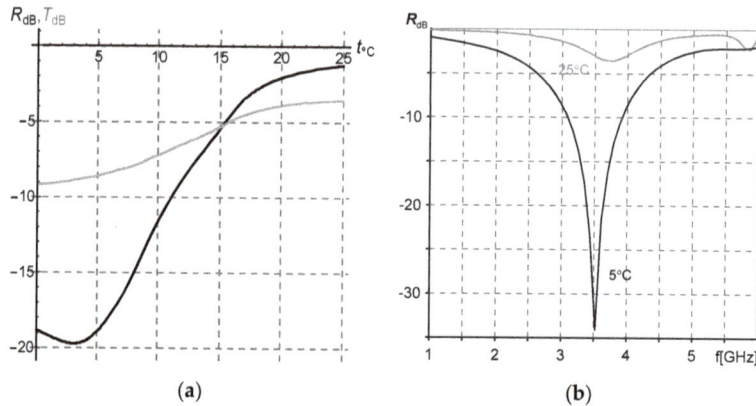

Figure 7. (a) Simulated dependence of reflection and transmission coefficients on temperature at f = 3.5 GHz. (b) Frequency dependence of reflection coefficient for shorted sample measured at 5 °C and 25 °C.

The calculated dependence of reflectivity on frequency and temperature is in agreement with the results of direct measurements (compare the maps in Figure 6 and the curves in Figure 7).

The composite under study may be applied as well as a tunable shield for electromagnetic interference (EMI) suppression. The shield efficiency is proportional to transmission attenuation [33] (see the grey curve in Figure 7a).

7. Conclusions

The mechanochemical technique to synthesize a composite filled with micrometer size gadolinium powder that is stable at room conditions is proposed.

The estimated dependence of the powder permeability on temperature is close to the data retrieved from the published magnetization curves for bulk gadolinium. The measured Curie temperature of the composite is about 5 °C lower that the reported temperature for the bulk metal; the difference may be related to the length limitation of the magnetized cluster by the particle size.

The composite quasistatic permeability decreases with temperature increase gradually; the main changes take place in the transition range about ±10° from the Curie point, and this range is much wider than the published theoretical estimation (Equation (2)). The effect is interpreted in terms of a cluster magnetization model (Equations (9)–(11)) developed from the Wiener mixing formula. The model explains the dependence of absorption peak frequency and intensity on temperature (Figure 4), and shows that the Hopkinson effect in Gd is masked by cluster magnetization; the effect reveals itself as the peak of intrinsic permeability (Figure 5) of magnetized clusters in the vicinity of the Curie temperature.

The treatment of the measured composite susceptibility in the frame of the cluster model shows that the cluster shape factor reaches the maximum in the vicinity of the Curie temperature; the maximal value is close to 0.5 (Figure 5), which corresponds to fiber-shaped clusters oriented perpendicular to the magnetic field; at the high- and low-temperature ends of the Curie transition range the shape factor is close to zero, which corresponds to the clusters oriented parallel to the magnetic field.

The model makes it possible to calculate the dependence of a microwave permeability spectrum on temperature and to estimate the performance of the composite layer as a microwave screen or an EMI suppressor shield with temperature tuned attenuation. The reflectivity measurements of a shorted sample are close to the simulated data (Figure 7).

Author Contributions: Conceptualization, methodology and writing S.N.S.; Investigation D.A.P., A.O.S. and S.F.L.; resources, review & editing K.N.R. All authors have read and agreed to the published version of the manuscript.

Funding: This research received no external funding.

Institutional Review Board Statement: Not applicable.

Informed Consent Statement: Not applicable.

Data Availability Statement: Not applicable.

Acknowledgments: Support for Russian Science Foundation (RSF) Grant No 21-19-00138 is acknowledged.

Conflicts of Interest: The authors declare no conflict of interest.

References

1. Kazemi, N.; Schofield, K.; Musilek, P. A high-resolution reflective microwave planar sensor for sensing of vanadium electrolyte. *Sensors* **2021**, *21*, 3759. [CrossRef]
2. Shahzad, W.; Hu, W.; Ali, Q.; Raza, H.; Abbas, S.M.; Ligthart, L.P. A Low-cost metamaterial sensor based on DS-CSRR for material characterization applications. *Sensors* **2022**, *22*, 2000. [CrossRef]
3. Tagantsev, A.K.; Sherman, V.O.; Astafiev, K.F.; Venkatesh, J.; Setter, N. Ferroelectric materials for microwave tunable applications. *J. Electroceramics* **2003**, *11*, 5–66. [CrossRef]
4. Starostenko, S.N.; Rozanov, K.N. Microwave screen with magnetically controlled attenuation. *Prog. Electromagn. Res.* **2009**, *99*, 405–426. [CrossRef]
5. Rozanov, K.N. Ultimate thickness to bandwidth ratio of radar absorbers. *IEEE Trans. Antennas Propag.* **2000**, *48*, 1230–1234. [CrossRef]
6. Dan'kov, S.Y.; Tishin, A.M.; Pecharsky, V.K.; Gschneidner, K.A., Jr. Magnetic phase transitions and the magnetothermal properties of gadolinium. *Phys. Rev. B* **1998**, *57*, 3478–3490. [CrossRef]
7. Egolf, P.; Rosensweig, R. Magnetic refrigeration at room temperature. *Ann. Telecommun.* **2003**, *58*, 45–50.
8. Quantum Design GmbH, Spectrum, Product & Application News, Magnetic Measurements on Gd and Dy. Available online: https://qd-europe.com/nl/en/news/product-application-news-spectrum/lanthanides-magnetic-measurements-on-gadolinium-and-dysprosium/ (accessed on 1 January 2020).
9. Landau, L.D.; Pitaevskii, L.P.; Lifshitz, E.M. *Electrodynamics of Continuous Media*; Elsevier Ltd.: Amsterdam, The Netherlands, 1984.
10. Starostenko, S.N.; Rozanov, K.N.; Lagar'kov, A.N. Electrical and magnetic properties of the binary heterogeneous mixture model. *Phys. Met. Metallogr.* **2021**, *122*, 323–344. [CrossRef]
11. Gottschall, T.; Kuz'min, M.D.; Skokov, K.P.; Skourski, Y.; Fries, M.; Gutfleisch, O.; Zavareh, M.G.; Schlagel, D.L.; Mudryk, Y.; Pecharsky, V.; et al. Magnetocaloric effect of gadolinium in high magnetic fields. *Phys. Rev. B* **2019**, *99*, 134429. [CrossRef]
12. Kaul, S.N.; Mathew, S.P. Anomalous resonant microwave absorption in nanocrystalline gadolinium. *Nanosci. Nanotechnol. Lett.* **2011**, *3*, 556–560. [CrossRef]
13. Fraga, G.L.F.; Pureur, P.; Cardoso, L.P. Impedance and initial magnetic permeability of gadolinium. *J. Appl. Phys.* **2010**, *107*, 053909. [CrossRef]
14. Khovailo, V.V.; Abe, T.; Takagi, T. Detection of weak-order phase transitions in ferromagnets by ac resistometry. *Condens. Matter* **2003**, *1*, 0307096. [CrossRef]
15. Coey, J.M.D.; Gallagher, K.; Skumryev, V. Alternating current susceptibility of a gadolinium crystal. *J. Appl. Phys.* **2000**, *87*, 7028–7030. [CrossRef]
16. Petinov, V.I.; Dremov, V. Magnetic resonance in Gd nanoparticles near the curie point. *Tech. Phys.* **2018**, *63*, 519–529. [CrossRef]
17. Petinov, V.I. Magnetism of Gd nanoparticles near Tc. *Tech. Phys.* **2017**, *62*, 882–889. [CrossRef]
18. Burgardt, P.; Seehra, M.S. Electron paramagnetic resonance in gadolinium near Tc. *Phys. Rev. B* **1977**, *16*, 1802–1807. [CrossRef]
19. Nigh, H.E.; Legvold, S.; Spedding, F.H. Magnetization and electrical resistivity of gadolinium single crystals. *Phys. Rev.* **1963**, *132*, 1092–1097. [CrossRef]

20. Cai, Y.; Chen, J.; Wang, C.; Xie, C. A second-order numerical method for Landau-Lifshitz-Gilbert equation with large damping parameters. *J. Comput. Phys.* **2022**, *451*, 110831, preprint. Available online: https://www.researchgate.net/publication/351476443 (accessed on 10 May 2021). [CrossRef]
21. Darnell, F.J.; Cloud, W.H. Magnetization of gadolinium near its curie temperature. *J. Appl. Phys.* **1964**, *35*, 935–937. [CrossRef]
22. Korolev, A.; Kurkin, M.I.; Sokolov, O.B.; Orlova, N.B. The 21/3 rule for magnetic susceptibility of gadolinium. *Phys. Solid State* **2010**, *52*, 561–567. [CrossRef]
23. Zverev, V.I. Magnetic and Thermomagnetic Properties of Gd, Tb and Ho in the Vicinity of Magnetic Phase Transitions. Ph.D. Thesis, Moscow State University, Moscow, Russia, 2012. (In Russian).
24. Ensign, B.; Choudhary, R.; Ucar, Y.; Paudyal, D. Electronic structure, magnetic properties, and exchange splitting of gadolinium intermetallics. *J. Magn. Magn. Mater.* **2020**, *509*, 166882. [CrossRef]
25. Bertelli, T.P.; Passamani, E.C.; Larica, C.; Nascimento, V.P.; Takeuchi, A.Y.; Pessoa, M.S. Ferromagnetic properties of fcc Gd thin films. *J. Appl. Phys.* **2015**, *117*, 203904. [CrossRef]
26. Gladun, A.D.; Igoshin, F.F.; Tsipenyuk, Y.M. Study of exchange interaction using magnetic properties of gadolinium. *Phys. High. Educ.* **2004**, *10*, 49–60.
27. Svalov, A.V.; Arkhipov, A.V.; Andreev, S.V.; Neznakhin, D.S.; Larrañaga, A.; Kurlyandskaya, G.V. Modified field dependence of the magnetocaloric effect in Gd powder obtained by ball milling. *Mater. Lett.* **2020**, *284*, 128921–128923. [CrossRef]
28. Causley, M.F.; Petropoulos, P.G. On the time-domain response of havriliak-negami dielectrics. *IEEE Trans. Antennas Propag.* **2013**, *61*, 3182–3189. [CrossRef]
29. Dos Santos, G.; Aparicio, R.; Linares, D.; Miranda, E.N.; Tranchida, J.; Pastor, G.M.; Bringa, E.M. Size and temperature dependent magnetization of iron nanoclusters. *Phys. Rev. B* **2020**, *102*, 184426. [CrossRef]
30. Furrer, A.; Waldmann, O. Magnetic cluster excitations. *Rev. Mod. Phys.* **2013**, *85*, 367–420. [CrossRef]
31. Sláma, J.; Ušáková, M.; Šoka, M.; Dosoudil, R.; Jančárik, V. Hopkinson effect in soft and hard magnetic ferrites. *Acta Phys. Pol. A* **2017**, *131*, 762–764. [CrossRef]
32. Queen, J.H. Saturation Moments of Gd Alloys. Ph.D. Thesis, Iowa State University, Ames, IA, USA, 1981.
33. Koledintseva, M.Y.; Rozanov, K.N.; Archambeault, B. Engineering of composite media for shields at microwave frequencies. In Proceedings of the IEEE International Symposium on Electromagnetic Compatibility, Chicago, IL, USA, 8–12 August 2005; Volume 1, pp. 169–174.

Article

Deposition of a SiO₂ Shell of Variable Thickness and Chemical Composition to Carbonyl Iron: Synthesis and Microwave Measurements

Arthur V. Dolmatov [1,2], Sergey S. Maklakov [1,*], Polina A. Zezyulina [1], Alexey V. Osipov [1], Dmitry A. Petrov [1], Andrey S. Naboko [1], Viktor I. Polozov [1,2], Sergey A. Maklakov [1], Sergey N. Starostenko [1] and Andrey N. Lagarkov [1]

[1] Institute for Theoretical and Applied Electromagnetics RAS, Izhorskaya St. 13, 125412 Moscow, Russia; dolmatov.av@phystech.edu (A.V.D.); zez-p@yandex.ru (P.A.Z.); avosipov@mail.ru (A.V.O.); dpetrov-itae@yandex.ru (D.A.P.); nas.webwork@gmail.com (A.S.N.); viktor.polozov@phystech.edu (V.I.P.); sergeymaklakov@yandex.ru (S.A.M.); snstar@mail.ru (S.N.S.); maklakov@itae.ru (A.N.L.)

[2] Moscow Institute of Physics and Technology, National Research University, 9 Institutskiy per., 141700 Dolgoprudny, Russia

* Correspondence: squirrel498@gmail.com

Abstract: Protective SiO₂ coating deposited to iron microparticles is highly demanded both for the chemical and magnetic performance of the latter. Hydrolysis of tetraethoxysilane is the crucial method for SiO₂ deposition from a solution. The capabilities of this technique have not been thoroughly studied yet. Here, two factors were tested to affect the chemical composition and the thickness of the SiO₂ shell. It was found that an increase in the hydrolysis reaction time thickened the SiO₂ shell from 100 to 200 nm. Moreover, a decrease in the acidity of the reaction mixture not only thickened the shell but also varied the chemical composition from $SiO_{3.0}$ to $SiO_{8.6}$. The thickness and composition of the dielectric layer were studied by scanning electron microscopy and energy-dispersive X-ray analysis. Microwave permeability and permittivity of the SiO₂-coated iron particles mixed with a paraffin wax matrix were measured by the coaxial line technique. An increase in thickness of the silica layer decreased the real quasi-static permittivity. The changes observed were shown to agree with the Maxwell Garnett effective medium theory. The new method developed to fine-tune the chemical properties of the protective SiO₂ shell may be helpful for new magnetic biosensor designs as it allows for biocompatibility adjustment.

Keywords: protective coating; soft magnetic powder; microwave permittivity; core–shell particles

1. Introduction

Iron powders are widely applied in power transformers, inductors, sensors [1], electromagnetic compatibility solutions [2,3], and materials designed to decrease electromagnetic pollution [4,5]. Commonly used is the carbonyl iron with spherical particles of 2 [5,6]–10 [4] micron in the mean diameter. Carbonyl iron possesses remarkable magnetic properties, although chemical stability and electromagnetic performance are still to be improved when embedded into composite materials. Surface modification of iron particles with a chemically inert non-conductive coating may solve these tasks. Inorganic and polymer coatings suit well: MnO_2, $BaTiO_3$, carbon, PMMA, polyaniline [4], parylene C [7], ZnO, Fe_3O_4 [8], etc. In addition, the SiO₂ is the most commonly used among others. Silica shell provides iron with oxidation resistance in the air [4,9] and corrosion resistance [6]. Modifying iron with these non-conductive shells prevents electric contact between particles and combines magnetic and dielectric losses in one material [8]. This may be used to fine-tune the electromagnetic properties of the latter [6,9,10].

Liquid-phase hydrolysis of silica precursors is frequently used due to its simplicity and effectiveness, although other techniques, even mechanical milling [11], also work. Recently publications on synthesizing iron powders coated with silica shells show the high importance of these studies. The primary requirement for the silica shell is uniformity. While simple tetraethyl orthosilicate (TEOS) hydrolysis in ammonia solution deposits uniformly coating to nano- and micro-particles of iron, modification of large, 200-μm-particles requires additional surface activation. Surface active agents were shown to successfully increase the uniformity of the SiO_2 coating (see references within [12]). Mechanical milling also improves the adhesion of the silica coating to coarse iron powders [12]. Another technique to increase the uniformity of the SiO_2 is the use of L-lysine instead of ammonia solution [4].

Despite all these studies, it is notable that the inherent properties of the silica shell, including chemical composition and dielectric constant, are rarely examined. However, it is known that the hydrolysis of organosilanes gives significantly different products depending on the acidity of the reaction mixture, which, in turn, is governed by ammonia concentration [13–16]. For example, two different final-stage thermal treatments may be applied for the Fe@SiO_2 drying. One is simply to keep the powder at 50–60 °C for a certain period [5,6,17]. The other is to anneal in an inert (N_2, [8]) or reducing atmosphere (H_2, [1]) at 500–800 °C for several hours. The latter decreases the oxygen content of the silica down to $SiO_{1.5}$ [8]. This is due to the volatilization of water that occurs at 100 °C and further thermal aging of the SiO_2 [8].

From pure silica gel studies, it is known that the real chemical composition affects the dielectric constant of the "SiO_2". Hydroxyl groups are those elements in the structure of the silica that are dealt with by polarization. There are at least two types of hydroxyls within the silica structure: intraglobular and surface hydroxyls (although the former may be further divided into subclasses, and some "free" hydroxyls are also distinguishable). The surface or "perturbed" hydroxyls are deemed to cause polarization via a constrained rotation; that is, a rotation from one hydrogen-bonded position to another [18]. It was also shown that annealing at a temperature up to 1000 °C gradually decreases the concentration of hydroxyl groups, increases density, decreases surface area, and decreases the dielectric constant of the silica. The maximum decrease is from 2.2 to 1.8, almost 20% of the initial value (see Table 2 from ref. [18]). The density of the silica was shown to influence the permittivity of the silica even in the GHz range [19].

These changes may not only be used to tune electromagnetic performance but to develop a desired biocompatibility of the Fe@SiO_2-based media. Particle size and surface charge of the silica are the key parameters in biocompatibility studies [20]. A method to optimize the surface charge of the carbonyl iron particles coated with the silica shell makes the powder a promising material for biomedical applications. The powder may be applied both in pure form or incorporated into some silicone matrix to form a magnetorheological elastomer [21]. The latter serves as a magnetic-field magnetostriction sensor (more precisely, a dual-mode magnetism/pressure sensor), which sensing performance is closely related to the matrix–polymer interactions that, in turn, are governed by surface modification of iron particles [21,22].

Here, the SiO_2 shell was deposited onto carbonyl iron of 3 μm mean diameter through hydrolysis of TEOS in water–ethanol solution. Duration of the hydrolysis reaction and ammonia concentration was studied to affect the thickness and chemical composition of the shell. The products were dried in the air under 60 °C for 6 h. Chemical composition was measured by EDX using scanning electron microscopy. Microwave measurements in the 0.5–15 GHz range were used to evaluate the permeability of the composite material comprised of the Fe@SiO_2 particles and paraffin wax matrix. The measured real permeability of the composite was shown to be in accordance with what was calculated following the Maxwell Garnett formula [23–25].

2. Materials and Methods

The deposition of the SiO_2 shell onto particles of carbonyl iron (CI) was carried out in a one-stage modified Stöber process. Pure carbonyl iron powder (≥97.0 mass.% Fe), tetraethyl orthosilicate (CAS 78-10-4, Aldrich №86578), and ammonia solution (25%, reagent grade) were used. First, 2 g of the metal powder was immersed in ethanol in a round-bottom 100-mL flask with a reflux condenser. The tetraethyl orthosilicate was added and ultrasonicated for 40 min with a power of 250 W and a frequency of 40 kHz. Then, ammonia was added to the mixture, and this was assigned as the start of the reaction.

Two different experiments were conducted. The first experiment studied the influence of the duration of the hydrolysis reaction on the properties of the product. The duration was 0.5, 1, 2, and 4 h. The ratio of volumes of TEOS and ammonia solution added to the reaction mixture, here and after [TEOS]/[$NH_3 \cdot H_2O$], was constant at 1. Another experiment studied the influence of concentrations of reagents on the properties of the product. The [TEOS]/[$NH_3 \cdot H_2O$] ratio was 0.75, 1, 1.5, and 4.5. The duration of the synthesis was 2 h in this experiment.

Whatever the process was, the product was separated by magnet-assisted decantation, rinsed multiple times in ethanol until the transparency of the liquid, and dried in air for 6 h at 60 °C.

Particle size and SiO_2 thickness were measured by scanning electron microscopy using the Zeiss Evo 50 VP microscope (Zeiss AG, Germany). Fifteen individual particles were studied and averaged to evaluate the thickness of the shell. Energy-dispersive X-ray (EDX) analysis (Si:O) was examined using the "Oxford instruments" platform (Oxford instruments, UK). A sample area of 15 × 15 µm was studied.

The composites for microwave measurements were made of Fe@SiO_2 particles mixed with molten paraffin wax with constant stirring during cooling [6]. The volume fraction of the Fe@SiO_2 powder was estimated at 35 vol.%. The volume fraction was calculated as follows. Initially, 30 mg of the filler was mixed with 100 mg of the wax. The shell thickness was set at 150 nm, since the real size distribution of iron particles and the presence of pure silica particles interferes with more precise calculations. The tabular density of iron was 7.8 g/cm^3, and the density for amorphous silica was 2.2 g/cm^3. The samples were formed in the torus and placed inside a standard 7/3 mm coaxial transmission line (Figure 1). Additionally, the sample inside the line was pressed slightly from both ends to force the composite material to fill the cross-section of the line and reduce possible air gaps between the sample and the metallic surface of the line.

Figure 1. Coaxial transmission line (on top) with a central conductor (in the middle) and a composite sample (at the bottom) for microwave measurements.

S-parameters of the composite samples placed in the airline were measured in the frequency range of 0.5 to 15 GHz with a vector network analyzer (VNA). Ports at the end of feeding coaxial cables were calibrated with standard TRL calibration procedure [26] with planes of phase reference at the ends of the measurement transmission line. The complex microwave permeability and permittivity were determined with the standard Nicolson–Ross–Weir (NRW) [27,28] method.

The quasi-static permittivity of the composites was determined at a frequency of 500 MHz. The frequency dependence of complex permittivity was fitted with the Havriliak–Negami empirical formula [29] to minimize possible errors. The measured data were fitted in the frequency range of 0.5 to 10 GHz to minimize low-frequency errors due to poor sensitivity of the microwave measurements at frequencies lower than 0.5 GHz. At frequencies higher than 10 GHz, the half-wavelength resonance on the sample thickness starts to affect, and NRW solution becomes inherently unstable.

Higher-order modes can also emerge on the sample boundary with typical resonance-like behavior of the calculated microwave permittivity and permeability [30,31]. Such resonance behavior cannot be attributed to the material properties and should be carefully considered. In our case, the sample length was chosen to force the possible emergence of higher-order modes beyond the frequency range of interest. However, for some samples, the effect of higher-order modes resonance on the sample length can be seen at frequencies higher than 10 GHz. The manifestation of this effect depends on the sample homogeneity and quality of the geometric shape and cannot be easily controlled.

The lower bound of the frequency region under study was chosen due to the following considerations. In the frequency range from 10 MHz to 500 MHz, TRL calibration performance starts to degrade due to the small phase difference in the transmission line calibration standard and direct-thru connection of the measurement ports. In addition, samples with small electric length, i.e., small permittivity, permeability, and geometric length, have low reflection coefficient in this frequency region (i.e., low contrast); thus, the NRW method performance also degrades.

3. Results and Discussion

3.1. Structure and Morphology

The increase in the duration of the deposition process from 0.5 to 4 h increased the thickness of the silica layer (Figures 2–4, Table 1). The minimum thickness obtained was 90 nm, and the maximum was 190 nm (Figure 5). Further prolongation of the process did not increase thickness. The distribution of the shell thickness in a sample narrowed with the reaction time (Table 1). Uniformity and final particle size of the silica are known to be dealt with by colloid interaction potentials. The reaction rate here is size-dependent and is governed by the competition between nucleation and aggregation [32]. Growth of the shell thicker than 200 nm was probably inhibited by the colloidal surface state of the product. The shell thickness evaluated from one-hour-deposited samples differed slightly from the monotonous trend from other data. This was probably a random error caused by a wide distribution of the shell thicknesses that were deposited with a duration of 1–1.5 h. No evidence was observed that the thickness of the shell might decrease with the reaction time in this synthesis.

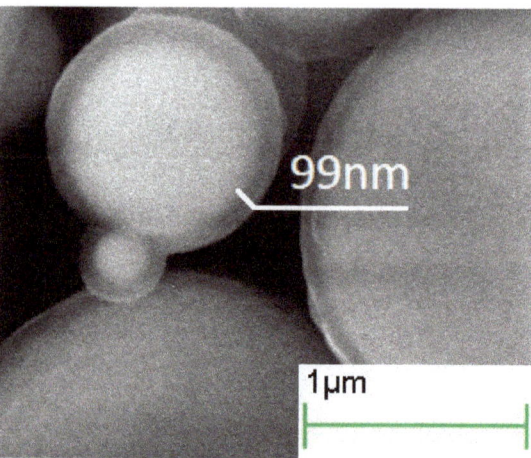

Figure 2. SEM images of core–shell particles obtained in the processes with 0.5 h hydrolysis duration.

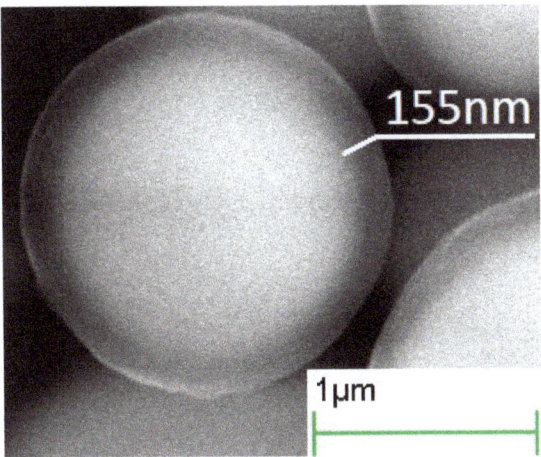

Figure 3. SEM images of core–shell particles obtained in the processes with 2 h hydrolysis duration.

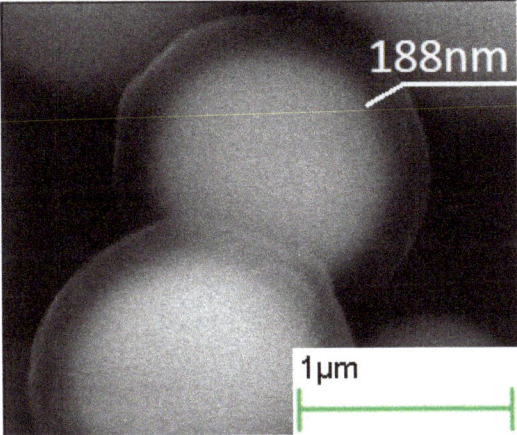

Figure 4. SEM images of core–shell particles obtained in the processes with 4 h hydrolysis duration.

Table 1. The mean thickness with the confidence interval ($p = 0.9$) of the silica shell, measured from the hydrolysis duration experiment.

Hydrolysis duration, h	0.5	1	1.5	2	3	4
Shell thickness, nm	95 ± 11	148 ± 22	125 ± 23	144 ± 13	172 ± 12	189 ± 14

According to EDX analysis, the changes in the duration of the reaction did not affect the stoichiometry of the "SiO$_x$". The atomic ratio Si:O remained at SiO$_{3.3}$. Excess of oxygen is in the form of hydrate water and surface hydroxyls, the real composition being SiO$_2 \cdot n$H$_2$O. An increase in atomic Si:Fe ratio was also observed, and it also showed an increase in the silica content within the samples subjected to the longer deposition process.

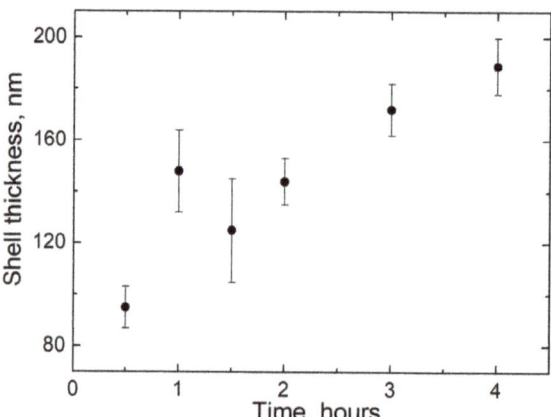

Figure 5. SiO$_2$ shell thickness vs. hydrolysis duration, "time".

Ammonia concentration was found to influence both the thickness and chemical composition of the shell (Figures 6–8, Table 2). Minimal thickness was estimated at ~60 nm, and it was deposited under low ammonia concentration. The thickest shell was deposited in an excess of ammonia, and it was ~220 nm (Figure 9). An increase in the [NH$_3$·H$_2$O] concentration resulted in a shell significantly enriched with oxygen, as much as SiO$_{8.6}$. An excess of ammonia also resulted in an enlarged fraction of individual SiO$_2$ nanoparticles with a mean size that was twice as high as the estimated thickness of the shell. The surface of particles obtained under the excess of ammonia was smooth, while those deposited under [TEOS]/[NH$_3$·H$_2$O] = 4.5 showed surface roughness of approximately 80 nm (Figure 6). These results are in agreement with the data on how ammonia influences the growth of the silica gel [33].

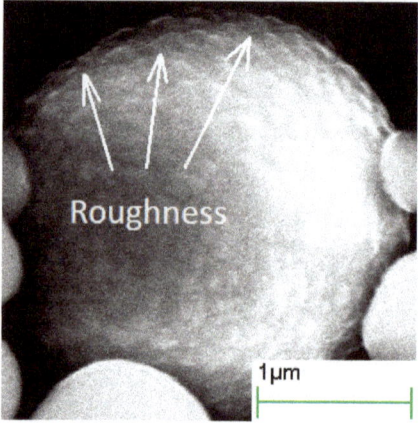

Figure 6. SEM images of core–shell particles obtained in the process with [TEOS]/[NH$_3$·H$_2$O] = 4.5 ammonia concentration.

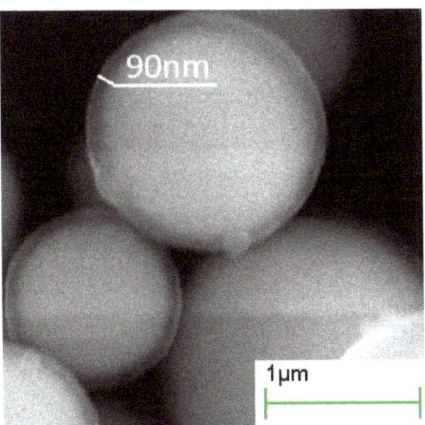

Figure 7. SEM images of core–shell particles obtained in the process with [TEOS]/[NH$_3$·H$_2$O] = 1.5 ammonia concentration.

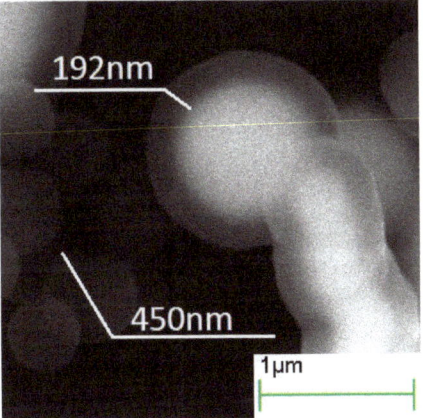

Figure 8. SEM images of core–shell particles obtained in the process with [TEOS]/[NH$_3$·H$_2$O] = 0.75 ammonia concentration.

Table 2. The mean thickness with the confidence interval ($p = 0.9$) of the silica shell, measured from the ammonia concentration experiment.

[TEOS]/[NH$_3$·H$_2$O] ratio	4.5	1.5	1	0.75
Shell thickness, nm	62 ± 9	90 ± 8	146 ± 12	218 ± 10
Si:O atomic ratio	SiO$_{3.0}$	SiO$_{3.3}$	SiO$_{3.8}$	SiO$_{8.6}$

It is interesting to note that the distribution of the thickness of the silica shell was narrower in those cases when the duration of hydrolysis was longer than 2 h (compare the confidence interval given in Table 1 when the duration time was 0.5, 1, and 1.5 h, and the rest of the confidence intervals provided in Tables 1 and 2). It may be assumed that during the first 1.5 h, the growth rate of the silica shell differs for iron particles of different sizes and the resulting thickness levels during the time interval between 1.5 and 2 h of the hydrolysis reaction. The deviation of the mean shell thickness coincides sufficiently with the deviations calculated in [32], where the monodisperse silica was grown and studied.

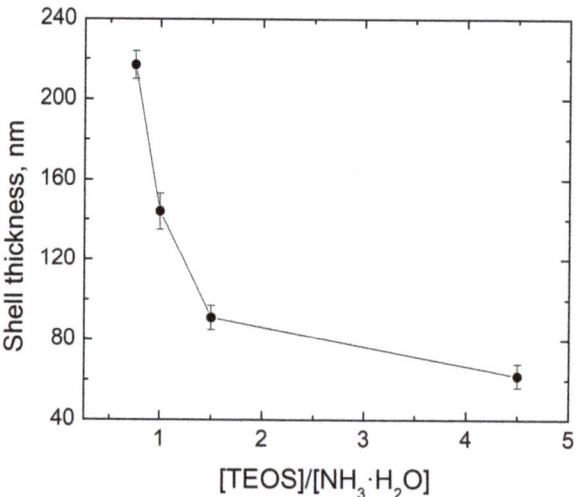

Figure 9. The thickness of the shell of the Fe@SiO$_2$ filler varied through *ammonia concentration* thickness of the shell vs. [TEOS]/[NH$_3$·H$_2$O] ratio.

3.2. Dielectric Permittivity: Theoretical Approach

In order to evaluate the dependence of the real part of the static relative permittivity of the core–shell particles on the thickness of the shell, the Maxwell Garnett effective medium theory was used. The theory allows calculating effective permittivity of the media comprised of two materials, one of which is a matrix and another one is inclusion. The Maxwell Garnett theory is valid in the case of a small concentration of the inclusion [34]. However, in a quasi-static regime, it is applicable for any inclusion concentration [35]. Effective permittivity of the composites in the quasi-static field was calculated as:

$$\varepsilon_{eff} = \varepsilon_h + 3f\varepsilon_h \frac{\varepsilon_i - \varepsilon_h}{\varepsilon_i + 2\varepsilon_h - f(\varepsilon_i - \varepsilon_h)} \quad (1)$$

where ε_h and ε_i were the relative permittivity of the matrix and inclusion, respectively, and f was a volume fraction of the inclusion. For the present Fe@SiO$_2$ composite $\varepsilon_h = \varepsilon_{SiO_2} = 3,9$ [36], $\varepsilon_i = \varepsilon_{Fe} = \infty$ and $f = \frac{R_{Fe}^3}{(R_{Fe}+t)^3}$, where R_{Fe} = 1500 nm was iron particle radius [6] and t was the thickness of the dielectric shell. The thickness varied from 30 to 210 nm. Taking ε_i as ∞ in (1) results in the following formula for ε_{eff}

$$\varepsilon_{eff} = \varepsilon_h + 3f\varepsilon_h \frac{1}{1-f} \quad (2)$$

The theoretical model showed a decrease in permittivity by 86%, with an increase in the thickness of the shell from 30 to 210 nm (Figure 10). The particle size distribution of the initial iron powder can be found in Figure 3 from [6]. Another calculation technique may be applied, which is first calculating the effective permeability of the paraffin and silicon shell, and then using that result as the environmental permittivity for the iron [37]. However, if the model applied in the manuscript may be derived rigorously with account for inclusion and shell shapes, another model is semi-empirical.

Measurements of the permittivity demanded blending of the Fe@SiO$_2$ particles into paraffin matrix. Consequently, it was necessary to evaluate the effective permittivity of the system composed of core–shell particles and paraffin in order to compare experimental and theoretical results. The permittivity was calculated using the Maxwell Garnett theory as well [38]. For the Fe@SiO$_2$ + Paraffin medium $\varepsilon_h = \varepsilon_{Paraffin} = 2.25$ [38], $\varepsilon_i = \varepsilon_{Fe@SiO_2}$, which was calculated previously, and $f = 0.35$. In the presence of the paraffin matrix dramatic decrease in the permittivity was smoothed, allowing fine tuning of the ε'. The permittivity dropped by 11%: at shell thickness of 30 nm, $\varepsilon_{Fe@SiO_2+Paraffin} = 5.69$, while at 150 nm $\varepsilon_{Fe@SiO_2+Paraffin} = 5.09$ (Figure 11).

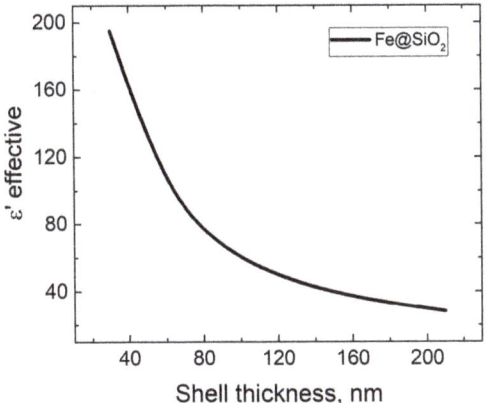

Figure 10. Calculated effective permittivity vs. SiO$_2$ shell thickness of the Fe@SiO$_2$ "composite" Fe@SiO$_2$–paraffin wax real composite.

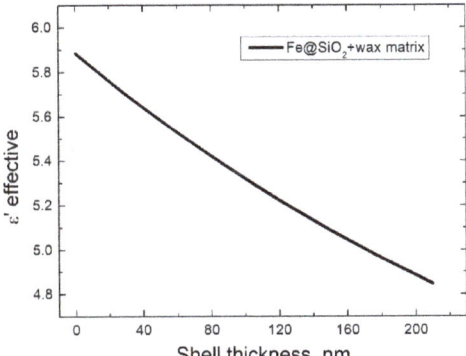

Figure 11. Calculated effective permittivity vs. SiO$_2$ shell thickness of the Fe@SiO$_2$–paraffin wax real composite.

The permittivity of the pure CI in the paraffin medium (shell thickness of 0 nm at Figure 11) was estimated according to (2) with $\varepsilon_m = \varepsilon_{Paraffin} = 2.25$ and $f = 0.35$.

3.3. Frequency Dispersions of Complex Permittivity and Permeability

Analysis of the reaction time variation allows assessing the impact of thickening of the dielectric shell on electromagnetic properties (ε and μ) of the composite. Samples acquired in more prolonged reactions tended to have a lower real part of permittivity (ε'), consistent with a thicker shell (Figure 12A). Values of the ε' in the quasi-static regime demonstrated a good agreement with the theoretical estimation based on the Maxwell Garnett effective medium theory (Figure 13). Frequency dispersions of the ε' of pure CI and the composite obtained in the half-hour reaction were indistinguishable, demonstrating that the shell was not uniform yet in a half-hour experiment. The shell can be considered uniform, starting with 90–100 nm thickness.

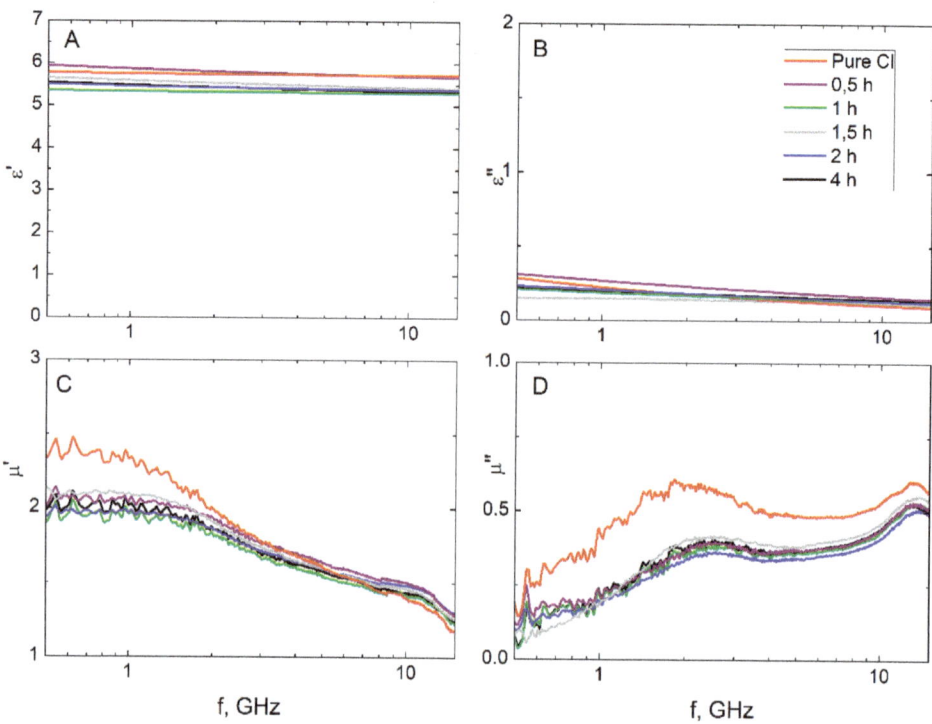

Figure 12. The measured frequency dispersions of complex permittivity, ε' (**A**) + $i\cdot\varepsilon''$ (**C**), and permeability, μ' (**B**) + μ'' (**D**) for the Fe@SiO$_2$–paraffin wax composite. The thickness of the shell of the Fe@SiO$_2$ filler varied through *hydrolysis duration*.

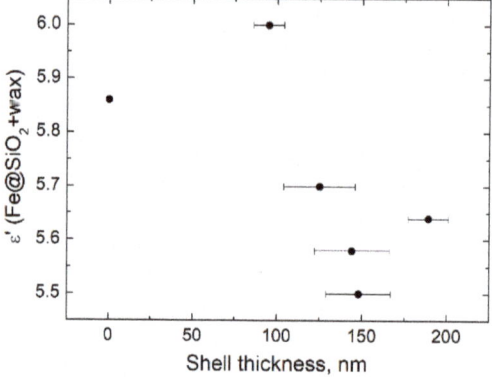

Figure 13. Quasi-static permittivity vs. thickness of the dielectric shell deposited at different durations of hydrolysis.

An increase in the shell thickness did not influence the imaginary part of the permittivity (ε''). The ε'' accounts for a loss in the medium. In the analyzed composites, conductive loss in the CI was a primary source of loss since SiO$_2$ and paraffin wax are low-loss materials (their dielectric loss tangents are ~0.002 [39] and ~0.007 [40]). Therefore, the absence of the changes in the ε'' dispersion demonstrated that the shell growth process does not affect the conductivity of the CI.

Both parts of the complex permeability (μ' and μ'') of the composite were lower than in the pure CI that indicates a larger volume fraction of magnetic material in the composite samples. The behavior of the frequency response curves of μ' and μ'' was the same for the pure CI and composites that implies that magnetic properties of the CI were unaffected by the shell.

The measured permittivity showed that a thin coating of 60 nm was not uniform: the ε' value was almost identical to that of pure iron powder-based composite (Figures 14 and 15). The amplitude of permeability also supported this proposition: thin SiO_2 shells did not significantly decrease this parameter. Further increase in thickness dropped quasi-static ε' value from 5.8 to 5.3, by 9%. This was also according to the Maxwell Garnett calculations, just as in the duration of the hydrolysis experiment. However, the increase in oxygen content within the SiO_x composition resulted in a slight increase in the ε'. This, in turn, was probably due to an increase in the ε' value of the dielectric shell itself, according to [18].

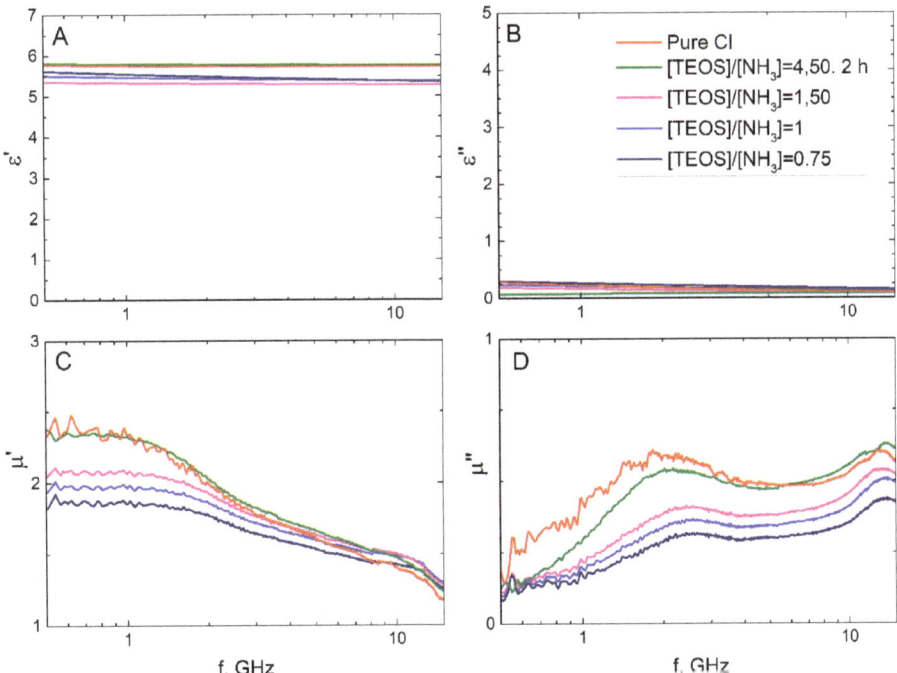

Figure 14. The measured frequency dispersions of complex permittivity, ε' (**A**) + $i \cdot \varepsilon''$ (**C**), and permeability, μ' (**B**) + μ'' (**D**) for the Fe@SiO$_2$–paraffin wax composite.

For the rest of the observations, the ε'', μ' and μ'' demonstrated the same dependencies on thickness as in the hydrolysis duration experiment (Figure 14). Changes in the chemical composition of the SiO_x did not vary the magnetic properties of the composite since SiO_x possesses no magnetic order. The ε'' did not depend on the shell thickness and composition, showing that SiO_x is a low-loss dielectric.

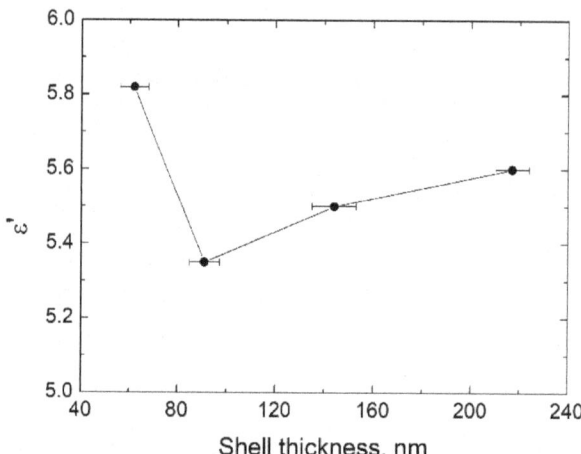

Figure 15. Quasi-static real part of permittivity vs. thickness of the shell that was obtained by varying ammonia concentration.

4. Conclusions

Hydrolysis of TEOS in the presence of ammonia of different concentrations gave uniform SiO_x coating on a surface of carbonyl iron micro-particles when the thickness of the coating was higher than 100 nm. Two techniques were found to increase the thickness of the shell up to approximately 200 nm. One was simply to prolong the duration of the deposition reaction to 4 h. Further prolongation was found to be ineffective for increasing the thickness. After deposition and drying at 60 °C, the shell composition was estimated at $SiO_{3.3}$. The other technique was to change the ammonia concentration in the reaction mixture: an increase in [$NH_3 \cdot H_2O$] concentration increased the thickness of the shell when the duration of the deposition was constant. Simultaneously, [$NH_3 \cdot H_2O$] was found to influence the composition of the shell. In relative terms, a decrease in [TEOS]/[$NH_3 \cdot H_2O$] ratio from 4.5 to 0.75 enriched the silica with oxygen from $SiO_{3.0}$ to $SiO_{8.6}$. Although it can be expected that the difference in chemical composition may vary the dielectric properties of the silica, the difference in the electromagnetic performance of the Fe@SiO_2 core–shell powders was found to be governed primarily by the thickness of the shell. This was estimated comparing the measured real permittivity values of the Fe@SiO_2–paraffin wax composites and theoretical values calculated following the Maxwell Garnett formula. The new method to easily fine-tune the chemical composition and thickness of the uniform silica shell deposited to carbonyl iron particles may be instructive for microwave performance and biocompatibility adjustment in a wide range of applications, including magnetic field sensors.

It can be expected, both from experience and the literature data given in the introduction section, that the dependencies reported here will remain the same when the size of the iron core is higher than 3 micrometers, up to at least 200–500 micrometers. With the decrease in the size of iron particles, an effective fraction of the SiO_2 shell will increase, which will undoubtedly affect the magnetic properties of the product. This effect may be expected to be the most obvious when protecting iron nanoparticles instead of microparticles. However, in general, the size of iron particles is deemed not to affect the mechanisms of the SiO_2 formation.

Author Contributions: Conceptualization, A.V.D. and S.S.M.; methodology, A.V.D., D.A.P., S.S.M., S.A.M. and P.A.Z.; validation, A.V.D., D.A.P., S.N.S., A.S.N. and A.N.L.; formal analysis, A.V.D., A.V.O., V.I.P. and S.N.S.; investigation, A.V.D., S.S.M.; writing—original draft preparation, A.V.D.; writing—review and editing, S.S.M.; visualization, A.V.D., A.S.N. and V.I.P.; supervision, S.S.M.;

project administration, A.N.L.; funding acquisition, A.N.L. All authors have read and agreed to the published version of the manuscript.

Funding: This study was financially supported by the Russian Science Foundation (RSF) under project No. 21-19-00138 (https://rscf.ru/en/project/21-19-00138/ access date 5 July 2021).

Institutional Review Board Statement: Not applicable.

Informed Consent Statement: Not applicable.

Data Availability Statement: Data is contained within the article.

Conflicts of Interest: The authors declare no conflict of interest. The funders had no role in the design of the study; in the collection, analyses, or interpretation of data; in the writing of the manuscript, or in the decision to publish the results.

References

1. Li, L.; Chen, Q.; Gao, Z.; Ge, Y.; Yi, J. Fe@SiO$_2$@(MnZn)Fe$_2$O$_4$ soft magnetic composites with enhanced permeability and low core loss for high-frequency applications. *J. Alloy. Compd.* **2019**, *805*, 609–616. [CrossRef]
2. Shukla, V. Review of electromagnetic interference shielding materials fabricated by iron ingredients. *Nanoscale Adv.* **2019**, *1*, 1640–1671. [CrossRef]
3. Sankaran, S.; Deshmukh, K.; Ahamed, M.B.; Khadheer Pasha, S.K. Recent advances in electromagnetic interference shielding properties of metal and carbon filler reinforced flexible polymer composites: A review. *Compos. Part A Appl. Sci. Manuf.* **2018**, *114*, 49–71. [CrossRef]
4. Wang, H.; Wang, M.; Zhang, X.; Lu, Z.; Fu, W.; Zhong, B.; Wang, C.; Zou, J.; Huang, X.; Wen, G. A new type of catalyst allows carbonyl iron powder to be coated with SiO$_2$ for tuned microwave absorption. *Surf. Interfaces* **2020**, *21*, 100755. [CrossRef]
5. Ge, C.; Wang, L.; Liu, G.; Wang, T. Enhanced electromagnetic properties of carbon nanotubes and SiO$_2$-coated carbonyl iron microwave absorber. *J. Alloy. Compd.* **2018**, *767*, 173–180. [CrossRef]
6. Maklakov, S.S.; Lagarkov, A.N.; Maklakov, S.A.; Adamovich, Y.A.; Petrov, D.A.; Rozanov, K.N.; Ryzhikov, I.A.; Zarubina, A.Y.; Pokholok, K.V.; Filimonov, D.S. Corrosion-resistive magnetic powder Fe@SiO$_2$ for microwave applications. *J. Alloy. Compd.* **2017**, *706*, 267–273. [CrossRef]
7. Wu, S.; Sun, A.; Lu, Z.; Cheng, C. Fabrication and properties of iron-based soft magnetic composites coated with parylene via chemical vapor deposition polymerization. *Mater. Chem. Phys.* **2015**, *153*, 359–364. [CrossRef]
8. Ge, J.; Cui, Y.; Liu, L.; Li, R.; Meng, F.; Wang, F. Enhanced electromagnetic wave absorption of hybrid-architectures Co@ SiOxC. *J. Alloy. Compd.* **2020**, *831*, 154442. [CrossRef]
9. Yuchang, Q.; Wancheng, Z.; Shu, J.; Fa, L.; Dongmei, Z. Microwave electromagnetic property of SiO$_2$-coated carbonyl iron particles with higher oxidation resistance. *Phys. B Condens. Matter* **2011**, *406*, 777–780. [CrossRef]
10. Li, J.; Feng, W.J.; Wang, J.S.; Zhao, X.; Zheng, W.Q.; Yang, H. Impact of silica-coating on the microwave absorption properties of carbonyl iron powder. *J. Magn. Magn. Mater.* **2015**, *393*, 82–87. [CrossRef]
11. Wu, Z.; Fan, X.; Wang, J.; Li, G.; Gan, Z.; Zhang, Z. Core loss reduction in Fe–6.5wt.%Si/SiO$_2$ core–shell composites by ball milling coating and spark plasma sintering. *J. Alloy. Compd.* **2014**, *617*, 21–28. [CrossRef]
12. Slovenský, P.; Kollár, P.; Mei, N.; Jakubčin, M.; Zeleňáková, A.; Halama, M.; Odnevall Wallinder, I.; Hedberg, Y.S. Mechanical surface smoothing of micron-sized iron powder for improved silica coating performance as soft magnetic composites. *Appl. Surf. Sci.* **2020**, *531*, 147340. [CrossRef]
13. Kim, S.-S.; Kim, H.-S.; Kim, S,G.; Kim, W.-S. Effect of electrolyte additives on sol-precipitated nano silica particles. *Ceram. Int.* **2004**, *30*, 171–175. [CrossRef]
14. Han, Y.; Lu, Z.; Teng, Z.; Liang, J.; Guo, Z.; Wang, D.; Han, M.-Y.; Yang, W. Unraveling the Growth Mechanism of Silica Particles in the Stöber Method: In Situ Seeded Growth Model. *Langmuir* **2017**, *33*, 5879–5890. [CrossRef]
15. Green, D.L.; Jayasundara, S.; Lam, Y.-F.; Harris, M.T. Chemical reaction kinetics leading to the first Stober silica nanoparticles—NMR and SAXS investigation. *J. Non-Cryst. Solids* **2003**, *315*, 166–179. [CrossRef]
16. Chen, S.-L.; Dong, P.; Yang, G.-H.; Yang, J.-J. Kinetics of Formation of Monodisperse Colloidal Silica Particles through the Hydrolysis and Condensation of Tetraethylorthosilicate. *Ind. Eng. Chem. Res.* **1996**, *35*, 4487–4493. [CrossRef]
17. Kosevich, A.; Petrusevich, E.; Maklakov, S.; Naboko, A.; Kolesnikov, E.; Petrov, D.; Zezyulina, P.; Pokholok, K.; Filimonov, D.; Han, M. Low Weight Hollow Microspheres of Iron with Thin Dielectric Coating: Synthesis and Microwave Permeability. *Coatings* **2020**, *10*, 995. [CrossRef]
18. Nichols, L.B.; Thorp, J.M. Dielectric constant of silica gel activated at different temperatures. *Trans. Faraday Soc.* **1970**, *66*, 1741–1747. [CrossRef]
19. Hotta, M.; Hayashi, M.; Nishikata, A.; Nagata, K. Complex Permittivity and Permeability of SiO$_2$ and Fe$_3$O$_4$ Powders in Microwave Frequency Range between 0.2 and 13.5 GHz. *ISIJ Int.* **2009**, *49*, 1443–1448. [CrossRef]
20. Malvindi, M.A.; Brunetti, V.; Vecchio, G.; Galeone, A.; Cingolani, R.; Pompa, P.P. SiO2 nanoparticles biocompatibility and their potential for gene delivery and silencing. *Nanoscale* **2012**, *4*, 486–495. [CrossRef]

21. Cvek, M.; Mrlík, M.; Ilčíková, M.; Mosnáček, J.; Münster, L.; Pavlínek, V. Synthesis of Silicone Elastomers Containing Silyl-Based Polymer-Grafted Carbonyl Iron Particles: An Efficient Way To Improve Magnetorheological, Damping, and Sensing Performances. *Macromolecules* **2017**, *50*, 2189–2200. [CrossRef]
22. Xu, J.; Pei, L.; Li, J.; Pang, H.; Li, Z.; Li, B.; Xuan, S.; Gong, X. Flexible, self-powered, magnetism/pressure dual-mode sensor based on magnetorheological plastomer. *Compos. Sci. Technol.* **2019**, *183*, 107820. [CrossRef]
23. Karkkainen, K.; Sihvola, A.; Nikoskinen, K. Analysis of a three-dimensional dielectric mixture with finite difference method. *IEEE Trans. Geosci. Remote Sens.* **2001**, *39*, 1013–1018. [CrossRef]
24. Polder, D.; van Santeen, J.H. The effective permeability of mixtures of solids. *Physica* **1946**, *12*, 257–271. [CrossRef]
25. Tuncer, E. Dielectric mixtures-importance and theoretical approaches. *IEEE Electr. Insul. Mag.* **2013**, *29*, 49–58. [CrossRef]
26. Engen, G.F.; Hoer, C.A. Thru-Reflect-Line: An Improved Technique for Calibrating the Dual Six-Port Automatic Network Analyzer. *IEEE Trans. Microw. Theory Tech.* **1979**, *27*, 987–993. [CrossRef]
27. Nicolson, A.M.; Ross, G.F. Measurement of the Intrinsic Properties of Materials by Time-Domain Techniques. *IEEE Trans. Instrum. Meas.* **1970**, *19*, 377–382. [CrossRef]
28. Weir, W.B. Automatic measurement of complex dielectric constant and permeability at microwave frequencies. *Proc. IEEE* **1974**, *62*, 33–36. [CrossRef]
29. Havriliak, S.; Negami, S. A complex plane analysis of α-dispersions in some polymer systems. *J. Polym. Sci. Part C Polym. Symp.* **1966**, *14*, 99–117. [CrossRef]
30. Petrov, D.A.; Rozanov, K.N.; Koledintseva, M.Y. Influence of Higher-order Modes in Coaxial Waveguide on Measurements of Material Parameters. In Proceedings of the 2018 IEEE Symposium on Electromagnetic Compatibility, Signal Integrity and Power Integrity (EMC, SI & PI), Long Beach, CA, USA, 30 July–3 August 2018; pp. 66–70.
31. LeFrancois, S.; Pasquet, D.; Maze-Merceur, G. A new model for microwave characterization of composite materials in guided-wave medium. *IEEE Trans. Microw. Theory Tech.* **1996**, *44*, 1557–1562. [CrossRef]
32. Greasley, S.L.; Page, S.J.; Sirovica, S.; Chen, S.; Martin, R.A.; Riveiro, A.; Hanna, J.V.; Porter, A.E.; Jones, J.R. Controlling particle size in the Stöber process and incorporation of calcium. *J. Colloid Interface Sci.* **2016**, *469*, 213–223. [CrossRef]
33. Rahman, I.A.; Vejayakumaran, P.; Sipaut, C.S.; Ismail, J.; Abu Bakar, M.; Adnan, R.; Chee, C.K. Effect of anion electrolytes on the formation of silica nanoparticles via the sol-gel process. *Ceram. Int.* **2006**, *32*, 691–699. [CrossRef]
34. Chettiar, U.K.; Engheta, N. Internal homogenization: Effective permittivity of a coated sphere. *Opt. Express* **2012**, *20*, 22976–22986. [CrossRef] [PubMed]
35. Gutierrez Vela, Y.; Ortiz, D.; Osa, R.; Saiz, J.; González, F.; Moreno, F. Electromagnetic Effective Medium Modelling of Composites with Metal-Semiconductor Core-Shell Type Inclusions. *Catalysts* **2019**, *9*, 626. [CrossRef]
36. Robertson, J. High dielectric constant oxides. *Eur. Phys. J. Appl. Phys.* **2004**, *28*, 265–291. [CrossRef]
37. Nazarov, R.; Zhang, T.; Khodzitsky, M. Effective Medium Theory for Multi-Component Materials Based on Iterative Method. *Photonics* **2020**, *7*, 113. [CrossRef]
38. Dionne, G.F.; Fitzgerald, J.F.; Aucoin, R.C. Dielectric constants of paraffin-wax–TiO2 mixtures. *J. Appl. Phys.* **1976**, *47*, 1708–1709. [CrossRef]
39. Mandal, A.K.; Sen, R. Microwave Absorption of Barium Borosilicate, Zinc Borate, Fe-Doped Alumino-Phosphate Glasses and Its Raw Materials. *Technologies* **2015**, *3*, 111. [CrossRef]
40. Ghassemiparvin, B.; Ghalichechian, N. Permittivity and dielectric loss measurement of paraffin films for mmW and THz applications. In Proceedings of the 2016 International Workshop on Antenna Technology (iWAT), Cocoa Beach, FL, USA, 29 February–2 March 2016; pp. 48–50.

Article

Adhesive and Magnetic Properties of Polyvinyl Butyral Composites with Embedded Metallic Nanoparticles

Tatyana V. Terziyan [1], Alexander P. Safronov [1,2], Igor V. Beketov [1,2], Anatoly I. Medvedev [2], Sergio Fernandez Armas [3] and Galina V. Kurlyandskaya [1,4,*]

1. Institute of Natural Sciences and Mathematics, Ural Federal University, 620002 Ekaterinburg, Russia; tatyana.terzyan@urfu.ru (T.V.T.); alexander.safronov@urfu.ru (A.P.S.); beketov@iep.uran.ru (I.V.B.)
2. Pulsed Processes Laboratory, Institute of Electrophysics UB RAS, 620016 Ekaterinburg, Russia; medtom@iep.uran.ru
3. SGIKER, Basque Country University UPV/EHU, 48940 Leioa, Spain; sergio.fernandez@ehu.eus
4. Department of Electricity and Electronics, Basque Country University UPV/EHU, 48940 Leioa, Spain
* Correspondence: kurlyandskaya.gv@ehu.eus or galinakurlyandskaya@urfu.ru; Tel.: +34-94-601-3237

Abstract: Magnetic metallic nanoparticles (MNPs) of Ni, Ni82Fe18, Ni50Fe50, Ni64Fe36, and Fe were prepared by the technique of the electrical explosion of metal wire. The average size of the MNPs of all types was in the interval of 50 to 100 nm. Magnetic polymeric composites based on polyvinyl butyral with embedded metal MNPs were synthesized and their structural, adhesive, and magnetic properties were comparatively analyzed. The interaction of polyvinyl butyral (supplied as commercial GE cryogenic varnish) with metal MNPs was studied by microcalorimetry. The enthalpy of adhesion was also evaluated. The positive values of the enthalpy of interaction with GE increase in the series Ni82Fe18, Ni64Fe36, Ni50Fe50, and Fe. Interaction of Ni MNPs with GE polymer showed the negative change in the enthalpy. No interfacial adhesion of GE polymer to the surface of Fe and permalloy MNPs in composites was observed. The enthalpy of interaction with GE polymer was close to zero for Ni95Fe5 composite. Structural characterization of the GE/Ni composites with the MNPs with the lowest saturation magnetization confirmed that they tended to be aggregated even for the materials with lowest MNPs concentrations due to magnetic interaction between permalloy MNPs. In the case of GE composites with Ni MNPs, a favorable adhesion of GE polymer to the surface of MNPs was observed.

Keywords: polyvinyl butyral; electric explosion of wire; metallic nanoparticles; magnetic nanoparticles; polymer filled composites

1. Introduction

Polyvinyl butyral (PVB) is a random terpolymer mainly composed of vinyl alcohol and vinyl butyral with relatively small amounts of vinyl acetate. A terpolymer is a copolymer in which two or more chemically distinct monomer units are alternating along linear chains in the irregular way. PVB is a colorless, amorphous thermoplastic resin [1], which is widely used in technological applications such as automotive laminated glass, paints, and adhesives due to its excellent flexibility, ability to form coatings in the film shape, good adhesion properties, and excellent UV resistance. Easy wettability and compatibility with various polar compounds (such as phenols, epoxies, isocyanates, etc.) make PVB an excellent candidate to be used in many functional applications. On its basis, composite materials with inorganic fillers of various chemical nature can be fabricated. Thus, the development of shape memory materials containing graphene oxide [2], photoactive materials with improved mechanical and heat-conducting properties with particles of TiO_2, CdS and other ceramic fillers can be mentioned [3–5].

Fillers of the metallic nature are also considered to be components of PVB-based systems. For instance, Angappan et al. [6] described the preparation of a composite based on PVB and a core-shell filler, where a thin nickel layer was used as a coating. The material was designed as a lightweight broadband microwave absorber for different applications. Metallic particles are often called "zero-valent" particles to distinguish them from metal oxide particles. In pure metals, atoms are zero-valent, while in oxides metal atoms have a positive oxidation number according to their valency. Although the term "metallic" presumably means elemental (zero-valent) metal, still in some biomedical references (and applications) it is sometimes attributed to metal oxides also.

In addition, PVB is a component of binders, in particular, a component of "GE varnish" (being not an abbreviation but a commercial product label) widely used to thermally anchor wires at cryogenic temperatures. It has fast track time and can be both air-dried or baked. Other important features of this adhesive are electrical insulating properties at cryogenic temperatures, suitable properties of a calorimeter cement, compatibility with a wide variety of materials, including cotton, nylon glass tapes, mica products, polyester products, vinyl products, plastics, and many others. In the research laboratories, GE varnish was widely used in the studies on the magnetic and microwave characteristics of the wide variety of materials [7–10]. In fact, when measuring the properties of magnetic particles, a filled polymer composition is usually prepared, where the particles under study act as a dispersed filler in a GE-varnish polymer matrix.

One of the major structural parameters of a polymer/filler composite material is the uniformity of the distribution of particles in the polymer matrix. However, it is difficult to achieve uniform distribution. Here it is important to mention that the techniques of characterization of the filler distribution are not developed yet at a satisfactory level. There are different parameters of the fillers themselves contributing to the uniformity of the filler distribution inside the composite. First, it is filler chemical composition and the shape of the filler particles (spherical, cubic, rod-like etc.). For the case of magnetic fillers, the interaction between the filler particles plays crucial role preventing their de-aggregation during polymer/filler composite fabrication. Second, practically all kinds of available fillers, especially those that can be obtained in the large quantities, have distribution of the shapes and sizes of the elements. This obstacle adds extra difficulties in the control of the uniformity of the distribution of particles in the polymer matrix. For example, it is well known that magnetic behavior of the ferrofluid critically depends on the presence of even a few particles of the large size in the ensemble [11].

Particle distribution affects thermal and electrical characteristics [5], mechanical, dielectric and microwave properties of composites based on PVB [3,7]. The uniformity of the particle distribution inside the composite is also affected by the adhesive interaction between the polymer matrix and the dispersed filler. The higher the adhesion of the polymer to the surface of the particles, the greater the likelihood of their disaggregation with distribution in the form of individual particles. In this regard, the use of GE varnish as a binder for fixing a certain distribution of magnetic filler requires an understanding of the degree of adhesive interaction between the components of the GE varnish and the magnetic particles.

When we refer to the bulk ferromagnets in thin film state, one of the most studied systems is the system of iron-nickel alloys starting from pure nickel and up to the pure iron [12–14]. The saturation magnetization evolution, magnetic anisotropy features, magnetic permeability and magnetostriction changes were widely discussed and comparatively analyzed [15–17]. Apart from the theoretical interest, this system is widely used in many practically important devices [18,19]. However, nanostructured FeNi alloys in the shape of nanoparticles and filled composites on their basis were studied to lesser extent and there is a gap or absence of the understanding to what extent the results obtained in the case of FeNi system in the bulk thin film state can be applicable to the MNPs related cases.

In this work, we have studied the structure, magnetic properties and interactions at the interface of the composite films based on polyvinyl butyral terpolymer (GE varnish) and magnetic nanoparticles of nickel, iron, and FeNi of various compositions.

2. Materials and Methods

GE varnish was a commercial product GE-7031-CT Thermal Varnish (CRYO-Technics, Büttelborn, Germany). It was a viscous liquid brownish 25% solution of a polymer in mixed organic solvent. GE varnish was used as received for the preparation of polymeric composites with embedded zero-valent metallic nanoparticles.

To minimize the contribution of the shape variation of the particles of the filler highly productive electrophysical technique of the electrical explosion of metal wire (EEW) was used for the synthesis of magnetic metallic nanoparticles (MNPs) [20]. The method itself was known long ago [21] but in recent years special attention was focused on it as a technique for MNPs fabrication [22]. Special advantage of the technique is the fabrication of the particles of close to spherical shape. MNPs were synthesized at the Laboratory of Pulsed Processes of the Institute of Electrophysics UB RAS via the electrical explosion of metal wire in argon. In EEW method a metallic wire is evaporated by a high voltage discharge, and the vapors condense in gas phase giving spherical non-agglomerated nanoparticles. The details of the method and the description of the experimental EEW setup can be found elsewhere [20,22]. Using EEW method magnetic nanoparticles of Ni, Ni82Fe18, Ni64Fe36, Ni50Fe50, and Fe were synthesized. The batches of Fe and Ni MNPs were approximately 500 g each and the batches of NiFe alloys were approximately 100 g each.

To prepare GE/MNPs composites, first, the weighted portion of MNPs of each type was grinded in the agate mortar with the addition of isopropanol to make a homogeneous suspension of MNPs. Then the weighted amount of GE varnish was added and vigorously mixed with the MNPs suspension. The resulted slurry was then cast onto the polished glass and left at ambient conditions for the evaporation of the solvent. The dried film of GE/MNPs composite was mechanically separated from the glass substrate and kept in a thermostat at 80 °C up to the reaching of the state of the constant weight when all residual solvents were eliminated. The wide range of polymer/MNPs compositions was selected to characterize extensively the enthalpy of adhesion of polymeric matrix to the surface of MNPs both at low MNPs content and at the highest obtainable content. In this respect mechanical properties of the compositions were not considered but they are certainly important for any practical application, e.g., for microwave adsorption. In this respect, compositions with MNPs content below 70% are preferable as they become brittle at the higher content of MNPs. Even so, for the present study in selected cases the MNPs content up to 88% was considered.

Transmission electron microscopy (TEM) images were obtained using a JEOL JEM2100 microscope operated at 200 kV (JEOL Ltd., Tokyo, Japan). The particles were dispersed in isopropanol under ultrasonic treatment and the resulted suspension was placed onto carbon coated copper grids and evaporated at ambient conditions. X-ray diffraction (XRD) studies were performed using a BrukerD8 Discover (Bruker, Billerica, MA, USA) instrument with Cu-Kα radiation (wavelength λ = 1.5418 Å), a graphite monochromator for a diffracted beam and a scintillation detector. Diffractograms were refined by Rietveld algorithm using TOPAS 3.0 program installed in the XRD instrument. The specific surface area of MNPs was measured by low-temperature adsorption of nitrogen using Micromeritics TriStar3000 automatic sorption analyzer (Micromeritics, Norcross, GA, USA).

Magnetic measurements of the magnetization value as a function of the applied magnetic field $M(H)$ (hysteresis loops) were performed at the room temperature by a vibrating sample magnetometer (Cryogenics Ltd. VSM, London, UK). Even though complete saturation was not always achieved in the field of 1.8 kOe, we designate the meaning of saturation magnetization (M_s) to the value of magnetization in H = 1.8 kOe. Both M_s and coercivity (H_c) value were calculated from the $M(H)$ hysteresis loops. The MNPs were measured in non-magnetic capsule (up to 5 mg of the sample weight) and GE/MNPs composites (up to 15 mg of the weight of the sample composite) were measured for in-plane configuration of the film.

Fourier transform infra-red (FTIR) spectra of composite films were obtained in the 4000–500 cm^{-1} range of wavenumbers using a Nicolet 6700 FTIR spectrometer equipped with an ATR add-on (Thermo Fisher Scientific, IN, USA)

Calorimetric measurements were performed at 25 °C using Calvet 3D differential microcalorimeter DAK-1-1 (EPSI, Chernogolovka, Russia) with ampoule cells. A portion (20–90 mg) of GE composite or air-dry MNPs was put into a thin glass ampoule ca.5 mL in volume and dried in an oven to a constant weight. After that the ampoule was sealed and placed in sliding holder, mounted to the top of a stainless steel tubular cell (7 cm^3). The cell was filled with 4 mL of isopropanol. Two assembled cells were positioned in the ports of microcalorimeter and it was thermostated at 25 °C for 2–3 h until the drift of the baseline fell below 0.01 mW in 30 min. Then the experiment in one of the cells (working cell) was initiated by breaking the glass ampoule in the solvent (isopropanol). The other cell was a reference. Heat evolution curve in the working cell was recorded for ca 60–90 min until the initial baseline was reestablished. The time dependence of heat flux was integrated using software program giving the enthalpy of dissolution for the sample in the ampoule. Then the experiment in the second cell was performed in a similar manner. Typical values of heat effects were within the range 0.1–5.0 J, depended on the load in the ampoule. The relative error of measurements was 5% for the heat effects ranging from 0.1 to 0.5 J and 2.0% for the heat effects in the 0.5–5.0 J range.

In addition, the structure of selected filled composites was studied by scanning electron JEOL JSM-640 microscope (JEOL Ltd., Tokyo, Japan) working at 20 kV accelerating voltage and equipped with energy dispersive X-ray (EDX) fluorescent detector for elemental analysis. As before [23] to avoid the charging of the non-conducting polymer surface, about 20 nm carbon layer was deposited onto the composite surface.

3. Results

Figure 1 shows transmission electron microscopy images of synthesized zero-valent metallic spherical EEW MNPs with different content of iron and nickel. As the shape of the MNPs was close to the spherical one and therefore their diameter was selected as characteristic geometrical parameter.

The average apparent characteristic diameters (d_S) in for the batches of all types were calculated based on the value of specific surface area according to following equation [24]:

$$d_S = \frac{6000}{\rho S_{sp}} \quad (1)$$

Here S_{sp} is the specific surface area of MNPs, which is conventionally measured using low-temperature adsorption of nitrogen (Brunauer–Emett–Teller (BET) method [16]); ρ—is the crystallographic density of MNPs. The calculated values of the apparent characteristic diameters of the MNPs are given in Table 1. One can see that all obtained batches can be considered to be the MNPs. The highest d_S value corresponded to the iron MNPs and the smallest one to the nickel MNPs. section may be divided by subheadings. It should provide a concise and precise description of the experimental results, their interpretation, as well as the experimental conclusions that can be drawn.

According to XRD data analysis phase composition of MNPs corresponded to solid solutions based on cubic crystal structure. Their specific parameters are given in Table 1 as well as the values of the saturation magnetization and the coercivities.

General comparative analysis of the d_S and M_s values shows good correlation between their values, and it is in accordance with the existing understanding of magnetic behavior of nanoparticles of these sizes and compositions [23,25–27]. In all cases the saturation magnetization was lower in comparison with bulk M_s values [28,29] with the difference of 15 to 25%. However, as to expect, the highest M_s was observed for iron MNPs and the lowest saturation magnetization was obtained for Ni MNPs.

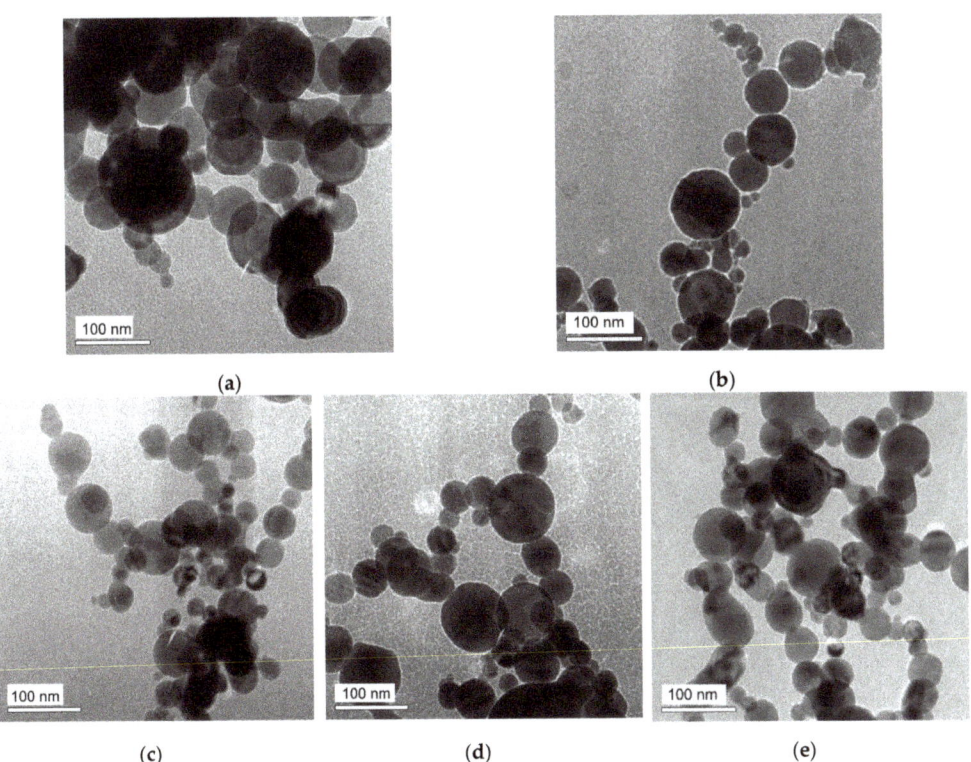

Figure 1. TEM images of zero-valent metal EEW MNPs of different types. (**a**)—Fe, (**b**)—Ni50Fe50, (**c**)—Ni64Fe36, (**d**)—Ni82Fe18, (**e**)—Ni.

The reduction of the value of the saturation magnetization in comparison with the bulk state can be assigned to different effects. The first one is the strongly pyrophoric features of EEW MNPs and the need for their surface passivation prior to the exposure to the atmosphere. In the previous studies [11,23,30] we have shown that the passivation oxide layer has a thickness of a few nm.

However, for the MNPs with d_S in the range of about 50 to 100 nm such a layer with lower magnetization can cause a reduction up to 15% of the total magnetization value (depending on the size of the MNPs). The second reason of the reduction of the saturation magnetization value is related to the concept of nanoscaling laws [31]. In the spherical MNPs at least three surface layers are not contributing to the ferromagnetic response not having the sufficient number of the nearest neighbors. For instance, in pure a-Fe one can obtain the reduction up to 10% in comparison with value of the bulk iron. Both abovementioned reasons for the M_s were given without taking into account the existence of the MNPs size distribution which makes the analysis even more difficult. Even so, fabricated batches of the metallic MNPs were used for comparative analysis of their adhesive and magnetic properties of GE varnish (polyvinyl butyral)-based composites.

Unfortutately, narrowing of the distribution of metallic MNPs is not achievable using conventional separation techniques such as filtering of separation. The basic reason for this is strong aggregation of metallic MNPs in their suspensions. Suspensions of metallic MNPs do not contain individual particles. These features of them were discussed in our recent paper by Shankar et al. [32]. Theoretical consideration by the extended DLVO approach favored strong magnetic interactions as a major reason for aggregation. Therefore, the batches of metal MNPs are to be used as synthesized.

Table 1. Selected characteristics of EEW metallic nanoparticles: Specific surface area—S_{sp}, density—ρ, apparent average diameter—d_S, phase composition (all phases are cubic); saturation magnetization M_s and coercivity H_c measured for nanoparticles at 20 °C.

Mark	Description	S_{sp} (m²/g)	ρ (g/cm³)	d_S (nm)	Phase Composition	M_s (emu/g)	H_c (Oe)
Fe	Iron	7.5	7.8	102 ± 8	98% (S.G: Fm-3m) fcc, 0.35927(2) 2% (S.G.: I m-3m) bcc a = 0.2862(3)	200 ± 5	300 ± 5
Ni50Fe50	Permalloy: 50% Ni, 50% Fe	12.5	8.4	62 ± 5	100% (S.G.: F m-3m) bcc, a = 0.3569(1)	140 ± 5	180 ± 2
Ni64Fe36	Permalloy: 64% Ni, 36% Fe	12.6	7.8	61 ± 5	90% (S.G: Fm-3m) fcc, a = 0.3592(2) 10% (S.G.: I m-3m) bcc, a = 0.2862(3)	110 ± 5	160 ± 2
Ni82Fe18	Permalloy: 82% Ni, 18% Fe	8.0	8.4	86 ± 7	100% (S.G: Fm-3m) bcc, a = 0.3548(1) nm	82 ± 5	100 ± 1
Ni	Nickel	12.6	8.9	53 ± 4	100% (S.G: Fm-3m) bcc, a = 0.3524(2) nm	48 ± 3	150 ± 2

GE varnish is a multi-component industrial product. Figure 2 presents FTIR spectrum in the range of wave numbers from 400 up to 4000 cm^{-1}, which was obtained aiming to clarify the chemical composition of the available product. The spectrum refers to the polymeric residue, which was obtained after the evaporation of the solvent from GE varnish. To mark it out further we will denote this polymeric residue as GE polymer or simply GE. The strongest bands in the GE spectrum were: a peak at 3370 cm^{-1}, which was attributed to the stretching of OH group, peaks at 2955 cm^{-1} and 2869 cm^{-1} due to the stretching of C-H bonds in aliphatic CH$_3$, CH$_2$ and CH groups, and a peak at 1712 cm^{-1} due to the stretching of carbonyl group. The peaks at 1434 cm^{-1} and 1129 cm^{-1} are attributed to the vibrations of CH$_2$ and C-O-C groups. Identification of the spectrum gave the 93% fit for poly(vinyl butyral) [33]. The difference in GE polymer and PVB IR spectra were observed only in the range of wave numbers 3600–2600 cm^{-1} corresponding to the vibrations of hydroxyl groups (Figure 2).

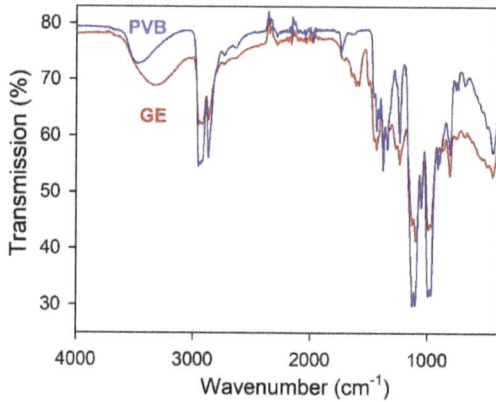

Figure 2. FTIR spectra of poly(vinyl butyral) (PVB) and GE polymer.

The frequency of OH group stretching in individual molecules is around 3600 cm^{-1} [25]. The broad bands in Figure 2 are shifted to lower frequencies for both GE and PVB. It means that the hydroxyl groups of polymers are linked by hydrogen bonds. In the case of the GE polymer, the shift is observed to a greater extent. FTIR spectrum of the GE shows very wide absorption peak at 3320 cm^{-1} due to self-associated OH groups.

The close identity of FTIR spectra for GE polymer and for PVB, except for the shape of the OH peak, indicated that both PVB and GE have the same basic chemical structure and differ in the degree of self-association by hydrogen bonds.

Interfacial interactions of GE polymer with zero-valent metallic MNPs were evaluated by microcalorimetry. For obvious reasons it is not possible to mix directly in a calorimeter solid nanoparticles with a solid polymer.

Therefore, the enthalpy of interaction among nanoparticles and polymeric matrix are calculated using thermochemical cycle [34,35] based on the measurable heat effects of several appropriate processes, which give in combination the desired enthalpy change. In the case of a GE polymeric composite the enthalpy of formation (ΔH_{comp}) refers to the process:

$$GE + MNPs => GE/MNPs\ composite + \Delta Hcomp \qquad (2)$$

As the components of a polymeric composite do not dissolve in each other, ΔH_{comp} solely depends on the interfacial interaction between solid particles and GE polymeric matrix. Note that ΔH_{comp} is a function of the content of MNPs in composite and it should better be written as $\Delta H_{comp}(\omega)$, with ω standing for the weight fraction of MNPs in GE composite.

The combination of processes that comprise Equation (2) and can be performed in calorimetric cell is given below:

$$GE + isopropanol => GE\ solution + \Delta H_{GE} \qquad (3)$$

$$MNPs + isopropanol => MNPs\ suspension + \Delta H_{MNPs} \qquad (4)$$

$$GE\ solution + MNPs\ suspension => MNPs\ suspension\ in\ GE\ solution + \Delta H_{mix} \qquad (5)$$

$$GE/MNPs\ composite + isopropanol => MNPs\ suspension\ in\ GE\ solution + \Delta H_{dis}(\omega) \qquad (6)$$

ΔH_{GE} is the enthalpy of dissolution of GE; ΔH_{MNPs} is the enthalpy of wetting of MNPs; ΔH_{mix} is the enthalpy of mixing suspension with solution; $\Delta H_{dis}(\omega)$ is the enthalpy of dissolution of a composite with weight fraction of MNPs equal to ω.

The combination of steps is: (2) = (3) + (4) + (5) − (6), and it gives the following equation for the enthalpy of composite formation:

$$\Delta H_{comp}(\omega) = \omega \times \Delta H_{GE} + (1 - \omega) \times \Delta H_{MNPs} + \Delta H_{mix} - \Delta H_{dis}(\omega) \qquad (7)$$

Typically, the term ΔH_{mix} is much lower than others. It falls within the experimental error of calorimetric measurements and can be neglected.

Figure 3a shows the typical view of concentration dependences of the enthalpy of dissolution for polymeric composites based on GE polymer with embedded MNPs. All experimentally measured thermal effects are expressed in Joules per gram of the samples used in the calorimetric experiment. Point on the left axis corresponds to the value of ΔH_{GE}, which was positive for the dissolution of GE in isopropanol. Points at the right axis correspond to ΔH_{MNPs} values for Ni and Fe MNPs. These values are small and negative. All other points in the plot correspond to $\Delta H_{dis}(\omega)$ of composites. These data were used for the calculation of the enthalpy of formation for GE/MNPs composites in the entire range of nanoparticles content. Concentration dependences are given in Figure 3b. According to Figure 3b, the enthalpy of formation of the GE composites with Fe and all marks of NiFe MNPs is endothermic over the whole range of compositions, i.e., during the formation of the composites the heat was absorbed. Concentration dependence of the enthalpy of formation for GE/Ni composite is negative over the entire composition range.

Concentration dependences for GE composites with NiFe MNPs lay between the plots for GE/Ni and GE/Fe composites.

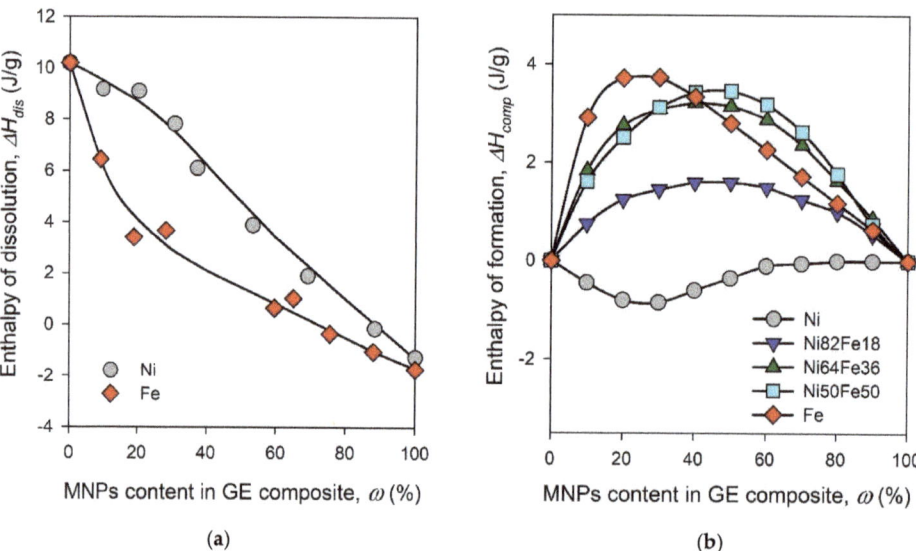

Figure 3. Concentration dependence of the enthalpy of dissolution in isopropanol for GE composites with Ni and Fe MNPs (a). Concentration dependences of the enthalpy of formation for GE/MNPs composites (b). T = 25 °C.

The reason for the existence of the concentration dependence of the enthalpy of formation of polymeric composites is not trivial, since the components in the composite do not dissolve in each other. In this case, the enthalpy change is due to the interfacial adhesion between polymer matrix and the surface of solid MNPs. Polymeric molecules are large, they can form a variety of conformations at the interface [27]. Therefore, the thickness of the interfacial layer at the solid surface in contact with polymer is likely extended compared to the interface with simple liquids. Thus, the enthalpy of adhesion of a polymer to a solid surface depends on the degree of saturation for interfacial polymeric layer, which is a function of the content of solids in polymeric composite. In Figure 3b ΔH_{comp} is zero at $\omega = 0$, i.e., for individual GE, because there are no interfacial layers in it. As the content of MNPs increases, the total area of the interface between MNPs and GE increases as well and so do the absolute values of ΔH_{comp}. At a certain content of MNPs in a composite all polymeric molecules would be involved in the formation of interfacial layers, and absolute values of ΔH_{comp} would reach their maximum. At a level of MNPs content above this threshold the interfacial layers would become progressively unsaturated and it would diminish the enthalpy of composite formation down to zero at $\omega = 0$ that corresponds to individual MNPs. Therefore, we might consider the maximum absolute value of the enthalpy of composite formation in Figure 3b as an indicative measure for the intensity of interfacial interactions (ΔH_{int}) of GE polymer with the surface of a certain type of MNPs.

Figure 4 shows a plot for these values for GE/MNPs composites. They were taken from $\Delta H_{comp}(\omega)$ dependences (Figure 3b) as maximum values at the plots for GE composites with Fe, Ni50Fe50, Ni64Fe36, Ni82Fe18 MNPs and minimum value at the plot for GE/Ni composite.

The trend in ΔH_{int} values is the same as the trend in concentration plots for ΔH_{comp} presented in Figure 3b. Composite GE/Ni had negative value of ΔH_{int}, while it was positive for GE/Fe composite. It is worthwhile noting that all the enthalpy changes by their definition are the difference between the enthalpy of a GE composite and the enthalpies of the components. Therefore, negative value of ΔH_{int} means that interaction in the composite become stronger than in components. Positive value of ΔH_{int} corresponds to the opposite. Thus, the embedding of Ni MNPs in GE composite resulted in the enhancement of interactions, which most likely occurred at the polymer/solid interface. On the contrary, positive values of ΔH_{int} for GE/Fe composite indicated the overall weakening of interactions compared to that in the components. The opposite sign of ΔH_{int} in GE/Ni and GE/Fe composites indicated that the mechanisms of interaction of GE polymer with the surface of Ni and Fe MNPs were different. In other words, there was a favorable adhesion of GE polymer to the surface of Ni MNPs and there was no adhesion of GE to the surface of Fe MNPs. Moreover, positive values of ΔH_{int} in GE/Fe composites were substantial, which meant that weak interactions at GE/Fe interface likely provided the weakening of interactions in the GE polymeric matrix. All permalloy MNPs: Ni82Fe18, Ni64Fe36, Ni50Fe50 also showed positive values of ΔH_{int}. The numerical value of ΔH_{int} increased with the Ni/Fe ratio in permalloy. Please note that even Ni82Fe18 with the major fraction of Ni had positive enthalpy of interaction with GE. It meant that surface properties of Fe are dominant in NiFe alloys.

Magnetic nanoparticles of Ni82Fe18 composition were obtained by EEW from the wire with Ni80Fe20 composition due to the slight change of the composition in the course of fabrication. The Ni80Fe20 alloy with maximum magnetic permeability and low magnetostriction (at about 79% nickel) is the most used in sensor applications [17–19]. High permeability in homogeneous magnetic material appears either due to the magnetization rotation in a condition of weak crystal anisotropy or due to the displacement of the domain walls. These results can be expected for the materials with close to zero magnetostriction value [12,15–17]. However, the most favorable composition and particular properties of the material depend in a complex way on the preparation conditions. We, therefore, outlined the interval around Ni80Fe20 (Figure 4) emphasizing the possibility of the shift toward either higher or lower Ni content.

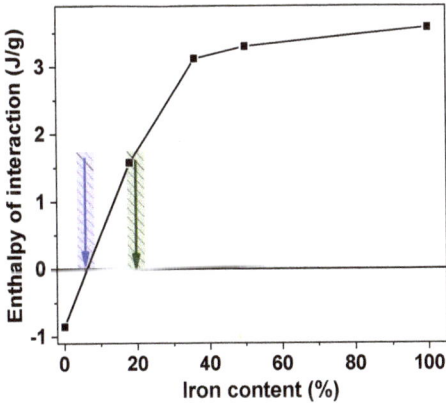

Figure 4. Indicative values of the enthalpy of interaction of GE polymer with zero-valent metal MNPs: Ni, Ni82Fe18, Ni64Fe36, Ni50Fe50, Fe. Blue arrow indicates zero-enthalpy of interaction composition and green arrow indicates the composition with maximum magnetic permeability (at about 79% nickel).

This experimental observation cannot for now find a reasonable explanation and further studies are needed to clarify this peculiar difference among zero-valent 3d metal nanoparticles. However, below we are making some additional comments, which could be useful for better understanding of the adhesion results.

4. Discussion

Let us now comparatively analyze some structural and magnetic properties of Ni and Ni82Fe18 GE-based composites. Backscattered electrons used for SEM evaluation [36] are high-energy electrons that are reflected or backscattered out of the volume of elastic scattering interaction with the atoms of the composite. Polymers consist of non-metallic elements with low atomic numbers and carbon is the most common element in polymer composition. Elements with the high atomic numbers scatter electrons stronger than elements with low atomic number. Therefore, the elements with the high atomic numbers appear brighter in the image offering the possibility to evaluate the contrast between areas with different chemical compositions. Figure 5 shows the surface properties of GE/Ni composites with selected concentrations of the filler.

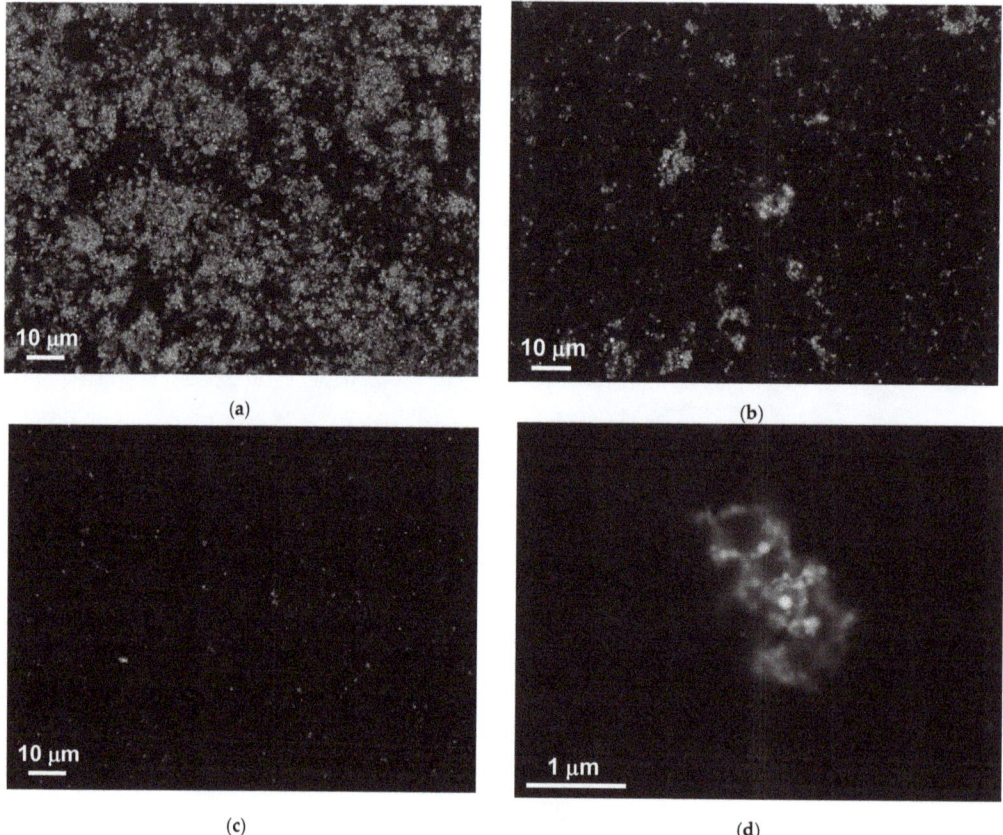

Figure 5. Surface properties of GE/Ni composites evaluated using scanning electron microscopy: (**a**) Ge/69 wt.% of Ni; (**b**) GE/31 wt.% of Ni; (**c**,**d**) GE/10 wt.% of Ni.

To analyze local distribution of the aggregates and MNPs in the aggregates step-by-step increase of the magnification method was used for the analysis of the structure of the composites with high and low particles concentrations (Figure 6). One can see that at high concentrations of the MNPs the structure of the composite can be described as sufficiently uniform re-distribution of large aggregates of the order of a few microns tending to be "star"-like units with many relatively short branches. Worth mentioning the presence of the "chain"-like structures formed by at least, 7–10 particles of the medium size (Figure 6e).

At low concentrations of the MNPs (even in the case of nickel MNPs with lowest saturation magnetization) the MNPs form aggregates, which are "star"-like units of the average size of the order of 1 micron or lower and turned to being spherical, i.e., the particles are not uniformly distributed over the composite even in the case when the composite preparation included all disaggregation steps. One of the reasons for such a behavior is magnetic interaction between the MNPs of this size and composition. It is also evident that "chain"-like aggregates are not a typical feature for GE composites with low concentrations of the MNPs.

Here we should come back to Table 1 and take into account available physical parameters for the comparison: d_S = 53 ± 4 nm and M_s = 48 ± 3 and H_c = 150 ± 2 Oe values for Ni MNPs. First, the substantial coercivity and saturation magnetization value quite close to the saturation magnetization of bulk nickel confirm that the MNPs are in the multidomain state, this means they have non-zero magnetic moment in zero applied field and tend to agglomerate due to dipole-dipole interactions. As the saturation magnetization increases in the set of the materials Ni–FeNi–Fe with the increase of the iron it is to expect that the level of the dipole-dipole interactions for the particles of the same size should also increase contributing to the elevation of the level of MNPs agglomeration.

However, it is very difficult to obtain the batches of MNPs of the same size, with the same particles size distributions and similar thickness of the passivating layer. In any case, due to the difference of the composition the surface properties can vary significantly, as they depend on the type of the oxides formed onto the MNPs surfaces [37].

Let us now analyze magnetic properties of selected composites. It is logical to take a close look on GE/Ni, and GE/Ni82Fe18 composites showing (see Figure 4) opposite signs of the enthalpy of interaction between the polymer matrix and the filler—these are representative cases. Figure 7 shows magnetic hysteresis loops of GE/Ni and GE/Ni82Fe18 composites with different filler concentrations (c) assigned in weight %. In both cases linear dependences of $M_S(c)$ were observed confirming fabrication of the composites of a good quality. Although the magnetic signal of GE is diamagnetic, in all fabricated composites, it is very small in comparison with ferromagnetic responses of the nanoparticles and therefore GE magnetic contribution should not be taken into account.

To understand to what extent the concentration of nanoparticles with different compositions and their interactions affect the magnetic properties of composites the magnetic hysteresis $M(H)$ loops were represented in M/M_s form (Figure 7e,f). One can see that in most general sense all M/M_s hysteresis loops of each type of the composites have very similar shape. Direct comparison of the M/M_s values for GE/Ni and GE/Fe18Ni82 composites with low concentration of the MNPs showed that they are quite similar. The same is true for the values of the remnant magnetizations, H_c values and even the field dependences of primary magnetization curves. Comparative analysis of M/M_s magnetic hysteresis loops of GE/Ni (c = 10 wt.%) and GE/Ni82Fe18 (c = 11 wt.%) composites shows their similarity. Even so, in the small magnetic fields M/M_s parameter increases much faster and coercivity is higher in the case of nickel composite. This can be a consequence in the average size of the particular batch (Table 1): magnetization of the larger nanoparticles includes processes that are more complex and for the whole ensemble requires an application of the higher magnetic field to start.

As mentioned before the Ni80Fe20 alloy in the shape of thin films, had found many technological applications in the area of inductors and magnetic field sensors [18,19]. In recent years, the FeNi films deposited onto flexible substrates attracted additional interest [37–41]. However, in most of the cases the functional properties of FeNi thin films or multilayered structures based on FeNi components are lower in comparison with the same structures deposited onto rigid substrates. One of the reasons for observed behavior is poor adhesion of the metallic film onto the surface of flexible substrate (such as Kapton, polyester, cyclo olefin copolymer, and others). One of the strategies to improve adhesion was the deposition of the appropriate buffer layers (Al, Cu, Ti) [36,38] in combination with usage the multilayered structures favoring the stress relaxation [39,42,43].

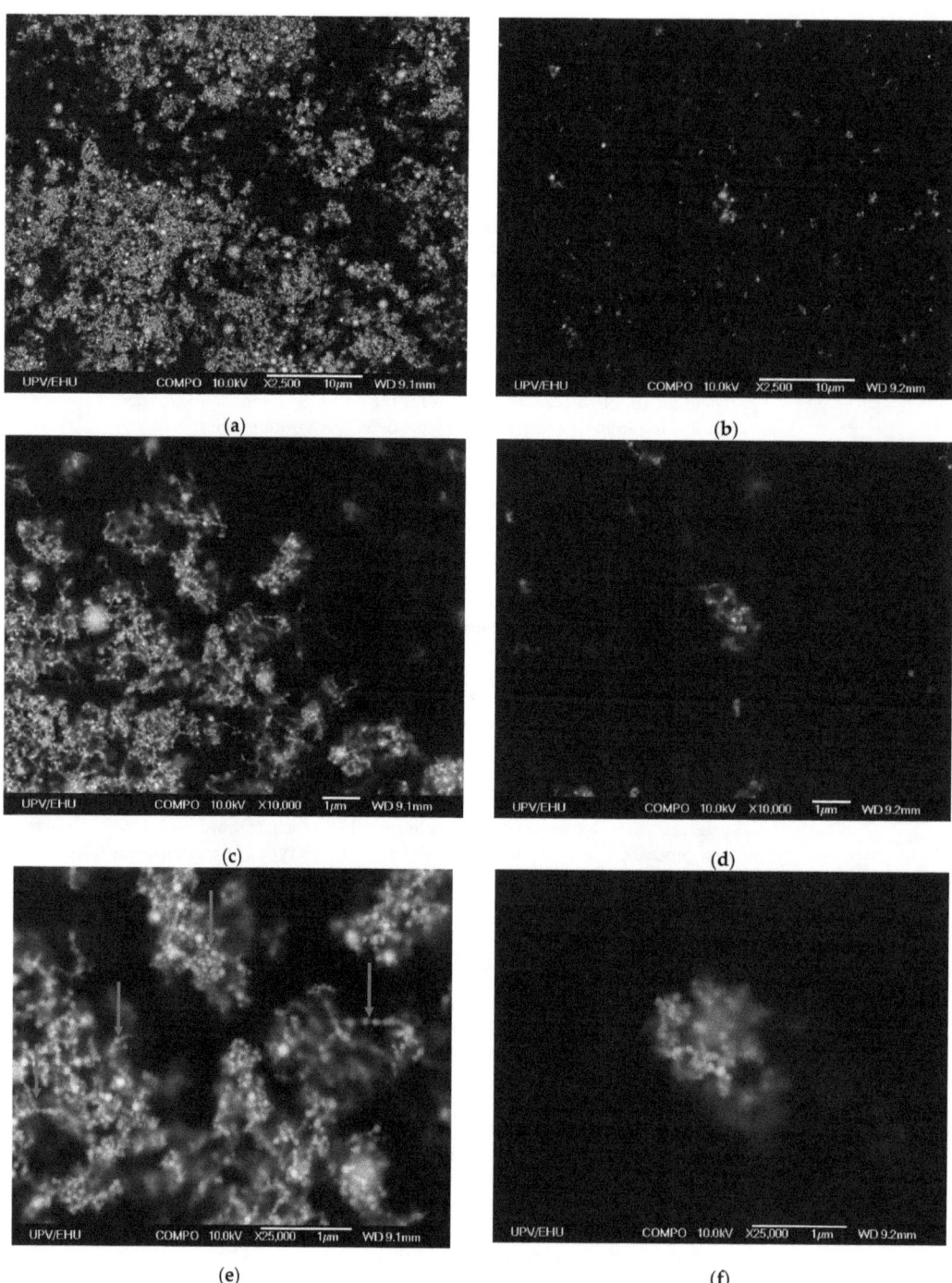

Figure 6. Surface properties of selected GE/Ni composites evaluated using scanning electron microscopy at different magnifications: (**a**,**c**,**e**)—GE/69 wt.% of Ni and (**b**,**d**,**f**)—GE/10 wt.% of Ni. "Chain" aggregates indicated by the red arrows.

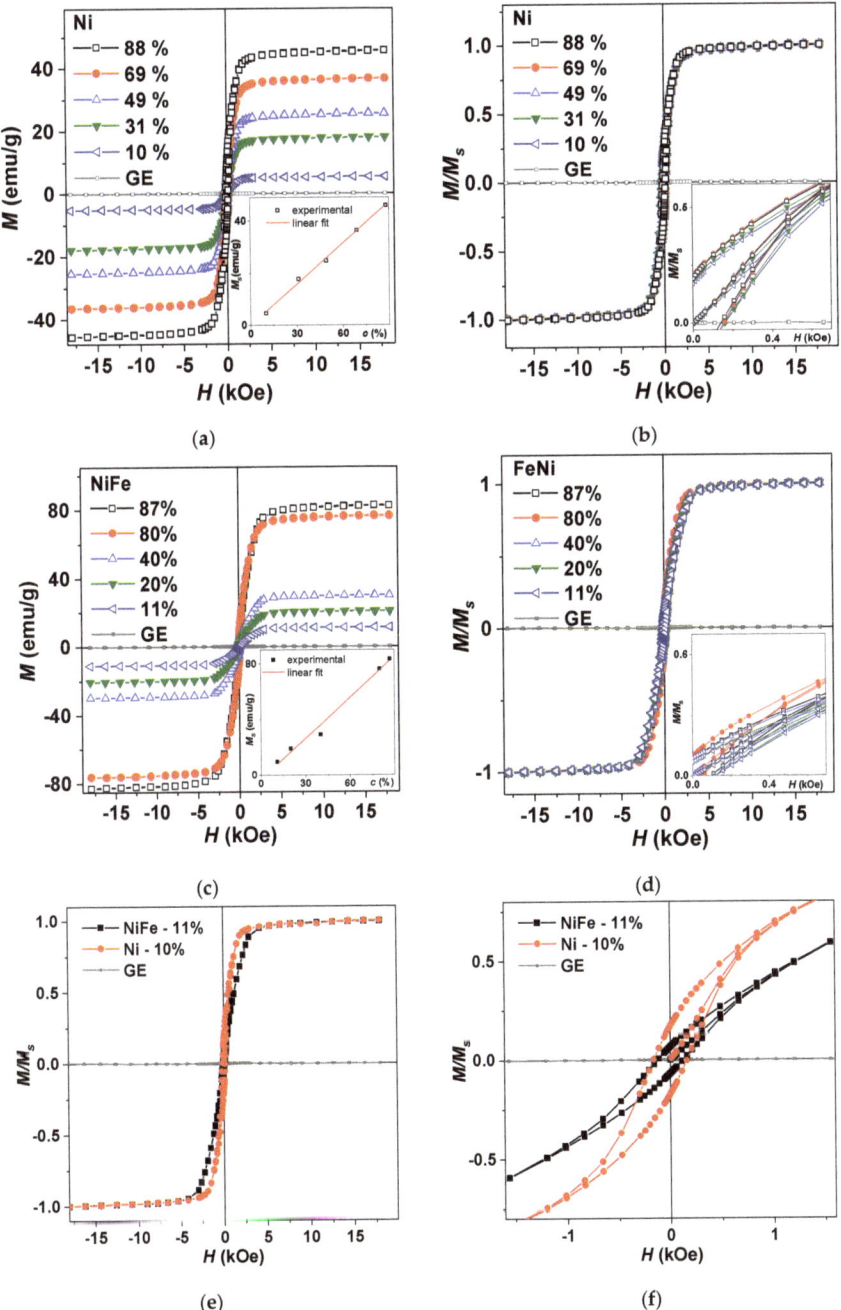

Figure 7. Magnetic hysteresis loops of GE/Ni (**a**,**b**) and GE/Ni82Fe18 (**c**,**d**) composites. Inset (**a**) shows the M_s dependence on the concentration of Ni MNPs; inset (**b**) shows the low field part of the hysteresis loop for GE/Ni; inset (**c**) shows the M_s dependence on the concentration c of NiFe MNPs; inset (**d**) shows the low field part of the $M(H)$ loop for GE/Ni82Fe18 composites. Hysteresis loops of relative magnetization for GE/Ni and GE/Ni82Fe18 composites (**e**,**f**).

In this study, in the case of GE/Ni composites with Ni MNPs as a filler, we had observed a favorable adhesion of GE polymer to the surface of Ni MNPs. This means that deposition of nickel layer onto polymer substrate might be the useful technological step and probably the solution of the well-known problem, at least this research direction seems to be interesting to follow. In addition, the microwave properties of Fe, FeNi, Ni, and NiCo MNPs of polymer-based composites can be mentioned. GE varnish was previously used for many nanostructured materials high frequency characterization due to very low contribution of the matrix itself and simplicity of the composite preparation [44,45]. Apart from usual applications as microwave absorbers [46,47] these composites were tested as model materials for the development of magnetic biosensors [48].

It is also important to attract attention to the strategy of the analysis of the surface properties of nanoparticles during their interactions with polymers for the prediction of the functional properties of thin films and multilayered structures deposited onto flexible substrates. As the effective surface available for the interaction of metallic film and polymer substrate is rather small, the only way to find appropriate combinations of thin film and polymer is multiple searches for the best deposition conditions. However, usage of the MNPs of requested composition and selected polymer for evaluation of their adhesive properties by existing methods of physical chemistry might be very useful complementary way to solve the problem.

5. Conclusions

Magnetic nanoparticles of various compositions were fabricated by EEW. The results of their structural and magnetic characterization were comparatively analyzed. The average size of all kinds of MNPs was in the interval of about 50 to 100 nm. All MNPs were in the multidomain state with the saturation magnetization slightly reduced in comparison with M_s value for corresponding composition. However, the observed reduction was explained in the framework of scaling laws for MNPs and based on the surface oxidation. Zero-valent nickel, iron, and permalloy MNPs were embedded into polymeric matrix based on GE varnish. Chemical nature of the GE varnish had been analyzed using IR spectroscopy. The main component of the GE polymer is self-associated poly(vinyl butyral). The interaction of GE polymer with zero-valent metal magnetic nanoparticles was studied by microcalorimetry. The positive values of the enthalpy of interaction with GE increase in the series Ni82Fe18, Ni64Fe36, Ni50Fe50, and Fe. Meanwhile, interaction of Ni MNPs with GE polymer was characterized by the negative change in the enthalpy. It meant that there is no interfacial adhesion of GE polymer to the surface of Fe and permalloy MNPs in composites.

Structural evaluation of the GE/Ni composites with the MNPs having the lowest saturation magnetization confirmed that despite the special efforts to separate MNPs in the course of fabrication of the composite they tended to be aggregated even for the materials with lowest MNPs concentrations. "Chain"-like aggregates were not typical for GE composites with low concentrations of the MNPs. At high concentrations of the Ni MNPs aggregates of the order of a few microns tending to be "star"-like units with the presence of the "chain"-like structures were observed. One of the reasons for such a behavior is a magnetic interaction between the MNPs.

To understand to what extent the concentration of MNPs and their interactions affect the magnetic properties M/Ms hysteresis loops were analyzed for GE/Ni and GE/NiFe composites with different concentration of the filler. Direct comparison of the M/Ms values for GE/Ni and GE/Fe18Ni82 composites with low concentration of the MNPs confirmed their similarity. It was shown that Ni95Fe5 composition might be interesting for future investigations, as the enthalpy of interaction with GE polymer was close to zero for it. In the case of GE/Ni composites with Ni MNPs as a filler, a favorable adhesion of GE polymer to the surface of Ni MNPs was observed. The deposition of nickel layer onto polymer substrate might be useful technological step for fabrication of FeNi films with enhanced functional properties when deposited onto flexible substrates.

Author Contributions: T.V.T., A.P.S. and G.V.K. conceived and designed the experiments; T.V.T., A.P.S., I.V.B., S.F.A., G.V.K. and A.I.M. performed the experiments; I.V.B., T.V.T., A.I.M., A.P.S., S.F.A. and G.V.K. analyzed the data; G.V.K., A.P.S., T.V.T.; visualization, T.V.T., A.P.S. and G.V.K. wrote the manuscript and participated in re-writing of the revised version. All authors discussed the results and implications, and commented on the manuscript at all stages. All authors read and approved the final version of the manuscript.

Funding: This research was supported in part by the Universidad del País Vasco/Euskal Herriko Unibertsitatea UPV/EHU Research Groups Funding.

Institutional Review Board Statement: Not applicable.

Informed Consent Statement: Not applicable.

Data Availability Statement: Data available from the corresponding author on reasonable request.

Acknowledgments: We thank A.M. Murzakaev, A. Larrañaga, A.V. Svalov, S.M. Bhagat, M.-L. Fdez-Gubieda and I. Orue for special support. Selected studies were done at SGIKER services of UPV-EHU.

Conflicts of Interest: The authors declare no conflict of interest.

References

1. Hallensleben, M.L. Polyvinyl Compounds, Others. In *Ullmann's Encyclopedia of Industrial Chemistry*; Wiley-VCH: Weinheim, Germany, 2000.
2. Bai, Y.; Jiwen Zhang, J.; Wen, D.; Gong, P.; Chen, X. A poly (vinyl butyral)/graphene oxide composite with NIR light-induced shape memory effect and solid-state plasticity. *Compos. Sci. Technol.* **2019**, *170*, 101–108. [CrossRef]
3. Georgopanos, P.; Eichner, E.; Filiz, V.; Handge, U.A.; Schneider, G.A.; Heinrich, S.; Abetz, V. Improvement of mechanical properties by a polydopamine interface in highly filled hierarchical composites of titanium dioxide particles and poly(vinyl butyral). *Compos. Sci. Technol.* **2017**, *146*, 73–82. [CrossRef]
4. Lin, P.-Y.; Wu, Z.-S.; Juang, Y.-D.; Fu, Y.-S.; Guo, T.-F. Microwave-assisted electrospun PVB/CdS composite fibers and their photocatalytic activity under visible light. *Microel. Engin.* **2016**, *149*, 73–77. [CrossRef]
5. Alva, G.; Lin, Y.; Fang, G. Thermal and electrical characterization of polymer/ceramic composites with polyvinyl butyral matrix. *Mater. Chem. Phys.* **2018**, *205*, 401–415. [CrossRef]
6. Angappan, M.; Bora, P.J.; Vinoy, K.J.; Vijayaraju, K.; Ramamurthy, P.C. Microwave absorption efficiency of poly (vinyl-butyral)/Ultra-thin nickel coated fly ash cenosphere composite. *Surf. Interfaces* **2020**, *19*, 100430.
7. Handge, U.A.; Wolff, M.F.H.; Abetz, V.; Heinrich, S. Viscoelastic and dielectric properties of composites of poly(vinyl butyral) and alumina particles with a high filling degree. *Polymer* **2016**, *82*, 337–348. [CrossRef]
8. Abdulvagidov, S.B.; Djabrailov, S.Z.; Abdulvagidov, B.S. Nature of novel criticality in ternary transition-metal oxides. *Sci. Rep.* **2019**, *9*, 19328. [CrossRef] [PubMed]
9. Jackson, E.M.; Liao, S.B.; Silvis, J.; Swihart, A.H.; Bhagat, S.M.; Crittenden, R.; Glover, R.E.; Manheimer, M.A. Initial susceptibility and microwave absorption in powder samples of $Y_1Ba_2Cu_3O_{6.9}$. *Phys. C Supercond.* **1988**, *152*, 125–129. [CrossRef]
10. Kurlyandskaya, G.V.; Bhagat, S.M.; Safronov, A.P.; Beketov, I.V.; Larranaga, A. Spherical magnetic nanoparticles fabricated by electric explosion of wire. *AIP Adv.* **2011**, *1*, 042122. [CrossRef]
11. Hubert, A.; Schäfer, R. *Magnetic Domains*; Springer: Berlin, Germany, 1998.
12. Bozorth, R.M. The permalloy problem. *Rev. Mod. Phys.* **1953**, *25*, 42–48. [CrossRef]
13. Sugita, Y.; Fujiwara, H.; Sato, T. Critical thickness and perpendicular anisotropy of evaporated permalloy films with stripe domains. *Appl. Phys. Lett.* **1967**, *10*, 229–231. [CrossRef]
14. Endo, Y.; Mitsuzuka, Y.; Shimada, Y.; Yamaguchi, M. Influence of magnetostriction on damping constant of Ni_xFe_{1-x} film with various Ni concentrations (x). *J. Appl. Phys.* **2011**, *109*, 07D336. [CrossRef]
15. Griffiths, J.H.E. A phenomenological theory of damping in ferromagnetic materials. *Nature* **1946**, *148*, 670. [CrossRef]
16. Svalov, A.V.; Aseguinolaza, I.R.; Garcia-Arribas, A.; Orue, I.; Barandiaran, J.M.; Alonso, J.; Fernandez-Gubieda, M.L.; Kurlyandskaya, G.V. Structure and magnetic properties of thin permaloy films near the "transcritical" state. *IEEE Trans. Magn.* **2010**, *16*, 333–336. [CrossRef]
17. Bonin, R.; Schneider, M.L.; Silva, T.J.; Nibarger, J.P. Dependence of magnetization dynamics on magnetostriction in NiFe alloys. *J. Appl. Phys.* **2005**, *98*, 123904. [CrossRef]
18. Gardner, D.S.; Schrom, G.; Paillet, F.; Jamieson, B.; Karnik, T.; Borkar, S. Review of on-chip inductor structures with magnetic films. *IEEE Trans. Magn.* **2009**, *45*, 4760–4766. [CrossRef]
19. Buznikov, N.A.; Safronov, A.P.; Orue, I.; Golubeva, E.V.; Lepalovskij, V.N.; Svalov, A.V.; Chlenova, A.A.; Kurlyandskaya, G.V. Modeling of magnetoimpedance response of thin film sensitive element in the presence of ferrogel: Next step toward development of biosensor for in tissue embedded magnetic nanoparticles detection. *Biosens. Bioelectron.* **2018**, *117*, 366–372. [CrossRef]

20. Kurlyandskaya, G.V.; Madinabeitia, I.; Beketov, I.V.; Medvedev, A.I.; Larrañaga, A.; Safronov, A.P.; Bhagat, S.M. Structure, magnetic and microwave properties of FeNi nanoparticles obtained by electric explosion of wire. *J. Alloys Compd.* **2014**, *615*, S231–S235. [CrossRef]
21. William, G. Chace exploding wires. *Phys. Today* **1964**, *17*, 19–24.
22. Sedoi, V.S.; Ivanov, Y.F. Particles and crystallites under electrical explosion of wires. *Nanotechnology* **2008**, *19*, 145710. [CrossRef]
23. Kurlyandskaya, G.V.; Safronov, A.P.; Terzian, T.V.; Volodina, N.S.; Beketov, I.V.; Lezama, L.; Marcano Prieto, L. Fe$_{45}$Ni$_{55}$ magnetic nanoparticles obtained by electric explosion of wire for the development of functional composites. *IEEE Magn. Lett.* **2015**, *6*, 38001041. [CrossRef]
24. Hiemenz, P.C.; Rajagopalan, R. *Principles of Colloid and Surface Chemistry*; Marcel Dekker: New York, NY, USA, 1997; p. 651.
25. Ucar, H.; Craven, M.; Laughlin, D.E.; McHenry, M.E. Effect of Mo additions on structure and magnetocaloric effect in γ-FeNi nanocrystals, J. *Electron. Mater.* **2014**, *43*, 137–141. [CrossRef]
26. Moghimi, N.; Rahsepar, F.R.; Srivastava, S.; Heinig, N.; Leung, K.T. Shape-dependent magnetism of bimetallic FeNi nanosystems. *J. Mater. Chem.* **2014**, *C2*, 6370–6375. [CrossRef]
27. Huber, D.L. Synthesis, Properties, and Applications of Iron Nanoparticles. *Small* **2005**, *1*, 482–501. [CrossRef]
28. O'Handley, R.S. *Modern Magnetic Materials*; John Wiley & Sons: New York, NY, USA, 1972; p. 740.
29. Kurlyandskaya, G.V.; Safronov, A.P.; Bhagat, S.M.; Lofland, S.E.; Beketov, I.V.; Marcano Prieto, L. Tailoring functional properties of Ni nanoparticles-acrylic copolymer composites with different concentrations of magnetic filler. *J. Appl. Phys.* **2015**, *117*, 123917. [CrossRef]
30. Jun, Y.; Seo, J.; Cheon, J. Nanoscaling laws of magnetic nanoparticles and their applicabilities in biomedical sciences. *Acc. Chem. Res.* **2008**, *41*, 179–189. [CrossRef]
31. Carini, G.; Bartolotta, A.; Carini, G.; D'Angelo, G.; Federico, M.; Di Marco, G. Water-driven segmental cooperativity in polyvinyl butyral. *Eur. Polym. J.* **2018**, *98*, 172–176. [CrossRef]
32. Safronov, A.P.; Istomina, A.S.; Terziyan, T.V.; Polyakov, Y.I.; Beketov, I.V. Influence of interfacial adhesion and the nonequilibrium glassy structure of a polymer on the enthalpy of mixing of polystyrene-based filled composites. *Polym. Sci. Ser. A.* **2012**, *54*, 214–223. [CrossRef]
33. Shankar, A.; Safronov, A.P.; Mikhnevich, E.A.; Beketov, I.V.; Kurlyandskaya, G.V. Ferrogels based on entrapped metallic iron nanoparticles in polyacrylamide network: Extended Derjaguin-Landau-Verwey-Overbeek consideration, interfacial interactions and magnetodeformation. *Soft Matter.* **2017**, *13*, 3359–3372. [CrossRef] [PubMed]
34. Rubinstein, M.; Colby, R.H. *Polymer Physics*; Oxford University Press: New York, NY, USA, 2003; p. 440.
35. Goldstein, J.; Echlin, P.; Joy, D.; Lifshin, E.; Lyman, C.E.; Michael, J.R.; Newbury, D.E.; Sawyer, L.C. *Scanning Electron Microscopy and X-Ray Microanalysis*; Kluwer Academic/Plenum Publishers: New York, NY, USA, 2003.
36. Kurlyandskaya, G.V.; Madinabeitia, I.; Murzakaev, A.M.; Sanchez-Ilarduya, M.B.; Beketov, I.V.; Medvedev, A.I.; Larrañaga, A.; Safronov, A.P.; Schegoleva, N.N. Core-shell fine structure of feni magnetic nanoparticles produced by electrical explosion of wire. *IEEE Trans. Magn.* **2014**, *50*, 2900304. [CrossRef]
37. Grimes, C.A. Sputter deposition of magnetic thin films onto plastic: The effect of undercoat and spacer layer composition on the magnetic properties of multilayer permalloy thin films. *IEEE Trans. Magn.* **1995**, *31*, 4109. [CrossRef]
38. Klopfer, M.; Cordonier, C.; Inoue, K.; Li, G.P.; Honma, H.; Bachman, M. Flexible, transparent electronics for biomedical applications. In Proceedings of the IEEE 63rd Electronic Components and Technology Conference (ECTC), Las Vegas, NV, USA, 28–31 May 2013; pp. 494–499.
39. Fernández, E.; Kurlyandskaya, G.V.; García-Arribas, A.; Svalov, A.V. Nanostructured giant magneto-impedance multilayers deposited onto flexible substrates for low pressure sensing. *Nanosc. Res. Lett.* **2012**, *7*, 230. [CrossRef]
40. Agra, K.; Mori, T.J.A.; Dorneles, L.S.; Escobar, V.M.; Silva, U.C.; Chesman, C.; Bohn, F.; Corrêa, M.A. Dynamic magnetic behavior in non-magnetostrictive multilayered films grown on glass and flexible substrates. *J. Magn. Magn. Mater.* **2014**, *355*, 136–141. [CrossRef]
41. Li, B.; Kavaldzhiev, M.N.; Kosel, J. Flexible magnetoimpedance sensor. *J. Magn. Magn. Mater.* **2015**, *378*, 499–505. [CrossRef]
42. Buznikov, N.A.; Svalov, A.V.; Kurlyandskaya, G.V. Influence of the Parameters of Permalloy-Based Multilayer Film Structures on the Sensitivity of Magnetic Impedance Effect. *Phys. Met. Metallogr.* **2021**, *122*, 223–229. [CrossRef]
43. Vas'kovskii, V.O.; Savin, P.A.; Volchkov, S.O.; Lepalovskii, V.N.; Bukreev, D.A.; Buchkevich, A.A. Nanostructuring Effects in Soft Magnetic Films and Film Elements with Magnetic Impedance. *Tech. Phys.* **2013**, *58*, 105–110. [CrossRef]
44. Kurlyandskaya, G.V.; Safronov, A.P.; Bhagat, S.M.; Larrañaga, A.; Bagazeev, A.V. Magnetic and microwave properties of Fe18Ni82 nanoparticles with close to zero magnetostriction. *J. Magn. Magn. Mater.* **2018**, *465*, 156–163. [CrossRef]
45. Qin, G.W.; Pei, W.L.; Ren, Y.P.; Shimada, Y.; Endo, Y.; Yamaguchi, M.; Okamoto, S.; Kitakami, O. Ni$_{80}$Fe$_{20}$ permalloy nanoparticles: Wet chemical preparation, size control and their dynamic permeability characteristics when composited with Fe micron particles. *J. Magn. Magn. Mater.* **2009**, *321*, 4057–4062. [CrossRef]
46. Iskhakov, R.S.; Chekanova, L.A.; Denisova, E.A. Width of the ferromagnetic resonance line in highly dispersed powders of crystalline and amorphous Co-P alloys. *Phys. Solid State* **1999**, *41*, 416–419. [CrossRef]

47. Lv, J.; Liang, X.; Ji, G.; Quan, B.; Liu, W.; Du, Y. Structural and carbonized design of 1D FeNi/C nanofibers with conductive network to optimize electromagnetic parameters and absorption abilities. *ACS Sustain. Chem. Eng.* **2018**, *6*, 7239–7249. [CrossRef]
48. Lodewijk, K.J.; Fernandez, E.; Garcia-Arribas, A.; Kurlyandskaya, G.V.; Lepalovskij, V.N.; Safronov, A.P.; Kooi, B.J. Magnetoimpedance of thin film meander with composite coating layer containing Ni nanoparticles. *J. Appl. Phys.* **2014**, *115*, 17A323. [CrossRef]

Article

Magnetoimpedance of CoFeCrSiB Ribbon-Based Sensitive Element with FeNi Covering: Experiment and Modeling

Stanislav O. Volchkov [1], Anna A. Pasynkova [1,2,*], Michael S. Derevyanko [3], Dmitry A. Bukreev [3], Nikita V. Kozlov [1], Andrey V. Svalov [1] and Alexander V. Semirov [3]

[1] Department of Magnetism and Magnetic Nanomaterials, Institute of Natural Sciences and Mathematics, Ural Federal University, 620002 Ekaterinburg, Russia; stanislav.volchkov@urfu.ru (S.O.V.); nikita.kozlov@urfu.ru (N.V.K.); andrey.svalov@urfu.ru (A.V.S.)
[2] Laboratory of Advanced Magnetic Materials, Institute of Metal Physics UD RAS, 620108 Ekaterinburg, Russia
[3] Department of Physics, Pedagogical Institute, Irkutsk State University, 664003 Irkutsk, Russia; mr.derevyanko@gmail.com (M.S.D.); da.bukreev@gmail.com (D.A.B.); semirov@mail.ru (A.V.S.)
* Correspondence: anna.chlenova@urfu.ru; Tel.: +7-343-389-9567

Abstract: Soft magnetic materials are widely requested in electronic and biomedical applications. Co-based amorphous ribbons are materials which combine high value of the magnetoimpedance effect (MI), high sensitivity with respect to the applied magnetic field, good corrosion stability in aggressive environments, and reasonably low price. Functional properties of ribbon-based sensitive elements can be modified by deposition of additional magnetic and non-ferromagnetic layers with required conductivity. Such layers can play different roles. In the case of magnetic biosensors for magnetic label detection, they can provide the best conditions for self-assembling processes in biological experiments. In this work, magnetic properties and MI effect were studied for the cases of rapidly quenched $Co_{67}Fe_3Cr_3Si_{15}B_{12}$ amorphous ribbons and magnetic $Fe_{20}Ni_{80}/Co_{67}Fe_3Cr_3Si_{15}B_{12}/Fe_{20}Ni_{80}$ composites obtained by deposition of $Fe_{20}Ni_{80}$ 1 μm thick films onto both sides of the ribbons by magnetron sputtering technique. Their comparative analysis was used for finite element computer simulations of MI responses with different types of magnetic and conductive coatings. The obtained results can be useful for the design of MI sensor development, including MI biosensors for magnetic label detection.

Keywords: magnetic field sensors; rapidly quenched amorphous ribbons; thin films; magnetic composites; computer simulation; finite elements method; thin film; magnetic field sensors

Citation: Volchkov, S.O.; Pasynkova, A.A.; Derevyanko, M.S.; Bukreev, D.A.; Kozlov, N.V.; Svalov, A.V.; Semirov, A.V. Magnetoimpedance of CoFeCrSiB Ribbon-Based Sensitive Element with FeNi Covering: Experiment and Modeling. *Sensors* **2021**, *21*, 6728. https://doi.org/10.3390/s21206728

Academic Editor: Andreas Hütten

Received: 17 August 2021
Accepted: 5 October 2021
Published: 10 October 2021

Publisher's Note: MDPI stays neutral with regard to jurisdictional claims in published maps and institutional affiliations.

Copyright: © 2021 by the authors. Licensee MDPI, Basel, Switzerland. This article is an open access article distributed under the terms and conditions of the Creative Commons Attribution (CC BY) license (https://creativecommons.org/licenses/by/4.0/).

1. Introduction

Magnetic materials are widely used in electronics and biomedical applications [1,2]. There are different types of ferromagnets designed for existing and proposed technological applications in in different fields, including medicine [3,4]. Among others, magnetic materials for detectors of small magnetic fields were considered in this research area [5,6]. The magnetic effect that ensures the highest sensitivity with respect to an applied magnetic field is the magnetoimpedance (MI) [7–9]. It can be used both for detection of the biomagnetic signals closely related to the functional activities of the living systems and magnetic label detection [3,6,10,11]. Special attention was paid to the development of the devices for the detection of the nanoparticles inside the living cells, in the blood stream or incorporated into a natural tissue [3,12,13]. The first prototype of MI biosensor for the detection of magnetic nanoparticles of iron oxide in commercial water-based stable suspension (ferrofluid) employed a Co-based amorphous ribbon as a magnetic sensitive element [14]. For the magnetic biosensing of the magnetizable labels, flat geometry is crucial because the biochemistry step includes various processes of self-assembling and washing.

Nanoparticles were not immobilized at the surface of the sensitive element. As the biomedical applications request nanoparticles provided as ferrofluid, all model experiments with ferrofluid detection have an additional value. Latter, this kind of testing was used for characterization of the properties of different kind of stable suspensions [15]. Cobased amorphous ribbon is a well-known material with a very low negative value of the magnetostriction coefficient, low coercivity, high magnetic permeability [5–10], and also high temperature stability due to chromium additions [16,17]. One of the most interesting and well-understood features is the possibility of creation of the induced magnetic anisotropy of the desired type [7,18].

The MI effect is associated with the high magnetic softness of the ferromagnetic conductor and the possibility to create a well-defined uniaxial magnetic anisotropy in the MI sensitive element [18]. The magnetoimpedance phenomenon consists of a change of the complex electrical impedance of ferromagnetic conductor under an influence of an external magnetic field [7,19,20]. The MI value in the amorphous cobalt-based ribbons and wires can reach hundreds of percent in weak magnetic fields of the order of a few Oersted [7–9], which distinguishes it favorably from other magnetic effects [21,22]. Therefore, their application in the area of the magnetic sensors [6,9], adapted for biodetection, is actively discussed, and proven in different compact analytical devices prototypes [11,14].

MI can be described in terms of the classic skin-effect [7–9,20], which consists of the inhomogeneous distribution of the current density over the cross section of the conductor. Due to the skin-effect, the density of the alternating current decreases in the direction from the surface toward the central part of the conductor. The skin-effect can be characterized by the penetration depth (δ) or the skin-depth:

$$\delta = c/(\pi f \sigma \mu_t)^{\frac{1}{2}}, \qquad (1)$$

where σ—the conductivity, f—the alternating current (AC) frequency, μ_t—the dynamic effective transverse magnetic permeability. It can be seen that the greater the magnetic permeability, electrical conductivity, and AC frequency, the smaller the depth of the skin layer and the more pronounced the skin-effect. The stronger the skin-effect, the stronger the difference between the effective and geometric cross-section of the conductor. Despite the fact that these representations are quite straight forward, a simple derivation of analytical expressions for the impedance is possible only for a limited number of idealized symmetric homogeneous cases [7,14,15]. Ref. [14] describes the first MI-based prototype of the biosensor for magnetic label detection. The last case not only requires sufficient sensitivity of the prototype with respect to applied field, but the flat geometry is also very important as the biofunctionalization and other related processes take place in a liquid and require manipulation of the biofluids.

One of the main tasks of MI-sensor development is optimization of their MI parameters in accordance with particular application request. In the case of the homogeneous MI element, the main task is the corresponding optimization of its magnetic properties [7,19,20,23,24]. Much wider possibilities for the optimization of MI responses are available in the case of the multilayered structures. Variations of the geometric parameters are also challenging. For example, in comparison with a continuous film, much higher MI values were achieved in multilayered structures, in which a conductive layer is located between layers with a high magnetic permeability [25,26].

More complex layered structures were also proposed and investigated. For example, a new type of MI structures has been proposed and studied theoretically for the case of a multilayered structure consisting of a highly conductive central layer and two outer ferromagnetic layers below and above the conductive one. The upper layer is a periodic structure, it consists of N multilayer elements and N + 1 regions in which there are no multilayer elements (the upper layer is profiled). An electrodynamic model has been created that allows one to find the values of the transverse magnetic permeability for the upper and lower layers of the MI structure. It is shown that for a profiled structure with a decrease in the deviation angle of the effective magnetic anisotropy axis from the transverse direction, the magnetic permeability of the upper layer increases, which leads to an increase in the skin effect and an increase in the MI effect [27].

Permalloy layers obtained by magnetron sputtering have been repeatedly investigated by the scientific community. Their composition and magnetic anisotropy can be very well controlled in stable deposition conditions [28–30]. FeNi-based MI prototypes were designed and developed [26,27,31,32].

On the other hand, the tendency of the development of the functional multilayered structures (for example, coated amorphous ribbons) is dictated by the goal of the functionalization of the surface of the MI element. So, it seems expedient to use gold, magnetite (Fe_3O_4), reduced graphene oxide (rGO), or iron coatings [33–36]. They are widely adapted in the self-assembling processes in biotechnologies in the tasks of the marker biodetection [5,11]. This area has recently become a very hot topic, especially when focused on carbon containing nanostructures with enhance functional properties [34,35].

An optimization of the MI response of a layered structure involves varying the material of the layers, induced magnetic anisotropy features and the geometric dimensions. A direct experimental search for a solution to this problem is very laborious, and obtaining of the analytical expressions is much more difficult than in the case of homogeneous sensitive elements. Therefore, the finite element method (FEM) has recently been used by researchers in the field of MI [37–39]. This approach allows us both to find the optimum configuration of the MI element and to simulate its operation in various conditions. A lot of work has been done on modeling the MI response in thin film structures [12,24,39] including those adapted for biodetection [40]. However, there are almost no studies on modeling the MI response of amorphous ribbons particularly for specialized configurations [41,42]. This applies to an even greater extent to the amorphous ribbons with coatings that functionalize their surface. However, the use of amorphous ribbons for biodetection is actively discussed (for both label-free and magnetic label detection) [4,11,15]. For this reason, the development of the computer models for the estimation of the properties of MI elements based on amorphous ribbons seems to be an important task.

This work presents the results of comparative studies of the magnetoimpedance effect of the amorphous CoFeCrSiB ribbons both with and without the FeNi magnetic layer coating. The experimental data for particular conditions are used for extended FEM analysis and prediction of the MI behavior of complex composites. In addition, the results of computer simulation are presented for MI ribbon-based materials with other coatings (Cu, Au, Fe, Fe_3O_4), which could be used in order to functionalize their surface in real biomedical devices.

2. Materials and Methods

2.1. Samples and Experimental Methods

Amorphous $Co_{67}Fe_3Cr_3Si_{15}B_{12}$ ribbons (IR—ribbons in the initial state) were prepared by the rapid quenching from the melt onto the rotating copper weal technique. Quenching then proceeded in the air. The thickness and the width of the ribbons were 20 μm and 2 mm respectively. The following composition of amorphous ribbons was selected on the basis of previous studies at the same laboratory [43]. All previously defined structural and magnetic parameters were checked by corresponding techniques and their values were confirmed with the accuracy above the 5%: the ribbons had the crystallization temperature 570 °C and saturation magnetostriction $\lambda \approx -0.2 \times 10^{-6}$.

The FeNi/CoFeCrSiB/FeNi composites (R-FeNi) were obtained by the deposition of $Fe_{20}Ni_{80}$ thin films at the bottom and on the top planes of the IR ribbon. The deposition was done at room temperature of the substrate by the magnetron sputtering method in Ar-atmosphere on ATC Orion Series Sputtering Systems (AJA International, North Scituate, MA, USA). The following parameters were applied: the background pressure was 3.0×10^{-7} mbar and working argon pressure 3.8×10^{-3} mbar. For every deposition onto the ribbon surface, we add additional piece of the substrate in the area of uniform deposition and the composition was specially checked afterwards by energy dispersive X-ray analysis technique: the permalloy film composition was very close to $Fe_{20}Ni_{80}$

corresponding to zero magnetostriction with an accuracy of about 1%. More details on the deposition conditions can be found elsewhere [4].

The thickness of the FeNi layers was 1 µm on each plane of the ribbon sample (Figure 1a).

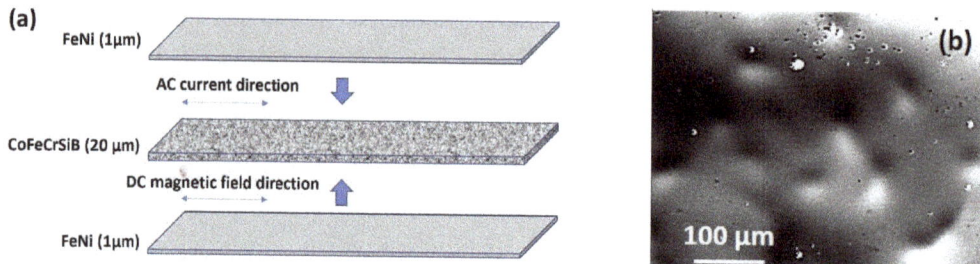

Figure 1. (a) Geometry of the model structure of FeNi/CoFeCrSiB/FeNi type. (b) Optical microscopy of the free surface of the amorphous CoFeCrSiB ribbon.

The optical microscopy photographs showed the surface features typical for amorphous ribbons of this composition. The most important point is that the FeNi deposited layer formed rough but continuous layers in both sides of the IR ribbons (Figure 1b). The observed roughness of the FeNi layer corresponded well to the features of the roughness of the ribbon itself. X-ray diffraction analysis of the IR was performed using a X'PERT PRO diffractometer (Philips, MX Amsterdam, The Netherlands) in Cu-Kα radiation.

Magnetic hysteresis loops of the IR samples were measured by the induction method by applying an external quasi-static magnetic field along the length of the sample (in-plane configuration). In addition, the amorphous ribbons were studied by the magneto-optical Kerr (MOKE) microscope and magnetometer (Evico magnetics GmbH, Dresden, Germany).

The magnetoimpedance measurements were carried out using the automatic system based on the Agilent impedance analyzer 4294A (Agilent/Keysight Technologies, Santa Rosa, CA, USA) using all necessary calibrations in order to extract the intrinsic impedance value corresponding to the signal of the magnetic sensitive element [9,42,43]. We studied the frequency range of an alternating current, f, of 0.1 to 70 MHz with an effective current value of 10 mA. The external magnetic field, H, was oriented along the long side of the sample, parallel to the direction of the flow of the alternating current, i.e., longitudinal MI configuration was employed. The maximum intensity of the external magnetic field, H_{max}, was as high as 150 Oe. The MI effect ratio was calculated as follows:

$$\Delta Z/Z(H) = \frac{Z(H) - Z(H_{max})}{Z(H_{max})} \times 100\%, \qquad (2)$$

where $Z(H)$ and $Z(H_{max})$ are the impedance modules in the magnetic fields H and H_{max}, respectively.

The magnetic field sensitivity of the impedance was determined by the expression:

$$S = \frac{\partial Z}{\partial H}, \qquad (3)$$

where ∂Z—is the impedance difference per magnetic field change discrete ∂H.

The samples for magnetic and MI measurements were 50 mm long.

2.2. Computer Simulations

The magnetodynamic behavior of the ribbons with and without covering was preliminary modeled using the computer simulation by the finite element method, which allows us to obtain a numerical solution of differential non-linear equations for a magnetoimpedance effect taking into account the geometry and physical-chemical parameters of the magnetic system [44]. The simulation was carried out using the Comsol Multiphysics 5.6 licensed software -core and AC/DC module (COMSOL AB, Stockholm, Sweden).

Typically, the shape and size of a finite element is determined by the geometrical parameters of the system in the conditions in which physical properties can be changed (for example, the presence of roughness, the size of the inhomogeneities of magnetic properties, etc.). In solving this problem, a tetrahedral non-structured mesh (network with uneven coupling) was used, since there is a large number of domains where the magnetic properties of the element are anisotropic. The solution uses a tetrahedral grid generator based on Delaunay algorithms [45].

The model of the amorphous CoFeCrSiB ribbon for computer simulation took into account the parallelogram geometry (coinciding with the shape of the sensitive element) (Figure 1a). Using software, the model was divided into tetrahedral sub-domains for 3D configuration. The size of the finite elements in the partition mesh depends on the wavelength of the electromagnetic excitation. A separate subsection in the form of hexahedral elements was established for the near-surface layer of the element (~20 nm), due to the prevailing current distribution in the condition of the significant skin-effect. The model with thin covering layers on top and at the bottom planes (Figure 1a) takes into account the program function of transient boundary conditions due to the presence of diffusion of the order of several nanometers between the covering and the ribbon and different propagation value of the electromagnetic wave in layers with different conductivity and magnetic permeability.

At the frequencies of alternating current above 1 MHz, the magnetization process is carried out only by the rotation of the magnetization vector. Therefore, the displacements of the magnetic domain boundaries cannot be taken into account as they are dumped by the eddy currents [7]. Therefore, the expression for the effective transverse magnetic permeability can be obtained using the procedure for minimizing the free energy functional, as described, for example, in [17]:

$$\mu_t = 1 + \frac{M_S \sin^2 \theta}{H \sin \theta + H_K \cos 2(\theta - \psi)}, \quad (4)$$

where M_S is the saturation magnetization, H_K is the transverse magnetic anisotropy field, ψ is the angle between the effective anisotropy axis and the transverse direction and θ is the angle between the magnetization and the direction of the external magnetic field H. In this case, the angle θ is related to H by the following expression:

$$H_K \sin(\theta - \psi) \cos(\theta - \psi) = H \cos \theta. \quad (5)$$

The values of M_S, H_K and ψ were set separately for each layer of the FEM model based on the results of work [19].

The models of the CoFeCrSiB ribbons with the following coatings were developed: FeNi, Cu, Au, Fe, Fe_3O_4. Copper and gold-based coatings have been selected as frequently used high conductivity materials. The designations for the samples analyzed in different models, as well as some properties of the coatings are given in Table 1. The thickness of the coatings in the FEM models varied from 10 to 1500 nm, which is the reasonable interval for continuous layers.

Table 1. Designation of the models (samples) of the CoFeCrSiB ribbons with various coatings with thickness 1 μm and some properties of coatings.

Designation of the Samples for Modelling	Structure	Covering	Magnetic Permeability (Calculated)	Electric Conductivity of Covering, S^{-1} (Constant)
IR	CoFeCrSiB	Absent	400–190,000	1.18×10^{16}
R-FeNi	Fe$_{20}$Ni$_{80}$/CoFeCrSiB/Fe$_{20}$Ni$_{80}$	Fe$_{20}$Ni$_{80}$	400–190,000	7.65×10^{17}
R-Cu	Cu/CoFeCrSiB/Cu	Cu	400–190,000	5.36×10^{15}
R-Au	Au/CoFeCrSiB/Au	Au	400–190,000	3.69×10^{15}
R-Fe	Fe/CoFeCrSiB/Fe	Fe	200–120,000	1.01×10^{7}
R-FeO	Fe$_3$O$_4$/CoFeCrSiB/Fe$_3$O$_4$	Fe$_3$O$_4$	1–2500	1.83×10^{17}

3. Results and Discussion

3.1. Experimental

According to X-ray analysis, the IR samples were amorphous. There was only an increase of intensity between the 2θ* angles 40°–55° corresponding to amorphous halo. No bright peaks corresponding to crystalline structures were observed.

Magnetometry studies in quasi-static conditions for the ribbon samples without and with FeNi covering have shown that only the magnetic saturation field and the saturation magnetization were slightly changed after FeNi covering of the initial CoFeCrSiB ribbon. The remnant magnetization and coercive force were unchanged (Figure 2a). Based on the shape of the magnetic hysteresis loops, it can be concluded that the effective magnetic anisotropy of both IR and R-FeNi samples is rather complex. $M(H)$ dependence is not linear for in-plane magnetization. On one hand, the coercivity is very small and magnetic hysteresis is almost negligible: $M(H)$ curves in increasing and decreasing magnetic fields are very close to each other. However, the slope is also small, and the saturation field is of the order of 1 Oe. Similar behavior was previously observed in Co-based rapidly quenched amorphous ribbons without additional heat treatments [46,47].

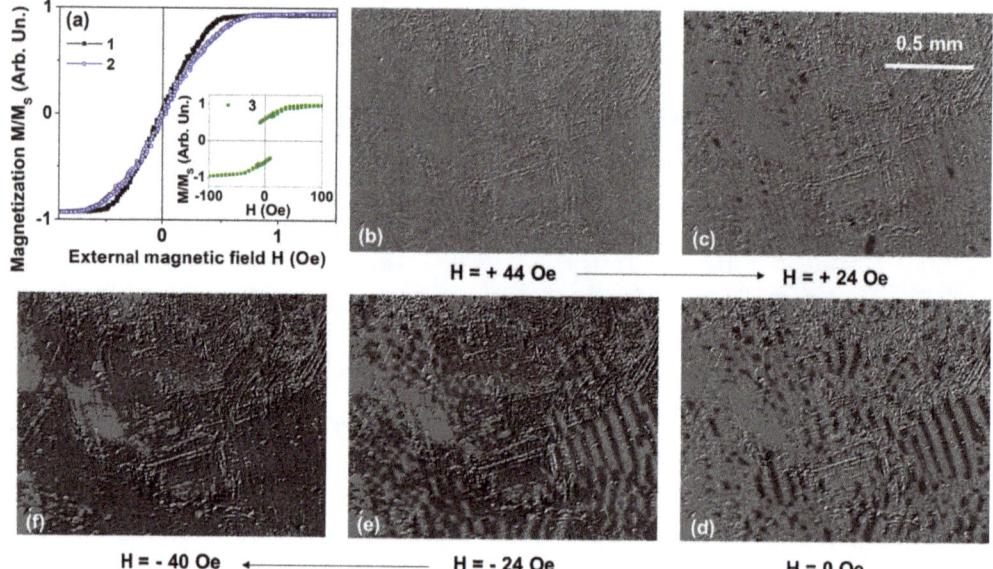

Figure 2. (a) Magnetoinductive magnetic hysteresis loops for the magnetic field applied in-plane of the ribbon and along the long side of it: (1)—the IR sample; (2)—the R-FeNi sample; and MOKE hysteresis loop (3)—Fe$_{20}$Ni$_{80}$ film 1 μm thick. (b–f) Magnetic domains of Co$_{67}$Fe$_3$Cr$_3$Si$_{15}$B$_{12}$ IR-samples starting from magnetic saturation state in the high positive magnetic field.

For the description of the effective magnetic anisotropy, one should consider the main contribution to be a longitudinal magnetic anisotropy with the anisotropy axis oriented along the long side of the ribbon. Figure 2 describes the surface magnetic domains revealed by the MOKE microscopy under applications of the external magnetic field in the plane of the sample along the long side of the ribbon starting with high positive magnetic field in the sequence order: (b)–(c)–(d)–(e)–(f).

One can see that magnetization process is indeed quite complex but expected for Co-based rapidly quenched amorphous ribbons without additional heat treatments [39]. It is very probable that surface domains reflect the domain structure inside the sample in a very rough way and that the surface anisotropy contribution is quite high. Magneto-optical studies of the FeNi layer of the R-FeNi sample showed that the shape of the magnetic hysteresis loops corresponds to the so-called "transcritical" state, due to the occurrence of anisotropy of the perpendicular surface [6,22,48]. It is associated with the formation of perpendicular magnetic anisotropy component, formation microstructure columnar during sputtering of the FeNi coating deposition, and the appearance of the stripe domains. Therefore, magnetron sputtering was chosen as a method for obtaining nanocrystalline permalloy with stable structural characteristics and well-defined magnetic anisotropy. However, different deposition techniques such as electrodeposition can also be effective [49,50].

As the volume of the permalloy film is small in comparison with the volume of the amorphous ribbon in the R-FeNi composite, parameters such as saturation magnetization, remanence magnetization and magnetic coercivity of the composite are close to the parameters of the ribbon.

Magnetoimpedance dependences for both IR and R-FeNi samples again have a complex shape. On one hand, the curve tends to approach a "two peaks" shape (Figure 3a) with two maxima near the field close to magnetic anisotropy fields for positive and negative H values. At the same time the $\Delta Z/Z(H = 0$ Oe$) \neq \Delta Z/Z(H = 150$ Oe$)$ and the observed difference is quite significant. This means the existence of two strong contributions to the effective magnetic anisotropy: the longitudinal (most probably corresponding to the bulk part of the ribbon [51,52]) and the transverse (related to the surface). This, like the hysteresis loops (Figure 2a), indicates predominantly longitudinal magnetic anisotropy but with very strong contribution of the transverse component [19,52]. In the case of the IR sample, the increase of the frequency from 1 to 10 MHz results in the significant decrease of the $\Delta Z/Z$ ratio and the appearance of the $\Delta Z/Z$ maximum in the lower external magnetic field. This is consistent with the supposition that the transverse anisotropy component is mostly associated with the surface anisotropy and a fairly large local anisotropy axis distribution near the surface. The last supposition is confirmed by the magnetic domains observations in the IR sample.

The MI of FeNi-coated ribbons is noticeably lower (Figure 3) than the initial ribbons in the entire frequency range. In part, this is due to a decrease in the transverse effective magnetic permeability due to the "transcritical" state of the FeNi coating.

However, the shape of the $\Delta Z/Z(H)$ curve for f = 1 MHz is much closer to the "one peak" shape corresponding to the longitudinal effective anisotropy ($\Delta Z/Z(H = 0$ Oe$) \sim \Delta Z/Zmax$) with small surface-related peak height anisotropy [53]. As the sputtered FeNi layer due to the shading effect has non-uniform thickness but at a time reduces the surface roughness and insures the better closing of a magnetic flux near the surface favoring the longitudinal anisotropy contribution. The effect of the coating is increased with the increase in the alternating current frequency, due to the skin-effect (see Expression (1)). So, at a frequency of 10 MHz and above, there is a sharp decrease in the MI of the R-FeNi samples, as well as a significant shift of the maximum to the region of large values of the magnetic fields. This phenomenon can be explained by the flow of alternating current mainly in the permalloy layer due to the skin effect. The presence of maxima of the $\Delta Z/Z(H)$ curves in the case of composites also indicates the presence of a transverse magnetic anisotropy component but the main decay of the $\Delta Z/Z$ ratio at high frequencies can be due to the flow of the alternating current mainly over the region with very high inhomogeneities in the structure (Inset Figure 3a).

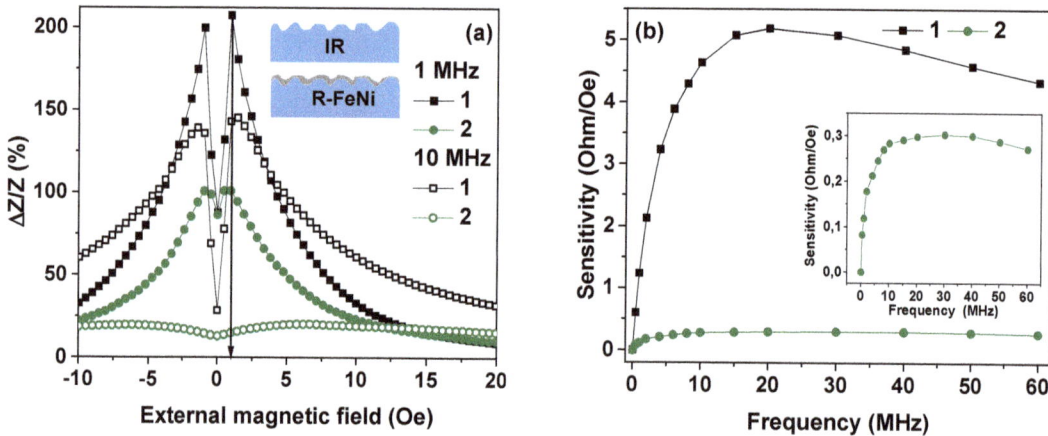

Figure 3. (a) Dependencies of the magnetoimpedance ratio on the value of the external magnetic field for different frequencies for the sample IR (black symbols, 1) and for the sample R-FeNi (green symbols, 2). The inset shows the description of the cross section of the IR ribbon and of the ribbon-based R-FeNi composite (not in the real scale) (b) Sensitivity dependencies of the MI ratio changes.

The magnetic field sensitivity of the impedance of both IR and R-FeNi structures, calculated using Expression (3), reaches a maximum, S_{max}, in the range of the magnetic fields from 0 to H, corresponding to $(\Delta Z/Z)_{max}$. Frequency dependences of S_{max} are shown in Figure 3b. It can be seen that S_{max} changes non-monotonically with AC frequency increase. In the case of uncoated ribbons, the highest sensitivity of about 5 Ohm/Oe was observed at the frequency of 20 MHz. For the coated ribbons, the highest sensitivity was about 0.3 Ohm/Oe and was observed at the frequency of 30 MHz. The sensitivity of the MI of the R-FeNi samples with respect to the applied magnetic field is an order of magnitude lower than the sensitivity of the IR samples in the entire investigated frequency range. However, the obtained result can be used for estimation of the behavior of different composites with variation of the type and the thickness of the covering. Such a usage of modern computer technologies is very useful for complex and time-consuming technological processes.

3.2. Computer Simulation

Figure 4 shows the results of the computer simulation of the MI responses of IR and R-FeNi ribbons. It is very important to mention that the values of M_S = 450 Gs, H_K = 0.75 Oe, and σ (Table 1) required for FEM modeling were determined from experimental data obtained in the present study (ψ = 0.1 rad). The choice of the angle of anisotropy was based on past parametric studies [19,44] in such a way that the maximum magnetic permeability of a sample with a transverse uniaxial magnetic anisotropy was achieved at a nonzero value of the linear magnetization component. The simulated Z(H) dependences have the same character as the experimental ones for the uncoated ribbons. For the AC frequencies below 10 MHz the experimental and simulated curves are very close to each other. In the case of the case of coated ribbons, their character repeats the character of the experimental dependences only in the current frequency range of 1 MHz and below. However, it was possible to reproduce the main experimental result for given thickness of the covering with the help of simulation. MI significantly decreases after the FeNi layer deposition onto both surfaces of the amorphous CoFeCrSiB ribbon.

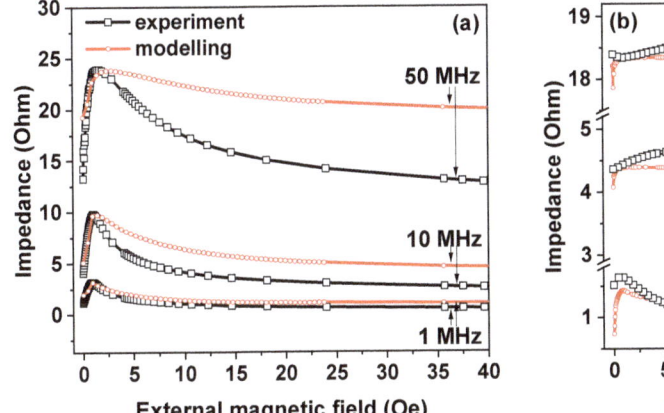 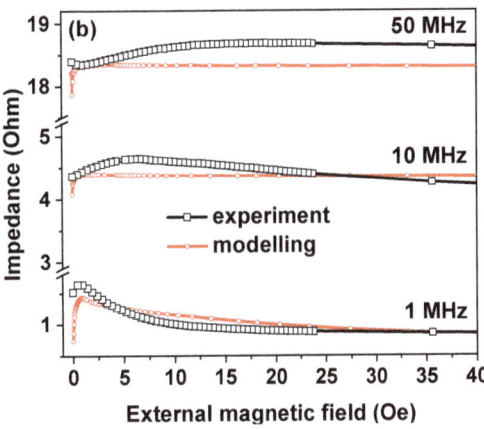

Figure 4. Difference between the experimental field dependence of the total impedance value and the same parameter obtained by the computer simulation: (**a**) for the IR (**b**) for the R-FeNi samples with a thickness of FeNi equal to 1 µm.

The main observations for the obtained experimental and calculated dependences with an increase in the AC frequency can be summarized as follows:

(a) The proposed FEM model is very simple, and it does not take into account the frequency dependence of the magnetic permeability, which can be a rather complex function for the frequency range under consideration (see Expressions (4) and (5)) [51].

(b) The non-flat morphology of the ribbon's surface is not considered (Figures 1b and 3a). Irregularities on the surface lead to a dispersion of the anisotropy [52,54], i.e., ψ (see Expressions (4) and (5)) takes on different values at different points of the ribbon's surface.

(c) The dispersion of the anisotropy over the cross section of the ribbon is not taken into account [52,54,55].

(d) The magnetic interaction peculiarities, including the features related to the variation of the additional layer thickness, for the amorphous ribbon and the FeNi coating are not considered.

Nevertheless, even the presented FEM solved model reproduces the experimental results at a most simple qualitative level.

Keeping in mind the fact that surface properties of the ribbons can be adjusted by the fabrication in different conditions, in addition to our experimental studies of IR and R-FeNi ribbons and their MI modeling, we also investigated the ribbons with different FeNi coating thicknesses. The results are shown in Figure 5. It can be seen that the larger the MI, the smaller the thickness of the magnetic coating layer. Improvement in the MI effect is possible for a deposition of the FeNi layer of 10 nm. In this case, rather satisfactory sensitivity with respect to the applied magnetic field of 25 Ohm/Oe was obtained.

Note that the obtained dependence of MI on the coating thickness can also be used for the development of the sensors for reactive chemical agents that cause coating dissolution (destructive detection scheme) [52].

As the multicomponent detection includes various requirements, we also simulated MI of the ribbons with such types of coatings as Cu, Au, Fe, and Fe_3O_4. Earlier it was shown that MI elements based on amorphous ribbons can be effectively used as a part of the biosensor prototype to detect functional properties of biofluids [52] confirming the idea that MI CoFeSiB amorphous ribbon-based elements can be used in chemical sensors. However, the goal of the present study is different—to obtain a stable MI response in chemically active biological medium for possible detection of magnetically labeled biocomponents of interest. Therefore, CoFeCrSiB ribbons were employed to insure this stability.

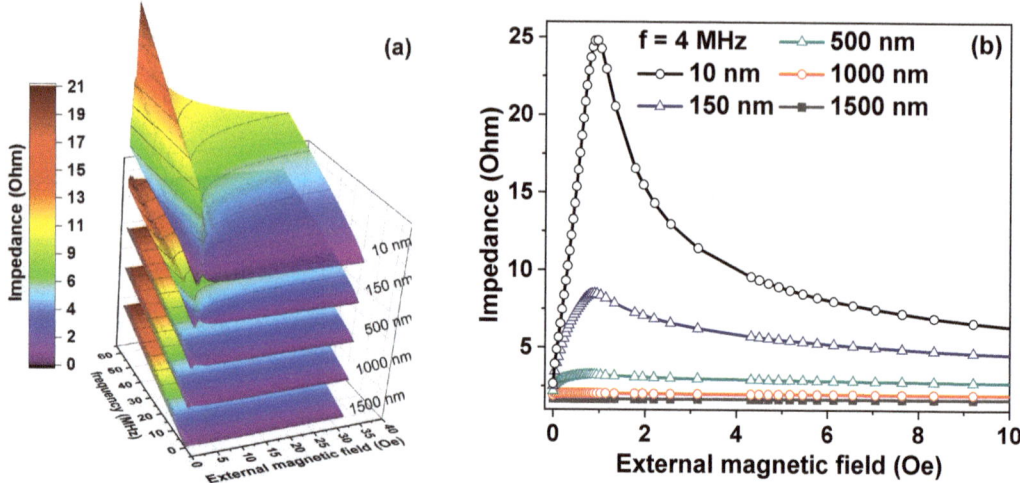

Figure 5. (**a**) Computing simulation of the dependence of the impedance modulus on the frequency and external magnetic field for different thickness of FeNi in the composite structure R-FeNi. (**b**) Computing modelling of absolute value of magnetoimpedance for frequency 4 MHz for different thicknesses of FeNi in the composite structure R-FeNi.

The results of our studies suggest that the deposition of a thin-film coating with high magnetic permeability or high electrical conductivity on an amorphous ribbon can contribute to a significant increase in the sensitivity of sensors for aggressive chemical media based on MI. At the same time, it is necessary for detected medium to dissolve the thin-film coating. Figure 6a shows the magnetic field dependences of the impedance of the composite on the basis of amorphous ribbon obtained for coating thickness of 1000 nm. It can be seen that the most significant decrease in MI is caused by coatings with high conductivity (Au, Cu) and high magnetic permeability (Fe, FeNi) (see Table 1). In the case of the ribbons coated with Fe_3O_4, the high MI is retained (Figure 6b).

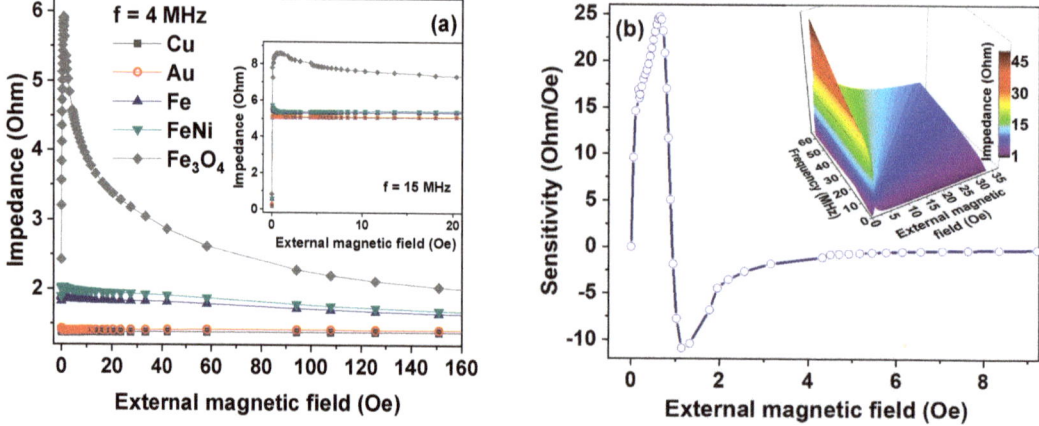

Figure 6. (**a**) Computer simulation of the field dependence of the impedance modulus at 4 MHz for the R-Cu, R-Au, R-Fe, R-FeNi and R-FeO samples (see Table 1) with the same covering thickness of the upper and lower layers of 1000 nm. The inset shows data for excitation current frequency f = 15 MHz. (**b**) Computer simulation of maximum sensitivity for structure R-FeO with thickness of upper and lower layers of 10 nm on 20 MHz. The inset shows dependence of absolute impedance value on frequency and external magnetic field values for this structure.

As noted in the introduction, the use of gold is considered in the development of biosensors of very different types, including magnetic biosensors based on MI as the gold coverings serve to create a self-assembling layers. The presented results of the modeling the effect of Au coating on MI can be useful for these purposes.

It is also important to emphasize that there is a large difference between the MI responses of Fe-coated ribbons and Fe_3O_4-coated ribbons because their maximum impedance differs by a factor of at least 6. As it is well known [56,57], the Fe_3O_4 layer is easily formed during the oxidation of Fe. Therefore, one can use the amorphous ribbon with a sufficient amount of iron in the composition or obtained required cover layer on the basis of Fe-coated ribbons, or depositthe Fe_3O_4 layer by sputtering technique or laser evaporation. In addition, one can compare the results of modeling with some experimental data available in the literature. The first attempt to obtain ribbon/film composite was made by Cerdeira et al. [33]. They reported on the experimental study of the CoFeMoSiB amorphous ribbons with the iron covering in the thickness interval of 0 to 240 nm. To some extent, the properties of CoFeCrSiB and CoFeMoSiB ribbons are similar and therefore comparison is valid and useful. However, the analysis of all obtained results indicates that lower thicknesses can be more effective, but the surface roughness can play very important role in contributing to the non-uniformity of the anisotropy of the amorphous ribbon-based composites. An additional iron layer can cause a modification of the roughness of the surface by a decrease of the depth of the Fe pores (see also Figure 3a).

As mentioned above, experimental studies have been undertaken into ribbon/film based composites. Ref. [58] reports a comparative study of uncoated (Fe50Ni50)81Nb7B12 nanocrystalline ribbons and ribbons coated with 120 nm of Co on both surfaces. The impact of the Co coating on the high frequency impedance of the (Fe50Ni50)81Nb7B12 ribbon was studied with techniques sensitive to surface magnetism. However, the thickness dependence was not discussed. We therefore propose to use the advantages of modeling techniques to make part of the studies via development of the appropriate models based on some experimental data.

In ref. [59] the authors studied copper oxide (CuO) film covering role on a surface of Co-based amorphous ribbon deposited using chemical successive ionic layer adsorption and reaction technique. The results showed that Co-based amorphous ribbons, which are coated CuO film, have a significant effect on the value and operation frequency for the MI effect as compared to the samples without coating. However, the overall MI ratio value was not very high.

In ref. [60] the authors describe a 50 nm-thick Co film grown either on the free surface or on the wheel-side surface of Co84.55Fe4.45Zr7B4 amorphous ribbons. They showed that the presence of the Co coating layer enhances both the MI ratio and its sensitivity. However, the maximum MI ratio was below 24%. This topic become very popular in recent years, and areas of interest include ideas for enhancing the GMI effect with both 3D metal and dielectric coatings [59–62]. However, this work is the first attempt to propose a methodology of the development of such a composite using modelization advantages which can be useful for the analysis of the experimental data of different authors.

4. Conclusions

The magnetoimpedance effect was experimentally investigated in amorphous $Co_{67}Fe_3Cr_3Si_{15}B_{12}$ ribbons, both coated with a 1 µm thick $Fe_{20}Ni_{80}$ layer and without coating. The experimental studies were combined with computer FEM modeling in the range of the alternating current frequencies of 0.1 to 70 MHz.

It was found experimentally that the presence of the FeNi layer leads to a significant decrease in MI and a decrease in its magnetic field sensitivity. Thus, in the case of uncoated ribbons, the sensitivity reaches 5 Ohm/Oe at the AC frequency of 20 MHz, while in the case of FeNi-coated ribbons, the maximum sensitivity is 0.3 Ohm/Oe at the frequency of 30 MHz. The significant decrease in MI and the decrease in its magnetic field sensitivity after coating is associated with a decrease in the transverse effective magnetic permeability due to its "transcritical" state. The "transcritical" state of the FeNi coating was demonstrated using magneto-optical Kerr effect studies.

During computer simulation, it was possible to reproduce the main experimental result—the significant decrease in MI in FeNi-coated ribbons. In addition, it was shown that the simulated magnetoimpedance dependences are very close to the experimental ones at AC frequencies below 10 MHz for the uncoated ribbons and at the frequencies of 1 MHz and below for FeNi-coated ribbons. It was found that the discrepancy between the simulated and experimental dependences increases with an increase in the frequency of the alternating current. We proposed various methods to refine the computer model to achieve a better agreement between the simulated and experimental dependences in a wider AC frequencies range, taking into account the frequency dependence of the magnetic permeability, anisotropy dispersion, and the magnetic interaction of the coating and ribbon.

In addition, models were developed for the ribbons with different thickness of $Fe_{20}Ni_{80}$ coating and for the ribbons with such coatings as: Cu, Au, Fe, Fe_3O_4.

Some of the simulation results can find application in the creation of chemical sensors. Thus, the discovered strong dependence of MI on the coating thickness can be used in the development of sensors for chemical agents that release the coating. At the same time, the large difference in the MI response of Fe-coated and Fe_3O_4-coated ribbons can be used as the basis for sensors of chemicals that oxidize iron. It was also noted that the results of modeling Au coated ribbons can be useful in the design of magnetic biosensors.

Author Contributions: Conceptualization, A.V.S. (Alexander V. Semirov) and A.V.S. (Andrey V. Svalov); methodology, M.S.D., D.A.B., A.V.S. (Alexander V. Semirov), A.V.S. (Andrey V. Svalov); software, S.O.V. and N.V.K.; validation, A.A.P., M.S.D., N.V.K. and A.V.S. (Andrey V. Svalov); formal analysis M.S.D., D.A.B., A.A.P.; investigation, M.S.D., D.A.B., A.V.S. (Andrey V. Svalov), N.V.K. and A.A.P.; writing—original draft preparation, M.S.D., D.A.B., S.O.V., A.A.P.; writing—review and editing, M.S.D., D.A.B., A.A.P.; visualization, M.S.D., D.A.B., S.O.V. and N.V.K.; supervision, A.V.S. (Alexander V. Semirov) and A.V.S. (Andrey V. Svalov). All authors have read and agreed to the published version of the manuscript.

Funding: This research was funded by Ministry of Science and Higher Education of the Russian Federation, grant number FEUZ-2020-0051, Ministry of Science and Higher Education of the Russian Federation, grant number AAAA-A19-119070890020-3, Act 211 Government of the Russian Federation, grant number 02. A03.21.0006.

Institutional Review Board Statement: Not applicable.

Informed Consent Statement: Not applicable.

Data Availability Statement: Data available from the corresponding author upon reasonable request.

Conflicts of Interest: The authors declare no conflict of interest.

References

1. Ferreira, H.A.; Graham, D.L.; Freitas, P.P.; Cabral, J.M.S. Biodetection using Magnetically Labeled Biomolecules and Arrays of Spin Valve Sensors. *J. Appl. Phys.* **2002**, *93*, 7281–7286. [CrossRef]
2. Coisson, M.; Barrera, G.; Celegato, F.; Martino, L.; Vinai, F.; Martino, P.; Ferraro, G.; Tiberto, P. Specific Absorption Rate Determination of Magnetic Nanoparticles through Hyperthermia Measurements in Non-Adiabatic Conditions. *J. Magn. Magn. Mater.* **2016**, *415*, 2–7. [CrossRef]
3. Zamani Kouhpanji, M.R.; Stadler, B.J.H. A Guideline for Effectively Synthesizing and Characterizing Magnetic Nanoparticles for Advancing Nanobiotechnology: A Review. *Sensors* **2020**, *20*, 2554. [CrossRef] [PubMed]
4. Svalov, A.V.; Aseguinolaza, I.R.; Garcia-Arribas, A.; Orue, I.; Barandiaran, J.M.; Alonso, J.; Fernández-Gubieda, M.L.; Kurlyandskaya, G.V. Structure and Magnetic Properties of Thin Permalloy Films Near the "Transcritical" State. *IEEE Trans. Magn.* **2010**, *46*, 333–336. [CrossRef]
5. Fal-Miyar, V.; Kumar, A.; Mohapatra, S.; Shirley, S.; Frey, N.A.; Barandiaránd, J.M.; Kurlyandskaya, G.V. Giant Magnetoimpedance for Biosensing in Drug Delivery. *Appl. Phys. Lett.* **2007**, *91*, 143902.
6. Uchiyama, T.; Mohri, K.; Honkura, Y.; Panina, L.V. Recent Advances of Pico-Tesla Resolution Magneto-Impedance Sensor Based on Amorphous Wire CMOS IC MI Sensor. *IEEE Trans. Magn.* **2012**, *48*, 3833–3839. [CrossRef]

7. Makhotkin, V.E.; Shurukhin, B.P.; Lopatin, V.A.; Marchukov, P.Y.; Levin, Y.K. Magnetic Field Sensors Based on Amorphous ribbons. *Sens. Actuators A Phys.* **1991**, *27*, 759–762. [CrossRef]
8. Beach, R.S.; Berkowitz, A.E. Giant Magnetic Field Dependent Impedance of Amorphous FeCoSiB Wire. *Appl. Phys. Lett.* **1994**, *64*, 3652–3654. [CrossRef]
9. Bukreev, D.A.; Derevyanko, M.S.; Moiseev, A.A.; Semirov, A.V.; Savin, P.A.; Kurlyandskaya, G.V. Magnetoimpedance and Stress-Impedance Effects in Amorphous CoFeSiB Ribbons at Elevated Temperatures. *Materials* **2020**, *13*, 3216. [CrossRef]
10. Chiriac, H.; Herea, D.D.; Corodeanu, S. Microwire Array for Giant Magnetoimpedance Detection of Magnetic Particles for Biosensor Prototype. *J. Magn. Magn. Mater.* **2007**, *311*, 425–428. [CrossRef]
11. Yang, Z.; Wang, H.; Guo, P.; Ding, Y.; Lei, C.; Luo, Y. A Multi-Region Magnetoimpedance-Based Bio-Analytical System for Ultrasensitive Simultaneous Determination of Cardiac Biomarkers Myoglobin and C-Reactive Protein. *Sensors* **2018**, *18*, 1765. [CrossRef]
12. Kurlyandskaya, G.V.; Fernández, E.; Safronov, A.P.; Svalov, A.V.; Beketov, I.; Burgoa Beitia, A.; García-Arribas, A.; Blyakhman, F.A. Giant Magnetoimpedance Biosensor for Ferrogel Detection: Model System to Evaluate Properties of Natural Tissue. *Appl. Phys. Lett.* **2015**, *106*, 193702. [CrossRef]
13. Blanc-Béguin, F.; Nabily, S.; Gieraltowski, J.; Turzo, A.; Querellou, S.; Salaun, P.Y. Cytotoxicity and GMI Bio-Sensor Detection of Maghemite Nanoparticles Internalized into Cells. *J. Magn. Magn. Mater.* **2009**, *321*, 192–197. [CrossRef]
14. Kurlyandskaya, G.V.; Sánchez, M.L.; Hernando, B.; Prida, V.M.; Gorria, P.; Tejedor, M. Giant-Magnetoimpedance-Based Sensitive Element as a Model for Biosensors. *Appl. Phys. Lett.* **2003**, *82*, 3053. [CrossRef]
15. Amirabadizadeh, A.; Lotfollahi, Z.; Zelati, A. Giant Magnetoimpedance Effect of Co68.15 Fe4.35 Si12.5 B15 Amorphous Wire in the Presence of Magnetite Ferrofluid. *J. Magn. Magn. Mater.* **2016**, *415*, 102–105. [CrossRef]
16. Kurlyandskaya, G.V.; Dmitrieva, N.V.; Lukshina, V.A.; Potapov, A.P. The thermomechanical treatment of an amorphous Co-based alloy with low Curie temperature. *J. Magn. Magn. Mater.* **1996**, *160*, 307–308. [CrossRef]
17. Nosenko, A.V.; Kyrylchuk, V.V.; Semen'ko, M.P.; Nowicki, M.; Marusenkov, A.; Mika, T.M.; Semyrga, O.M.; Zelinska, G.M.; Nosenko, V.K. Soft Magnetic Cobalt Based Amorphous Alloys with Low Saturation Induction. *J. Magn. Magn. Mater.* **2020**, *515*, 167328. [CrossRef]
18. Kurlyandskaya, G.V.; Garcia-Arribas, A.; Barandiaran, J.M.; Kisker, E. Giant Magnetoimpedance Stripe and Coil Sensors. *Sens. Actuators A* **2001**, *91*, 116–119. [CrossRef]
19. Kraus, L. Theory of Giant Magneto-Impedance in the Planar Conductor with Uniaxial Magnetic Anisotropy. *J. Magn. Magn. Mater.* **1999**, *195*, 764–778. [CrossRef]
20. Kraus, L. GMI Modeling and Material Optimization. *Sens. Actuators A Phys.* **2003**, *106*, 187–194. [CrossRef]
21. Catalan, G. Magnetocapacitance without Magnetoelectric Coupling. *Appl. Phys. Lett.* **2006**, *88*, 102902. [CrossRef]
22. Knobel, M.; Vázquez, M.; Kraus, L. Giant Magnetoimpedance. In *Handbook of Magnetic Materials*; Buschow, K.H.J., Ed.; North-Holland Elsevier Science, B.V.: Amsterdam, The Netherlands, 2003; Volume 5, pp. 497–563.
23. Derevyanko, M.S.; Bukreev, D.A.; Moiseev, A.A.; Kurlyandskaya, G.V.; Semirov, A.V. Effect of Heat Treatment on the Magnetoimpedance of Soft Magnetic Co68.5Fe4Si15B12.5 Amorphous Ribbons. *Phys. Met. Metallogr.* **2020**, *121*, 28–31. [CrossRef]
24. Semirov, A.V.; Derevyanko, M.S.; Bukreev, D.A.; Moiseev, A.A.; Kurlandskaya, G.V. High Frequency Impedance of Cobalt-Based Soft Magnetic Amorphous Ribbons Near the Curie Temperature. *Bull. Russ. Acad. Sci. Phys.* **2014**, *78*, 81–84. [CrossRef]
25. Xiao, S.; Liu, Y.; Yan, S.; Dai, Y.; Zhang, L.; Mei, L. Giant Magnetoimpedance and Domain Structure in FeCuNbSiB Films and Sandwiched Films. *Phys. Rev. B* **2000**, *61*, 5734–5739. [CrossRef]
26. García-Arribas, A.; Fernández, E.; Svalov, A.; Kurlyandskaya, G.V.; Barandiaran, J.M. Thin-Film Magneto-Impedance Structures with Very Large Sensitivity. *J. Magn. Magn. Mater.* **2016**, *400*, 321–326. [CrossRef]
27. Buznikov, N.A.; Kurlyandskaya, G.V. Magnetic Impedance of Periodic Partly Profiled Multilayered Film Structures. *Phys. Met. Metall.* **2021**, *122*, 755–760. [CrossRef]
28. Wang, Z.; Dai, B.; Zhang, Y.; Ren, Y.; Tan, S.; Zeng, L.; Ni, J.; Li, J. High Resonance Frequencies Induced by In-Plane Antiparallel Magnetization in NiFe/FeMn Bilayer. *J. Magn. Magn.Mater.* **2020**, *514*, 167139. [CrossRef]
29. Yang, X.; Zhang, S.; Li, Q.; Zhao, G.; Li, S. The Abnormal Damping Behavior Due to the Combination between Spin Pumping and Spin Back Flow in Ni80Fe20/Rut Bilayers. *J. Magn. Magn.Mater.* **2020**, *502*, 166495. [CrossRef]
30. Li, X.; Sun, X.; Wang, J.; Liu, Q. Magnetic Properties of Permalloy Films with Different Thicknesses Deposited onto Obliquely Sputtered Cu Under Layers. *J. Magn. Magn. Mater.* **2015**, *377*, 142–146. [CrossRef]
31. Kurlyandskaya, G.V.; de Cos, D.; Volchkov, S.O. Magnetosensitive Transducers for Nondestructive Testing Operating on the Basis of the Giant Magnetoimpedance Effect: A Review. *Russ. J. Nondestruct. Test.* **2009**, *45*, 377–398. [CrossRef]
32. Corrêa, M.A.; Bohn, F.; Chesman, C.; da Silva, R.B.; Viegas, A.D.C.; Sommer, R.L. Tailoring the Magnetoimpedance Effect of NiFe/Ag Multilayer. *J. Phys. D Appl. Phys.* **2010**, *43*, 295004. [CrossRef]
33. Cerdeira, M.A.; Kurlyandskaya, G.V.; Fernandez, A.; Tejedor, M.; Garcia-Miquel, H. Giant Magnetoimpedance Effect in Surface Modified CoFeMoSiB Amorphous Ribbons. *Chin. Phys. Lett.* **2003**, *20*, 2246–2249. [CrossRef]
34. Yang, Z.; Lei, C.; Sun, X.C.; Zhou, Y.; Liu, Y. Enhanced GMI Effect in Tortuous-Shaped Co-Based Amorphous Ribbons Coated with Graphene. *J. Mater. Sci.-Mater. Electron.* **2016**, *27*, 3493–3498. [CrossRef]
35. Chen, Y.; Zou, J.; Shu, X.; Song, Y.; Zhao, Z. Enhanced Giant Magneto-Impedance Effects in Sandwich FINEMET/rGO/FeCo Composite Ribbons. *Appl. Surf. Sci.* **2021**, *545*, 149021. [CrossRef]

36. Mukherjee, D.; Devkota, J.; Ruiz, A.; Hordagoda, M.; Hyde, R.; Witanachchi, S.; Mukherjee, P.; Srikanth, H.; Phan, M.H. Impacts of Amorphous and Crystalline Cobalt Ferrite Layers on the Giant Magneto-Impedance Response of a Soft Ferromagnetic Amorphous Ribbon. *J. Appl. Phys.* **2014**, *116*, 123912. [CrossRef]
37. Li, B.; Kosel, J. Three Dimensional Simulation of Giant Magneto-Impedance Effect in Thin Film Structures. *J. Appl. Phys.* **2011**, *109*, 07E519. [CrossRef]
38. Yabukami, S.; Mawatari, H.; Horikoshi, N.; Murayama, Y.; Ozawa, T.; Ishiyama, K.; Arai, K.I. A Design of Highly Sensitive GMI Sensor. *J. Magn. Magn. Mater.* **2005**, *290*, 1318–1321. [CrossRef]
39. García-Arribas, A. The Performance of the Magneto-Impedance Effect for the Detection of Superparamagnetic Particles. *Sensors* **2020**, *20*, 1961. [CrossRef]
40. Mansourian, S.; Bakhshayeshi, A.; Taghavi Mendi, R. Giant Magneto-Impedance Variation in Amorphous CoFeSiB Ribbons as a Function of Tensile Stress and Frequency. *Phys. Lett. A* **2020**, *384*, 126657. [CrossRef]
41. Chlenova, A.A.; Kozlov, N.V.; Lukshina, V.A.; Neznakhin, D.S.; Kurlyandskaya, G.V. Giant Magnetoimpedance in the Ferromagnetic Amorphous Alloys. In *AIP Conference Proceedings*; AIP Publishing LLC: Melville, NY, USA, 2020; Volume 2313, p. 030073.
42. Gazda, P.; Nowicki, M.; Szewczyk, R. Comparison of Stress-Impedance Effect in Amorphous Ribbons with Positive and Negative Magnetostriction. *Materials* **2019**, *12*, 275. [CrossRef]
43. Dmitrieva, N.V.; Kurlyandskaya, G.V.; Lukshina, V.A.; Potapov, A.P. The Recovery Kinetics of the Magnetic Anisotropy Induced by Stress Annealing of the Amorphous Co-Based Alloy with Low Curie Temperature. *J. Magn. Magn. Mater.* **1999**, *196*, 320–321. [CrossRef]
44. Kozlov, N.V.; Chlenova, A.A.; Volchkov, S.O.; Kurlyandskaya, G.V. The Study of Magnetic Permeability and Magnetoimpedance: Effect of Ferromagnetic Alloy Characteristics. In *AIP Conference Proceedings*; AIP Publishing LLC: Melville, NY, USA, 2020; Volume 2313, p. 030050.
45. Hurtado, F.; Noy, M.; Urrutia, J. Flipping Edges in Triangulations. *Discret. Comput. Geom.* **1999**, *22*, 333–346. [CrossRef]
46. Hubert, A.; Schäfer, R. *Magnetic Domains*; Springer: Berlin, Germany, 1998.
47. Saito, N.; Fujiwara, H.; Sugita, Y. A New Type Magnetic Domain in Negative Magnetostriction Ni-Fe Films. *J. Phys. Soc. Jpn.* **1964**, *19*, 1116–1125. [CrossRef]
48. Svalov, A.V.; Kurlyandskaya, G.V.; Hammer, H.; Savin, P.A.; Tutynina, O.I. Modification of the "Transcritical" State in Ni75Fe16Cu5Mo4 Films Produced by RF Sputtering. *Tech. Phys.* **2004**, *49*, 868–871. [CrossRef]
49. Zubar, T.I.; Sharko, S.A.; Tishkevich, D.I.; Kovaleva, N.N.; Vinnik, D.A.; Gudkova, S.A.; Trukhanov, E.L.; Trofimov, E.A.; Chizhik, S.A.; Panina, L.V.; et al. Anomalies in Ni-Fe Nanogranular Films Growth. *J. All. Comp.* **2018**, *748*, 970–978. [CrossRef]
50. Torabinejad, V.; Aliofkhazraei, M.; Assareh, S.; Allahyarzadeh, M.H.; Sabour Rouhaghdam, A. Electrodeposition of Ni-Fe Alloys, Composites, and Nano Coatings-A Review. *J. All. Comp.* **2017**, *691*, 841–859. [CrossRef]
51. Chen, D.; Muñoz, J.; Hernando, A.; Vázquez, M. Magnetoimpedance of Metallic Ferromagnetic Wires. *Phys. Rev. B* **1998**, *57*, 10699–10704. [CrossRef]
52. Kurlyandskaya, G.V.; Fal Miyar, V. Surface Modified Amorphous Ribbon Based Magnetoimpedance Biosensor. *Biosens. Bioelectron.* **2007**, *22*, 2341–2345. [CrossRef]
53. Vazquez, M.; Sinnecker, J.P.; Kurlyandskaya, G.V. Hysteretic Behavior and Anisotropy Fields in the Magneto-Impedance Effect. *Mat. Sci. Forum* **1999**, *302*, 209–218. [CrossRef]
54. Tsukahara, S.; Satoh, T.; Tsushima, T. Magnetic Anisotropy Distribution Near the Surface of Amorphous Ribbons. *IEEE Trans. Magn.* **1978**, *14*, 1022–1024. [CrossRef]
55. Kraus, L.; Tomáš, I.; Keatociivílová, E.; Speingmann, B.; Müller, K. Magnetic Anisotropy Caused by Oriented Surface Roughness of Amorphous Ribbons. *Phys. Status Solidi.* **1987**, *100*, 289–299. [CrossRef]
56. Coey, J.M.D. *Magnetism and Magnetic Materials*; Cambridge University Press: New York, NY, USA, 2010; p. 628.
57. Pankhurst, Q.A.; Connolly, J.; Jones, S.K.; Dobson, J. Applications of Magnetic Nanoparticles in Biomedicine. *J. Phys. D Appl. Phys.* **2003**, *36*, R167. [CrossRef]
58. Eggers, T.; Lama, D.S.; Thiabgoha, O.; Marcin, J.; Švec, P.; Huong, N.T.; Škorvánek, I.; Phan, M.H. Impact of the Transverse Magnetocrystalline Anisotropy of a Co Coating Layer on the Magnetoimpedance Response of FeNi-Rich Nanocrystalline Ribbon. *J. All. Comp.* **2018**, *741*, 1105–1111. [CrossRef]
59. Taysioglu, A.A.; Peksoz, A.; Kaya, Y.; Gazilrez, N.D.; Kaynak, G. GMI Effect in CuO Coated Co-Based Amorphous Ribbons. *J. All. Comp.* **2009**, *487*, 38–41. [CrossRef]
60. Laurita, N.; Chaturvedi, A.; Bauer, C.; Jayathilaka, P.; Leary, A.; Miller, C.; Phan, M.H.; McHenry, M.E.; Srikanth, H. Enhanced giant Magnetoimpedance Effect and Field Sensitivity in Co-Coated Soft Ferromagnetic Amorphous Ribbons. *J. Appl. Phys.* **2011**, *109*, 07C706. [CrossRef]
61. Jamilpanah, L.; Hajiali, M.R.; Mohseni, S.M.; Erfanifam, S.; Houshiar, M.; Roozmeh, S.E. Magnetoimpedance Exchange Coupling in Different Magnetic Strength Thin Layers Electrodeposited on Co-Based Magnetic Ribbons. *J. Phys. D Appl. Phys.* **2017**, *50*, 155001. [CrossRef]
62. Dadsetan, A.; Almasi Kashi, M.; Mohseni, S.M. ZnO thin Layer/Fe-Based Ribbon/ZnO Thin Layer Sandwich Structure: Introduction of a New GMI Optimization Method. *J. Magn. Magn. Mater.* **2020**, *4934*, 165697. [CrossRef]

Article

Directional Field-Dependence of Magnetoimpedance Effect on Integrated YIG/Pt-Stripline System

Arthur L. R. Souza [1], Matheus Gamino [1], Armando Ferreira [2], Alexandre B. de Oliveira [1], Filipe Vaz [2], Felipe Bohn [1] and Marcio A. Correa [1,2,*]

1 Departamento de Física, Universidade Federal do Rio Grande do Norte, Natal 59078-900, Brazil; arthur_souza77@yahoo.com.br (A.L.R.S.); mgamino@fisica.ufrn.br (M.G.); abo1980@gmail.com (A.B.d.O.); felipebohn@fisica.ufrn.br (F.B.)
2 Centro de Física, Universidade do Minho, 4710-057 Braga, Portugal; armando.f@fisica.uminho.pt (A.F.); fvaz@fisica.uminho.pt (F.V.)
* Correspondence: marciocorrea@fisica.ufrn.br

Abstract: We investigated the magnetization dynamics through the magnetoimpedance effect in an integrated YIG/Pt-stripline system in the frequency range of 0.5 up to 2.0 GHz. Specifically, we explore the dependence of the dynamic magnetic behavior on the field orientation by analyzing beyond the traditional longitudinal magnetoimpedance effect of the transverse and perpendicular setups. We disclose here the strong dependence of the effective damping parameter on the field orientation, as well as verification of the very-low damping parameter values for the longitudinal and transverse configurations. We find considerable sensitivity results, bringing to light the facilities to integrate ferrimagnetic insulators in current and future technological applications.

Keywords: magnetization dynamics; magnetoimpedance; YIG

1. Introduction

The giant magnetoimpedance (MI) effect corresponds to the strong variation of the electrical impedance of a soft magnetic material when submitted to an external magnetic field [1–9]. Since the discovery of the effect, MI has attracted attention due to its versatility. In the context of fundamental physics, the effect is as an exciting alternative, with some potential advantages, to the traditional ferromagnetic resonance (FMR) effect. The fields configuration in the MI experiment is quite similar to that verified in the traditional FMR measurements [10,11]. However, while FMR makes use of a resonant cavity having a fixed frequency, the MI effect, in turn, allows us to explore the evolution of the ferromagnetic resonance with frequency and field strength in both saturated and unsaturated magnetic states. Hence, from the technological point of view, the MI effect arises as a sharp tool to detect small magnetic field changes. In addition, materials exhibiting MI appear as sensitive field sensor elements that can be integrated into a wide variety of electronic devices [11–16].

In recent years, the interest in the MI effect has increased, which at first glance may be based on the increasing demand for biosensors [12,17–21]. Within this field, different groups have recently reported interesting results. For instance, Kurlyandskaya and colleagues [17] have explored the MI response of a Co-based alloy ribbon composing a MI-sensitive element. In this case, the studied ribbon is quite thick, favoring the obtainment of high MI variations at the low frequency regime, an essential feature to the integration of the MI-sensor elements into devices. Following a different line, Yu and coworkers [12] have investigated the MI effect in thin films as a sensor to detect magnetic particles in blood vessels. Remarkably, the analyzed multilayer films present significant MI sensitivity despite the smaller thickness of the samples, reaching up to 41%/Oe for low field variations at the moderate frequency regime.

However, in the meantime, the search for electronic devices characterized by low energy consumption and low cost of production has been the main reason for the renewed attention to the MI effect.

Owing to this area, the nanostructured systems engineered to these ends have enabled us to scrutinize numerous spintronics effects. Among the several materials taken into account to the building of the spintronic nanostructures, ferrimagnetic insulators (FMI) as $Y_3Fe_5O_{12}$ (YIG) alloy deserves notability, bringing unique features that are advantageous in this context over the ones of other materials, including magnetic conductors [22–25]. Specifically, the electrical nature, magnetic moment, and very-low damping parameter make YIG the ideal playground for investigations in which pure spin currents are considered, suppressing the charge current [26,27].

The magnetization dynamics in YIG films have been extensively probed through broadband ferromagnetic resonance in recent years; but the link between FMI materials and MI experiments remained elusive until recently [28–31]. With this spirit, connecting YIG and MI, Kang and coworkers [28], for instance, have addressed the dynamics in YIG spheres and a single-crystalline YIG film deposited by using liquid phase epitaxy. The authors have disclosed very interesting results by employing a vector network analyzer (VNA) to perform MI measurements in a wide range of frequencies, with MI ratio values of around 256% and MI sensitivity of ∼8.8%/Oe for such systems. Next, Madwal and colleagues [29] have investigated a sputtered YIG single-layer thin film with a thickness of 45 nm. In this case, the magnetoimpedance experiments have been performed using the inductive method, in which a signal coil is wound around the film and surprisingly, despite the reduced dimension of the film, the FMR effect is found even at the low-frequency regime, from 0.5 up to 2.0 GHz.

Remarkably, the studies aforementioned deal with techniques in which the sample is electrically disconnected from the measurement system, which complicates the integration of the samples as MI-sensor elements in an electronic device. It is worth remarking that part of the difficulty in obtaining MI results in YIG samples resides in the high resistivity of the material. Nevertheless, recently, it has been shown that the growth of YIG/NM heterostructures, where NM is a non-ferromagnetic metal such as Ag, Cu, and W [30,31], may act as a way to circumvent this adversity and perform investigations on magnetization dynamics through the MI effect in the low frequency regime. In these previous works, the YIG/NM heterostructures were entirely produced by using the Magnetron Sputtering technique. This experimental procedure limits the reach of YIG with high thicknesses. Moreover, the NM materials in these previous studies present low spin–orbit coupling when compared with Pt material. Therefore, the modification of the YIG deposition technique (allowing for increased thickness) and the use of Pt material can bring interesting results, mainly in the MI response at moderate- and high-frequency regime.

For these heterostructures, beyond obtaining from the MI measurements important magnetic parameters such as effective magnetization M_{eff} and effective damping parameters α_{eff}, fingerprints of the FMR effect have been identified. Consequently, it has been demonstrated that these YIG/NM-stripline systems are promising candidates as MI-sensor elements to integrate magnetic devices.

This article reports a systematic investigation of the magnetization dynamics through the MI effect in an integrated YIG/Pt-stripline system. Specifically, we explored the dependence of dynamic magnetic behavior on field orientation. We went beyond the traditional longitudinal MI (LMI) effect, in which magnetic field and probe current are parallel, and also acquired the MI response for the transverse (TMI) and perpendicular (PMI) setups. From the results, we disclose the strong dependence of the effective damping parameter on the field orientation, as well as verify very-low damping parameter values for the LMI and TMI configurations. The observed high MI sensitivity and the experimental setup employed here turn easy to integrate ferrimagnetic insulators in current and future technological applications.

2. Experiment

To engineer the integrated YIG/Pt-stripline system, we considered a $Y_3Fe_5O_{12}$/Pt bilayer grown in two steps. First, we produced a YIG film with a thickness of 6 µm by Liquid Phase Epitaxy (LPE) onto a (111) Gadolinium Gallium Garnet (GGG) substrate with dimensions of 5×5 mm. After, we covered the YIG with a 6-nm-thick Pt layer deposited by magnetron sputtering. The deposition process was carried out using a Pt target (99.99% of purity) with the following parameters: base pressure of 5×10^{-8} Torr, deposition pressure of 3×10^{-3} Torr, with 99.99% pure Ar at 20 sccm constant flow, and using a DC source with a current of 50 mA.

The structural features of the sample were verified through X-ray diffraction (XRD). The XRD experiment was performed in the $\theta - 2\theta$ geometry, with Cu-K_α radiation ($\lambda = 1.54060$ Å), using a Rigaku Miniflex II system.

The quasi-static magnetic behavior was obtained through magnetization curves at room temperature, acquired using a Vibrating Sample Magnetometer (VSM) Lakeshore model 7407. In particular, the magnetization curves were taken with the magnetic field applied along distinct orientations, corresponding to the very same field configurations employed to the MI experiments.

The MI measurements were performed using a RF-impedance analyzer Agilent model E4991, with an E4991A test head connected to the integrated YIG/Pt-stripline system, in which the bilayer is the central conductor, and it is separated from the ground plane by the substrate. The electrical contacts between the YIG/Pt bilayer and the stripline system were made with 24 h cured low-resistive silver paint. To avoid propagative effects and acquire just the sample contribution to MI, the RF impedance analyzer was calibrated at the end of the connection cable by performing open, short, and load (50 Ω) measurements using reference standards. The probe current is fed directly to one side of the sample, while the other side is in a short circuit with the ground plane. We went beyond the traditional longitudinal MI (LMI) effect, in which external magnetic field and probe current are parallel, and also acquired the MI response for the transverse (TMI) and perpendicular (PMI) setups. Specifically, in the TMI configuration, the external magnetic field and probe current are transverse, and the field remains in the plane of the film; in the PMI one, in turn, the field is perpendicular to the film plane. A schematic representation of the MI setups explored here is depicted in Figure 1a. For the LMI case, we employed a solenoid as a source for the magnetic field, with a maximum amplitude of ± 300 Oe, while for the TMI and PMI cases, we used an electromagnet, thus reaching ± 1500 Oe. While the external magnetic field was varied for all experiments, a 0 dBm (1 mW) constant power was applied to the sample, characterizing a linear regime of driving signal. MI measurements were taken over a frequency range between 0.5 and 2.0 GHz. The frequency sweep was made for each field value, and the real R and imaginary X components of the impedance Z are simultaneously acquired. In the meantime, in our RF-impedance analyzer, the test head may estimate the Z from the ratio between the electrical voltage and current, both acquired as sketched in Figure 1b. Then, its maximum amplitude for a given field strength and frequency is obtained by

$$Z = \frac{V}{I} = \frac{V}{V_R} R. \tag{1}$$

where V is the peak voltage provided by the test head, V_R is the potential difference between the terminals of a reference resistor R, and I is the peak current flowing through the sample. In this sense, such a configuration allows us to wonder a simple circuit with an MI-sensor element to be embarked in a magnetic device. In order to make easier a direct comparison between the measurements, we show here discounted values of the impedance, given by

$$\Delta Z = Z(H) - Z(H_{max}), \tag{2}$$

where $Z(H)$ is the electrical impedance for a given external magnetic field value and $Z(H_{max})$ is the impedance value for the maximum external magnetic field, where the sample is saturated magnetically. It is worth pointing out that similar definitions of variation were also taken for the real R and imaginary X components of the impedance, i.e.,

$$\Delta R = R(H) - R(H_{max}) \tag{3}$$

and

$$\Delta X = X(H) - X(H_{max}). \tag{4}$$

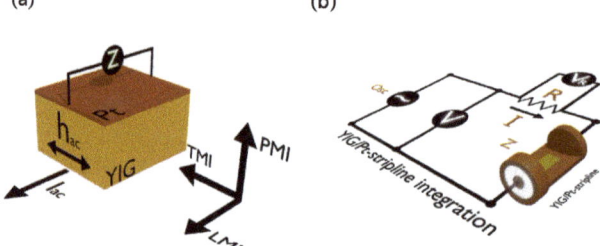

Figure 1. Schematic representation of the MI experiments. (**a**) LMI, TMI, and PMI setups employed in the dynamic magnetic characterization. (**b**) YIG/Pt-stripline system integrated in a circuit as a MI-sensor element.

3. Results and Discussion

Figure 2 shows the XRD result for our YIG/Pt bilayer. The diffractogram reveals two peaks at $2\theta \approx 51.1°$, which are associated with the GGG (111) substrate (ICSD 9237) and ascribed to the coexistence of the $K_{\alpha 1}$ and $K_{\alpha 2}$ contributions. In addition, a low-intensity peak is found at $2\theta = 50.9°$, assigning the YIG (444) preferential growth (ICSD 80139). The peaks of YIG (444) and GGG (444) overlap due to their good lattice match, and no evidence of polycrystalline YIG is verified, meaning that textured YIG layer is formed in the growth direction of [111]. Our findings are in accordance with results previously reported in the literature for similar heterostructures [32–35].

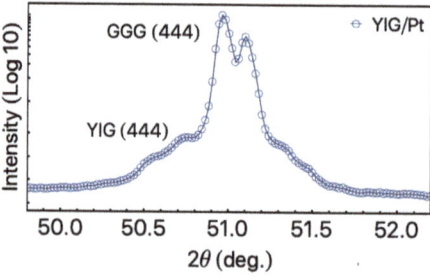

Figure 2. X-ray diffraction result for the YIG/Pt bylayer grown onto a (111) GGG substrate. The peaks are indexed considering the ICSD cards 9237 and 80139 for the GGG and YIG, respectively.

Although the XRD pattern shows evidence of the YIG crystallization, which is also corroborated from the comparison with the literature, we yet may employ quasi-static magnetic technique to further characterize the YIG phase after the annealing. Figure 3 shows the normalized magnetization curves acquired with the field along different orientations, LMI, TMI, and PMI (see Figure 1a). Notice that only the ferrimagnetic response of the YIG is seen here, given the paramagnetic contribution of the GGG substrate is removed from each curve. Remarkably, the YIG/Pt bilayer has quite weak anisotropic in-plane magnetic properties [36], depicted by the similar magnetization curves acquired for the LMI and TMI configurations. The results suggest soft magnetic properties, with low saturation field H_s, low coercive field H_c, and high magnetic permeability. Specifically, we find H_s values of 4.2 Oe and 4.5 Oe, while H_c ones of 0.5 Oe, from the LMI and TMI experiments, respectively. On the other hand, for the PMI one, the shape of the curve is completely modified. The change is attributed to the shape magnetic anisotropy, which leads to a significant increase of the coercive and saturation fields, reaching to $H_c \approx 3.0$ Oe and $H_s \approx 14$ Oe. In addition, it is worth highlighting that we observe a drastic reduction of the effective magnetization in the PMI measurement, a fact not identified in the plot since the curve is normalized, but that can be inferred here from the decrease in the signal-to-noise ratio.

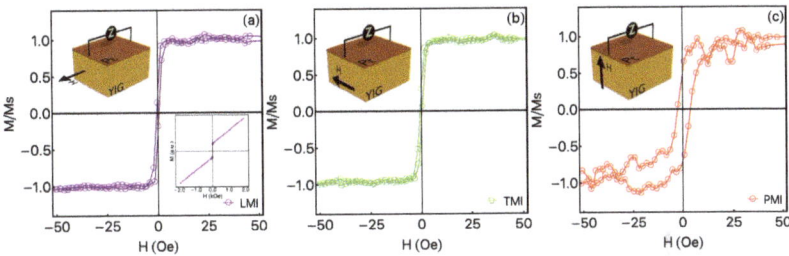

Figure 3. Normalized magnetization curves for the YIG/Pt bilayer acquired with the field along different orientations, (**a**) LMI, in which the external magnetic field is parallel to the drive current on the MI experiment. (**b**) TMI, here the external magnetic field is applied in the film-plane, and perpendicular to the drive current (MI experiment). (**c**) PMI configurations, which the external magnetic field is perpendicular to the film-plane. It is worth mentioning that the paramagnetic contribution of the GGG substrate is removed from each curve and, therefore, only the ferrimagnetic response of the YIG is seen here. The inset in (**a**) shows a representative example of the magnetization curves before the remotion of the paramagnetic contribution.

The ferrimagnetic behavior and the coercive field value verified here from the magnetization curves, when combined with the XRD result, are indicators of the quite-good quality of our YIG/Pt heterostructure. However, the general features of the whole film, including its magnetic and electrical properties, are essential issues for the MI response. As a consequence, in order to make it feasible to carry out MI experiments in magnetic insulators and place them as MI-sensor elements for magnetic devices, we overcome any experimental adversity due to the high electrical resistivity of the material by capping the YIG with a non-magnetic conductor Pt layer.

It is well known that the magnetic properties of our YIG/Pt bilayer are reflected in the magnetization dynamics, including the magnetoimpedance effect [2,10]. These features establish the limits in which distinct mechanisms command the MI response. Here, the soft magnetic behavior and the integration between the YIG/Pt bilayer and the stripline system allow us to observe FMR contributions even in the low-frequency regime. Consequently, we can induce substantial MI modifications, making the integrated system a promising candidate for sensor elements.

Figure 4 shows the evolution of ΔR as a function of the external magnetic field with the frequency. The curves were acquired over a complete magnetization loop and present hysteretic behavior. However, here we show just part of the curves, when the field goes from the negative to maximum positive value.

At first glance, we notice the curves have similar general shapes. Nevertheless, a closer inspection reveals that the ones acquired for the PMI configuration occur in a field range dissimilar to the one verified for the LMI and TMI ones. This feature is due to the substantial modification in the shape anisotropy and, consequently, in the anisotropy field. In addition, for all experiments, we observe the amplitude of the peaks is dependent on the field orientation. More specifically, the peak amplitude and the peak's position in field are a result of the orientation between field and magnetization, and of the interplay of the effects associated with the magnetic anisotropy field and damping parameter [37].

From the general point of view, our samples have all classical features of the magnetoimpedance observed in conducting ferromagnetic systems. Specifically, the curves exhibit a double peak behavior, symmetrical around $H = 0$, for the whole frequency range, irrespectively of the field configuration. An interesting feature related to the ΔR behavior resides in the dependence of the position peaks with probe current frequency. We observe the displacement of the peaks towards higher fields as the frequency increases, even for the smallest frequency values. Such peak behavior is a fingerprint of the FMR effect controlling the magnetization dynamics and the MI variations, in a response similar to that obtained through the broadband FMR technique.

From the ΔR results, the resonance field H_r and the linewidth ΔH were estimated by fitting the peaks using a Lorentzian function, as shown in Figure 4. Such quantities are key parameters for obtaining the effective magnetization M_{eff} and the effective damping parameter α_{eff}. Specifically, the dependence of f_r with H_r provides M_{eff} through the Kittel equation

$$f_r = \frac{\gamma}{2\pi}\sqrt{(H_r + H_k)(H_r + H_k + 4\pi M_{eff})}, \qquad (5)$$

in which $\gamma/2\pi$ is the gyromagnetic ratio and H_k is the anisotropy field. In addition, α_{eff} is achieved from the relation between ΔH and f_r,

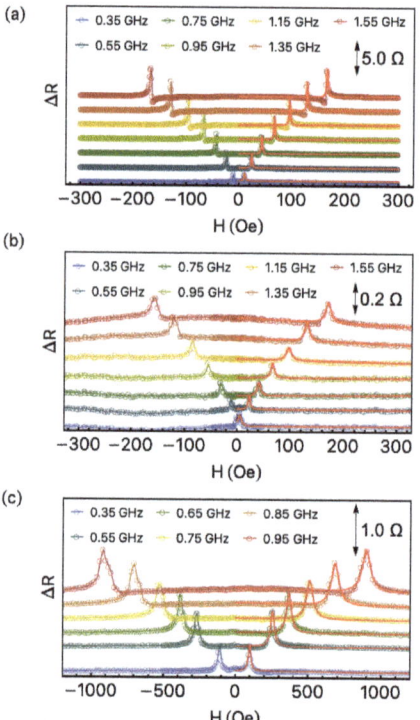

Figure 4. Variation of the real component of the impedance, ΔR, as a function of the external magnetic field for selected frequency values. Response of the YIG/Pt-stripline system for the (**a**) LMI, (**b**) TMI, and (**c**) PMI configurations. The curves are shifted on a vertical scale in order to make the visualization clearer. The symbols correspond to the experimental data, while the red lines are the fit from which the resonance field H_r and linewidth ΔH are obtained. The arrows indicate the scales of each measured configuration with the appropriate values.

$$\Delta H = \Delta H_\circ + \frac{2\pi f \alpha_{eff}}{\gamma}, \qquad (6)$$

where $\Delta H_\circ$ is the extrinsic inhomogeneous contribution to ΔH and f is the frequency, i.e., f_r.

Figure 5 brings both plots, f_r vs. H_r and ΔH vs. f_r. From Figure 5a, we verify the dependences of f_r with H_r for the LMI and TMI configurations are similar, as expected due to the quite-weak anisotropic in-plane magnetic properties. Assuming $\gamma/2\pi = 2.8$ GHz/kOe and $H_k = 0.5$ Oe, this latter close to the H_c values obtained from the quasi-static magnetization curves, we infer $4\pi M_{eff} \approx 1696$ G using Equation (5). The $4\pi M_{eff}$ value is in concordance with results previously reported in the literature [38]. For the PMI configuration, not shown here, the effective magnetization is significantly smaller, which is attributed to the shape anisotropy due to the reduced thickness.

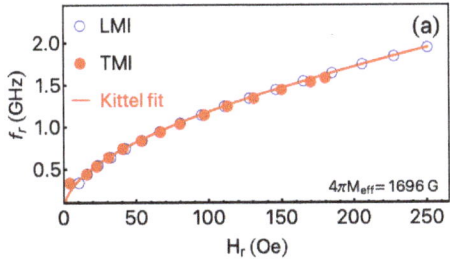

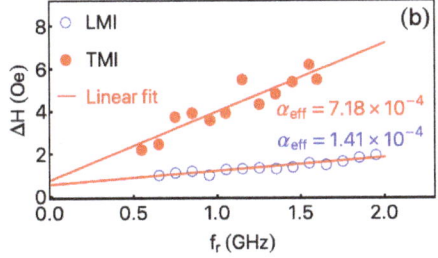

Figure 5. (a) Resonance frequency f_r as a function of resonance field H_r of the YIG/Pt-stripline system for the LMI and TMI configurations. The symbols are the experimental data and the red lines correspond to the fit obtained with the Kittel equation, Equation (5), then inferring the effective magnetization M_{eff}. (b) FMR linewidth ΔH as a function of the f_r in the LMI and TMI configurations. The symbols are the experimental data and the red lines correspond to the possible fit obtained with Equation (6), then providing an estimate of the effective damping parameters α_{eff}.

From Figure 5b, we identify the relations between ΔH and f_r for the LMI, TMI. Generally, our system consists of a ferrimagnetic insulator capped by a non-magnetic metallic layer. Given Pt is a metal with high spin–orbit coupling, the effective damping parameter in our sample has contributions of distinct mechanisms. The first one is the well-known Gilbert damping parameter. Such contribution consists of representing the relaxation mechanisms by a torque that pulls the magnetization toward the equilibrium direction [39]. Moreover, considering that the longitudinal and transverse components of the magnetization are stirred through different relaxation rates, the Bloch–Bloembergen phenomenology [39] is present in our sample. The second contribution to the effective damping parameter comes from the two-magnon and spin pumping mechanisms, especially due to the bilayer structure of our sample. There are numerous interesting studies playing with the mechanisms influencing α_{eff} and bringing the theoretical background to understand contributions for a given system [39–43]. Here, we do not address such issue in detail and instead focus our efforts in the fit using Equation (6) to infer the effective damping parameters and the extrinsic inhomogeneous contribution to ΔH. Once the effective damping (α_{eff}) is the parameter considered for the future sensor applications.

For the LMI and TMI configurations, we find α_{eff} of 1.41×10^{-4} and 7.18×10^{-4}, respectively. Remarkably, such values are very low and are in agreement with results for YIG films [43,44]. Moreover, the in-plane inhomogeneous contribution to ΔH seems to be the same for both, as expected. It is well known α_{eff} has a central role in the magnetic response of the sample. For instance, samples with low a α_{eff} reach the magnetic stabilization quickly, an important parameter for a sensor in which the fast magnetic response is primordial, as is the case of biosensors. For the PMI one, in turn, we are not able to fit the α_{eff} due to the mechanisms associated with the non-uniform excitation modes and domain contribution what leads to a considerable increase in the ΔH as observed in Figure 4c. Nevertheless, the α_{eff} value for the PMI setup seems to be comparable with those found for other interesting ferromagnetic systems, such as thin films of Co_2FeAl full-Heusler alloy [45,46].

With the straight potential for sensor applications, we focus on the MI performance as a function of the frequency and field strength. Figure 6 shows the maximum ΔZ value, ΔZ_{max}, as a function of the frequency for the LMI, TMI, and PMI configurations. Such analysis allows us to infer the frequency range in which the sensor element has the best MI efficiency. For LMI, the highest ΔZ_{max} takes place at around 1.45 GHz and is found for an external magnetic field of 150 Oe, as can see in Figure 6a,b. We observe a monotonic rise of ΔZ_{max} up to frequencies close to 1.0 GHz, where the curve reaches a constant value within the experimental error. This behavior is interesting since the frequency may be modified without losing the MI efficiency of the sensor. The ΔZ_{max} behavior for the TMI configuration is quite similar to that discussed for the LMI one. However, as we can confirm from Figure 6c, the MI efficiency is very low, becoming negligible if compared to the results acquired for the other experimental setups. Although we observe low α_{eff} value for this field configuration, the alignment between the external magnetic field, alternating magnetic field, and magnetization seem to affect the MI variations drastically, vanishing ΔZ_{max}. At last, Figure 6d,e shows the results for the PMI configuration. In this case, the MI efficiency presents an initial increase up to 0.5 GHz, achieving a roughly constant value between 0.5 and 1.0 GHz, followed by a decrease above 1.0 GHz. Then, the highest ΔZ_{max} is found at around 0.96 GHz. It is worth highlighting such result is found at 900 Oe, in a sense the PMI configuration allows the identification of interesting MI efficiency values at the high-field regime.

These findings bring to light an exciting way to promote the integration of insulating ferrimagnetic materials in sensor elements, and modify the field range in which the optimal MI response is achieved, i.e., simply by changing the orientation of the magnetic field in the experiment.

The α_{eff} values verified for our integrated YIG/Pt-stripline reveal fingerprints of a low-damping dynamical system. Within this context, the narrow ΔZ peaks as a function of the field provides insights on the MI sensitivity. To quantify the sensitivity as a function of the frequency, we calculate the magnitude of the impedance change

$$\text{Sens.} = \frac{\Delta Z_{max}(H) - \Delta Z(H-10)}{10}, \qquad (7)$$

where $\Delta Z_{max}(H)$ corresponds to the maximum values of the ΔZ that is observed at the field H, and $\Delta Z(H-10)$ is the ΔZ observed at $H-10$ (see inset in Figure 7b).

Figure 7 shows the sensitivity as a function of frequency for the LMI, TMI, and PMI configurations. For LMI, Figure 7a, we observe the maximum value reaches ~ 415 mΩ/Oe at 1.5 GHz. However, we point out there is a broad range of frequencies, between 0.5 and 2.0 GHz, in which the sensitivity has significant values, in a sense, we may vary the frequency without loss of sensitivity. For the TMI configuration, we find a narrow range of frequencies with a sensitivity of 10 mΩ/Oe. In particular, such behavior becomes interesting since high sensitivity values are found in a range of 0.2 GHz, taking place at the low frequency regime, below 0.35 GHz. At last, for the PMI configuration, we verify a maximum sensitivity of 7.5 mΩ/Oe at around 0.65 GHz. In this case, it is worth emphasizing such value takes place at high-field values, as previously mentioned (see Figure 6e).

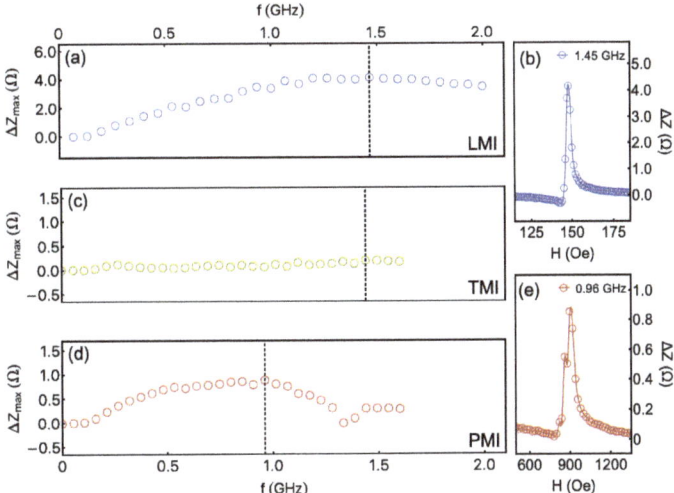

Figure 6. (**a**) Maximum ΔZ value, ΔZ_{max} as a function of the frequency for the LMI configuration. The maximum efficiency is found for $f = 1.45$ GHz, as indicated by the dashed line. (**b**) The ΔZ vs. H at $f = 1.45$ GHz, as a representative example of the analyzed curves. The ΔZ_{max} is achieved from the difference between the Z value at the peak and the Z at the maximum magnetic field. (**c**) ΔZ_{max} as a function of the frequency for the TMI configuration, whose maximum at 1.4 GHz is indicated by the dashed line. (**d**) A similar plot for the PMI configuration. (**e**) The ΔZ vs. H at 0.96 GHz for the PMI setup.

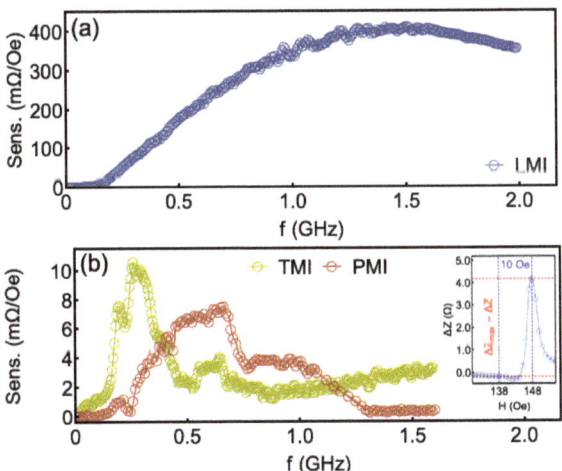

Figure 7. (**a**) Sensitivity as a function of frequency for the LMI field configuration. The maximum value is 415 mΩ/Oe at 1.5 GHz. (**b**) A similar plot for the TMI and PMI configurations. The inset depicts a representative example of the experimental procedure employed to calculate the sensitivity.

4. Conclusions

In summary, we have investigated herein the magnetization dynamics through the magnetoimpedance effect in an integrated YIG/Pt-stripline system. Specifically, we have explored the dependence of dynamic magnetic behavior on field orientation. To this end, we have analyzed the magnetoimpedance response for the traditional longitudinal configuration, as well as for the transverse and perpendicular ones. From the experimental results, we have estimated magnetic parameters that are fundamental for a sensor element, such as the effective magnetization and the effective damping parameter. We have observed low α_{eff} values of 1.41×10^{-4} and 7.18×10^{-4} for the LMI and TMI configurations, respectively. Moreover, we have found a significant increase of α_{eff} for the PMI configuration, as expected. From the technological perspective, we have obtained the efficiency ΔZ_{max} as a function of the frequency and have estimated the sensitivity for our system. Within this context, the change of field configuration suggests the integrated YIG/Pt-stripline system may be used in different magnetic devices as a sensor element. In particular, the LMI configuration reveals high sensitivity for a wide frequency range, in which a change of frequency does not yield any loss of sensitivity, while the TMI and PMI ones disclose high sensitivity in a limited frequency interval. However, for the PMI field configuration, the higher sensitivity happens at high fields, bringing the possibility to explore distinct field ranges with a single sensor. The verified high MI sensitivity brings to light the facilities to integrate ferrimagnetic insulators in current and future technological applications, as ultra-fast sensors.

Author Contributions: Conceptualization, A.L.R.S., M.A.C., M.G., and F.B.; methodology, M.A.C., A.L.R.S., A.F., and A.B.d.O.; software, M.A.C.; validation, F.V., F.B., M.G., A.B.d.O., and M.A.C.; formal analysis, A.L.R.S, M.A.C., F.B., M.G.; investigation, A.L.R.S., A.F., M.A.C., M.G.; data curation, A.B.d.O., M.A.C.; writing—original draft preparation, A.L.R.S., M.A.C., M.G.; writing—review and editing, F.B., F.V., A.F.; supervision, M.A.C. and F.B.; project administration, M.A.C.; funding acquisition, M.A.C., F.B. and F.V. All authors have read and agreed to the published version of the manuscript.

Funding: This research was funded by CNPq grand numbers 304943/2020-7 and 407385/2018-5, Capes grand number 88887.573100/2020-00 and FCT grant number CTTI-31/18-CF(2).

Institutional Review Board Statement: Not applicable.

Informed Consent Statement: Not applicable.

Acknowledgments: The authors thank the Brazilian agencies CNPq and CAPES for the financial support. From Portugal side, the authors thank the Portuguese Foundation for Science and Technology (FCT) for the strategic funding UID/FIS/04650/2020. Armando Ferreira thanks the FCT for the contract under the Stimulus of Scientific Employment (CTTI-31/18–CF (2) junior researcher contract).The work reported in this paper was supported by On-Surf Mobilizar Competencias Tecnologicas em Engenharia de Superficies, Project POCI-01-0247-FEDER-024521

Conflicts of Interest: The authors declare no conflict of interest.

References

1. Kraus, L. GMI modeling and material optimization. *Sens. Actuators A Phys.* **2003**, *106*, 187–194. [CrossRef]
2. Kraus, L. Theory of giant magneto-impedance in the planar conductor with uniaxial magnetic anisotropy. *J. Magn. Magn. Mater.* **1999**, *195*, 764–778. [CrossRef]
3. Hika, K.; Panina, L.V.; Mohri, K. Magneto-impedance in sandwich film for magnetic sensor heads. *IEEE Trans. Magn.* **1996**, *32*, 4594–4596. [CrossRef]
4. Panina, L.V.; Makhnovskiy, D.P.; Mapps, D.J.; Zarechnyuk, D.S. Two-dimensional analysis of magnetoimpedance in magnetic/metallic multilayers. *J. Appl. Phys.* **2001**, *89*, 7221–7223. [CrossRef]
5. Panina, L.V.; Mohri, K. Magneto-impedance effect in amorphous wires. *Appl. Phys. Lett.* **1994**, *65*, 1189–1191. [CrossRef]
6. Sommer, R.L.; Chien, C.L. Role of magnetic anisotropy in the magnetoimpedance effect in amorphous alloys. *Appl. Phys. Lett.* **1995**, *67*, 857. [CrossRef]
7. Sommer, R.L.; Chien, C.L. Longitudinal, transverse, and perpendicular magnetoimpedance in nearly zero magnetostrictive amorphous alloys. *Phys. Rev. B* **1996**, *53*, R5982–R5985. [CrossRef] [PubMed]

8. Sommer, R.L.; Chien, C.L. Longitudinal and transverse magneto-impedance in amorphous Fe73.5Cu1Nb3Si13.5B9 films. *Appl. Phys. Lett.* **1995**, *67*, 3346. [CrossRef]
9. Morikawa, T.; Nishibe, Y.; Yamadera, H.; Nonomura, Y.; Takeuchi, M.; Sakata, J.; Taga, Y. Enhancement of giant magneto-impedance in layered film by insulator separation. *IEEE Trans. Magn.* **1996**, *32*, 4965–4967. [CrossRef]
10. Yelon, A.; Menard, D.; Britel, M.; Ciureanu, P.; Menard, D.; Britel, M.; Ciureanu, P. Calculations of giant magnetoimpedance and of ferromagnetic resonance response are rigorously equivalent. *Appl. Phys. Lett.* **1996**, *69*, 3084. [CrossRef]
11. Menard, D.; Yelon, A. Theory of longitudinal magnetoimpedance in wires. *J. Appl. Phys.* **2000**, *88*, 379–393. [CrossRef]
12. Melnikov, G.Y.; Lepalovskij, V.N.; Svalov, A.V.; Safronov, A.P.; Kurlyandskaya, G.V. Magnetoimpedance Thin Film Sensor for Detecting of Stray Fields of Magnetic Particles in Blood Vessel. *Sensors* **2021**, *21*, 3621. [CrossRef] [PubMed]
13. Rouhani, A.A.; Matin, L.F.; Mohseni, S.M.; Zoriasatain, S. A Domain Dynamic Model Study of Magneto-impedance Sensor in the Presence of Inhomogeneous Magnetic Fields. *J. Supercond. Nov. Magn.* **2021**, *34*, 571–580. [CrossRef]
14. Nakai, T. Nondestructive Detection of Magnetic Contaminant in Aluminum Casting Using Thin Film Magnetic Sensor. *Sensors* **2021**, *21*, 4063. [CrossRef]
15. Li, B.; Kavaldzhiev, M.N.; Kosel, J. Flexible magnetoimpedance sensor. *J. Magn. Magn. Mater.* **2015**, *378*, 499–505. [CrossRef]
16. Yamaguchi, M.; Takezawa, M.; Ohdaira, H.; Arai, K.I.; Haga, A. Directivity and sensitivity of high-frequency carrier type thin-film magnetic field sensor. *Sens. Actuators A Phys.* **2000**, *81*, 102–105. [CrossRef]
17. Kurlyandskaya, G.V.; Sánchez, M.L.; Hernando, B.; Prida, V.M.; Gorria, P.; Tejedor, M. Giant-magnetoimpedance-based sensitive element as a model for biosensors. *Appl. Phys. Lett.* **2003**, *82*, 3053–3055. [CrossRef]
18. Wang, T.; Yang, Z.; Lei, C.; Lei, J.; Zhou, Y. An integrated giant magnetoimpedance biosensor for detection of biomarker. *Biosens. Bioelectron.* **2014**, *58*, 338–344. [CrossRef]
19. Zhu, Y.; Zhang, Q.; Li, X.; Pan, H.; Wang, J.; Zhao, Z. Detection of AFP with an ultra-sensitive giant magnetoimpedance biosensor. *Sens. Actuators B Chem.* **2019**, *293*, 53–58. [CrossRef]
20. Kurlyandskaya, G.; Levit, V. Magnetic Dynabeads® detection by sensitive element based on giant magnetoimpedance. *Biosens. Bioelectron.* **2005**, *20*, 1611–1616. [CrossRef]
21. Yang, Z.; Liu, Y.; Lei, C.; Sun, X.c.; Zhou, Y. A flexible giant magnetoimpedance-based biosensor for the determination of the biomarker C-reactive protein. *Microchim. Acta* **2015**, *182*, 2411–2417. [CrossRef]
22. Kang, Y.M.; Wee, S.H.; Baik, S.I.; Min, S.G.; Yu, S.C.; Moon, S.H.; Kim, Y.W.; Yoo, S.I. Magnetic properties of YIG ($Y_3Fe_5O_{12}$) thin films prepared by the post annealing of amorphous films deposited by rf-magnetron sputtering. *J. Appl. Phys.* **2005**, *97*, 10A319. [CrossRef]
23. Evelt, M.; Safranski, C.; Aldosary, M.; Demidov, V.E.; Barsukov, I.; Nosov, A.P.; Rinkevich, A.B.; Sobotkiewich, K.; Li, X.; Shi, J.; et al. Spin Hall-induced auto-oscillations in ultrathin YIG grown on Pt. *Sci. Rep.* **2018**, *8*, 1269. [CrossRef] [PubMed]
24. Serga, A.A.; Chumak, A.V.; Hillebrands, B. YIG magnonics. *J. Phys. D Appl. Phys.* **2010**, *43*, 264002. [CrossRef]
25. Čermák, J.; Abrahám, A.; Fabián, T.; Kaboš, P.; Hyben, P. YIG based LPE films for microwave devices. *J. Magn. Magn. Mater.* **1990**, *83*, 427–429. [CrossRef]
26. Lotze, J.; Huebl, H.; Gross, R.; Goennenwein, S.T. Spin Hall magnetoimpedance. *Phys. Rev. B-Condens. Matter Mater. Phys.* **2014**, *90*, 174419. [CrossRef]
27. Hahn, C.; De Loubens, G.; Klein, O.; Viret, M.; Naletov, V.V.; Ben Youssef, J. Comparative measurements of inverse spin Hall effects and magnetoresistance in YIG/Pt and YIG/Ta. *Phys. Rev. B* **2013**, *87*, 174417. [CrossRef]
28. Kang, C.; Wang, T.; Jiang, C.; Chen, K.; Chai, G. Investigation of the giant magneto-impedance effect of single crystalline YIG based on the ferromagnetic resonance effect. *J. Alloys Compd.* **2021**, *865*, 158903. [CrossRef]
29. Medwal, R.; Chaudhuri, U.; Vas, J.V.; Deka, A.; Gupta, S.; Duchamp, M.; Asada, H.; Fukuma, Y.; Mahendiran, R.; Rawat, R.S. Magnetoimpedance of Epitaxial $Y_3Fe_5O_{12}$ (001) Thin Film in Low-Frequency Regime. *ACS Appl. Mater. Interfaces* **2020**, *12*, 41802–41809. [CrossRef]
30. Santos, J.; Silva, E.; Rosa, W.; Bohn, F.; Correa, M. Role of the spin-orbit coupling on the effective damping parameter in $Y_3Fe_5O_{12}$/(Ag,W) bilayers explored through magnetoimpedance effect. *Mater. Lett.* **2019**, *256*, 126662. [CrossRef]
31. Correa, M.; Rosa, W.; Melo, A.; Silva, E.; Della Pace, R.; Oliveira, A.; Chesman, C.; Bohn, F.; Sommer, R. Magnetoimpedance effect in ferrimagnetic insulator yttrium iron garnet films capped by copper. *J. Magn. Magn. Mater.* **2019**, *480*, 6–10. [CrossRef]
32. Cao Van, P.; Surabhi, S.; Dongquoc, V.; Kuchi, R.; Yoon, S.G.; Jeong, J.R. Effect of annealing temperature on surface morphology and ultralow ferromagnetic resonance linewidth of yttrium iron garnet thin film grown by rf sputtering. *Appl. Surf. Sci.* **2018**, *435*, 377–383. [CrossRef]
33. Shang, T.; Zhan, Q.F.; Ma, L.; Yang, H.L.; Zuo, Z.H.; Xie, Y.L.; Li, H.H.; Liu, L.P.; Wang, B.M.; Wu, Y.H.; et al. Pure spin-Hall magnetoresistance in Rh/$Y_3Fe_5O_{12}$ hybrid. *Sci. Rep.* **2015**, *5*, 17734. [CrossRef] [PubMed]
34. Jin, L.; Jia, K.; He, Y.; Wang, G.; Zhong, Z.; Zhang, H. Pulsed laser deposition grown yttrium-iron-garnet thin films: Effect of composition and iron ion valences on microstructure and magnetic properties. *Appl. Surf. Sci.* **2019**, *483*, 947–952. [CrossRef]
35. Gamino, M.; Silva, E.; Alves Santos, O.; Mendes, J.; Rodríguez-Suárez, R.; Machado, F.; Azevedo, A.; Rezende, S. The role of metallic nanoparticles in the enhancement of the spin Hall magnetoresistance in YIG/Pt thin films. *J. Magn. Magn. Mater.* **2018**, *466*, 267–272. [CrossRef]
36. Castel, V.; Vlietstra, N.; van Wees, B.J.; Youssef, J.B. Frequency and power dependence of spin-current emission by spin pumping in a thin-film YIG/Pt system. *Phys. Rev. B* **2012**, *86*, 134419. [CrossRef]

37. Correa, M.; Santos, J.; Silva, B.; Raza, S.; Della Pace, R.; Chesman, C.; Sommer, R.; Bohn, F. Exploring the magnetization dynamics, damping and anisotropy in engineered CoFeB/(Ag, Pt) multilayer films grown onto amorphous substrate. *J. Magn. Magn. Mater.* **2019**, *485*, 75–81. [CrossRef]
38. Wu, M. Nonlinear Spin Waves in Magnetic Film Feedback Rings. In *Solid State Physics*; Academic Press: Cambridge, MA, USA, 2010; Volume 62, pp. 163–224. [CrossRef]
39. Rezende, S.M.; Rodríguez-Suárez, R.L.; Azevedo, A. Magnetic relaxation due to spin pumping in thick ferromagnetic films in contact with normal metals. *Phys. Rev. B* **2013**, *88*, 014404. [CrossRef]
40. Maier-Flaig, H.; Harder, M.; Gross, R.; Huebl, H.; Goennenwein, S.T.B. Spin pumping in strongly coupled magnon-photon systems. *Phys. Rev. B* **2016**, *94*, 054433. [CrossRef]
41. Schoen, M.A.W.; Lucassen, J.; Nembach, H.T.; Koopmans, B.; Silva, T.J.; Back, C.H.; Shaw, J.M. Magnetic properties in ultrathin 3d transition-metal binary alloys. II. Experimental verification of quantitative theories of damping and spin pumping. *Phys. Rev. B* **2017**, *95*, 134411. [CrossRef]
42. Noack, T.B.; Vasyuchka, V.I.; Bozhko, D.A.; Heinz, B.; Frey, P.; Slobodianiuk, D.V.; Prokopenko, O.V.; Melkov, G.A.; Kopietz, P.; Hillebrands, B.; et al. Enhancement of the Spin Pumping Effect by Magnon Confluence Process in YIG/Pt Bilayers. *Phys. Status Solidi* **2019**, *256*, 1900121. [CrossRef]
43. Chang, H.; Praveen Janantha, P.A.; Ding, J.; Liu, T.; Cline, K.; Gelfand, J.N.; Li, W.; Marconi, M.C.; Wu, M. Role of damping in spin Seebeck effect in yttrium iron garnet thin films. *Sci. Adv.* **2017**, *3*, e1601614. [CrossRef]
44. Jungfleisch, M.B.; Chumak, A.V.; Kehlberger, A.; Lauer, V.; Kim, D.H.; Onbasli, M.C.; Ross, C.A.; Kläui, M.; Hillebrands, B. Thickness and power dependence of the spin-pumping effect in $Y_3Fe_5O_{12}$/Pt heterostructures measured by the inverse spin Hall effect. *Phys. Rev. B* **2015**, *91*, 134407. [CrossRef]
45. Hait, S.; Husain, S.; Barwal, V.; Gupta, N.K.; Pandey, L.; Svedlindh, P.; Chaudhary, S. Comparison of high temperature growth versus post-deposition in situ annealing in attaining very low Gilbert damping in sputtered Co2FeAl Heusler alloy films. *J. Magn. Magn. Mater.* **2021**, *519*, 167509. [CrossRef]
46. Husain, S.; Akansel, S.; Kumar, A.; Svedlindh, P.; Chaudhary, S. Growth of Co_2FeAl Heusler alloy thin films on Si(100) having very small Gilbert damping by Ion beam sputtering. *Sci. Rep.* **2016**, *6*, 28692. [CrossRef] [PubMed]

Article

Advanced Characterization of FeNi-Based Films for the Development of Magnetic Field Sensors with Tailored Functional Parameters

Sergey V. Komogortsev [1,2,*], Irina G. Vazhenina [1,2], Sofya A. Kleshnina [1,2], Rauf S. Iskhakov [1], Vladimir N. Lepalovskij [3], Anna A. Pasynkova [3,4] and Andrey V. Svalov [3]

1. Kirensky Institute of Physics, Federal Research Center KSC SB RAS, 660036 Krasnoyarsk, Russia; irina-vazhenina@mail.ru (I.G.V.); sofya.antipckina@yandex.ru (S.A.K.); rauf@iph.krasn.ru (R.S.I.)
2. Institute of Physics, Siberian Federal University, 660041 Krasnoyarsk, Russia
3. Department of Magnetism of Solid State, Institute of Natural Sciences and Mathematics, Ural Federal University, 620002 Ekaterinburg, Russia; vladimir.lepalovsky@urfu.ru (V.N.L.); pasynkova_a@imp.uran.ru (A.A.P.); andrey.svalov@urfu.ru (A.V.S.)
4. Laboratory of Advanced Magnetic Materials, Institute of Metal Physics UD RAS, 620108 Ekaterinburg, Russia
* Correspondence: komogor@iph.krasn.ru

Abstract: Magnetometry and ferromagnetic resonance are used to quantitatively study magnetic anisotropy with an easy axis both in the film plane and perpendicular to it. In the study of single-layer and multilayer permalloy films, it is demonstrated that these methods make it possible not only to investigate the average field of perpendicular and in-plane anisotropy, but also to characterize their inhomogeneity. It is shown that the quantitative data from direct integral and local measurements of magnetic anisotropy are consistent with the direct and indirect estimates based on processing of the magnetization curves. The possibility of estimating the perpendicular magnetic anisotropy constant from the width of stripe domains in a film in the transcritical state is demonstrated. The average in-plane magnetic anisotropy field of permalloy films prepared by magnetron sputtering onto a Corning glass is almost unchanged with the thickness of a single-layer film. The inhomogeneity of the perpendicular anisotropy field for a 500 nm film is greater than that for a 100 nm film, and for a multilayer film with a total permalloy thickness of 500 nm, it is greater than that for a homogeneous film of the same thickness.

Keywords: magnetic field sensors; thin films; multilayered structures; magnetic anisotropy; anisotropy distribution; ferromagnetic resonance; magnetoimpedance; high frequency applications

1. Introduction

The functionality of a magnetic sensor is largely determined by magnetic anisotropy or difference in the magnetic response at different directions of the applied field. In a magnetic film sensor, both the magnetic shape anisotropy (characterized by the easy magnetization plane and anisotropy constant equal to $\mu_0 M_s^2/2$) and the contribution to the magnetic anisotropy associated with the material of the magnetic film are important. In addition to the shape anisotropy, two major contributions to the macroscopic magnetic anisotropy of the film are made by the in-plane magnetic anisotropy with an easy magnetization axis (EA) in the film plane ($K_{inplane} \equiv K_{ip}$, $H_{ip} = 2K_{ip}/\mu_0 M_s$, where M_s is the saturation magnetization) and out-of-plane magnetic anisotropy ($K_{out\,of\,plane} \equiv K_{op}$, $H_{op} = 2K_{op}/\mu_0 M_s$). Both the in-plane and out-of-plane magnetic anisotropies are usually induced by the deposition of a thin film or a multilayered structure in the presence of a constant magnetic field [1,2], by inclined sputtering [3,4] due to the anisotropic substrate surface [5,6], film texture [7] or due to anisotropic stresses via magnetoelastic phenomena [8].

The H_{ip} field value is almost the upper limit of fields, which can be detected by the sensor. The values of K_{op} or H_{op} limit the film thicknesses of the sensitive element components. This limitation is associated with the transition to the "transcritical" state with the thickness L of the uniform film greater than the critical thickness $L_{cr} = 2\pi\sqrt{(A/K_{op})}$ (where A is the exchange stiffness). The magnetic hysteresis of the film in this state ($L > L_{cr}$) sharply increases, and the in-plane magnetic anisotropy becomes weakly pronounced due to the appearance of rotatable magnetic anisotropy [9,10]. The transition into the "transcritical" state is accompanied by the formation of stripe domains oriented parallel to the external magnetic field previously applied to the film in the film plane.

The easy magnetization axis of the rotatable anisotropy is parallel to the stripes. The stripe domains can be arranged by a certain applied field in any direction in the film plane, which implies that the easy magnetization axis along the stripes is rotated by the applied field as well [2]. Usually, the rotatable anisotropy in soft magnetic films exceeds the induced magnetic anisotropy, and thus, the total magnetic anisotropy in the film plane becomes weakly pronounced. The high magnetic hysteresis and weakly pronounced magnetic anisotropy strongly decrease the sensitivity of the thin film element with respect to the external magnetic field, making the element unsuitable for the magnetic field sensing. This limitation is sometimes in conflict with the technical requirements for the sensor parameters, in particular, concerning the requirementsfor particular applications (such as GMI) demanding films with a thickness exceeding L_{cr}. An example of resolving such a contradiction is a multilayer design of a magnetic thin film-based element [2]. Magnetic sensors with sensitive elements containing single-layered or multilayered film structures are of interest for applications in electronic devices and biomedicine [11–13].

Iron–nickel $Fe_{20}Ni_{80}$ (permalloy)-based films are the best historically proven choice for magnetic nanostructured-sensitive elements with high magnetic permeability and fairly stable characteristics in different media, wide temperature ranges and various radiation levels (up to airspace conditions) [14–16]. Permalloy-based multilayers are becoming more and more important in sensor design due to a number of advantages, including the developed methodologies to avoid the transition into the "transcritical" state for rather thick films [9,10,17,18], which are required for many high-frequency applications, as well as efficient sensors working on the principle of the giant magnetoimpedance effect (GMI) [14–16].

New requirements for the functional properties of thin film-based sensors call for the development of measurement and characterization approaches leading to the deeper understanding both of integral and local properties of thin films and multilayered structures. The approaches traditionally used in magnetic anisotropy characterization (measurement of magnetization curves and ferromagnetic resonance—FMR) are continuously improved, both through the development of standard magnetometers and spectrometers with enhanced sensitivities and resolutions. The development of new approaches to data analysis also contributes to this progress. The latter is important in the research related to the inhomogeneity of magnetic anisotropy (both the variation value of the anisotropy constant K and the variation of the easy magnetization axis direction). The inhomogeneity of the magnetic anisotropy is due both to the variation of the technological parameters during the thin film fabrication and to the natural features of the process of thin film growth. The above-mentioned contributions are quite common for many types of magnetic films and they should be taken into account during the fabrication of the sensitive element of magnetic films. Such an inhomogeneity can narrow the range of linear and reversible responses of the film element. In this work, we discuss magnetic anisotropy measurements using ferromagnetic resonance and magnetometry and compare various approaches to the data processing using thin films and multilayered permalloy-based systems.

In this study, we propose, discuss and develop several approaches for the detailed evaluation of magnetic anisotropy using static and dynamic magnetic measurements, including ferromagnetic resonance, as a tool for the comprehensive characterization of thin films and multilayered permalloy-based structures.

2. Experiment

FeNi-based films including single-layer and multilayer films (Fe$_{20}$Ni$_{80}$ (100 nm)/Cu(3 nm))$_5$ were prepared by magnetron sputtering onto Corning glass substrates at room temperature. The background pressure was 3×10^{-7} mbar, and the working argon pressure was 3×10^{-3} mbar. Permalloy films (Py) were deposited using a Fe$_{20}$Ni$_{80}$ alloy target. The thickness control of the layers was carried out through the deposition time based on the previously calibrated deposition rates. The single-layer FeNi film thickness was varied in a range of 50–500 nm. A constant magnetic field of 20 kA/m was applied parallel to the film plane during the deposition in order to induce a well-defined uniaxial magnetic anisotropy. In some cases, a Ta buffer layer was used in order to improve the properties of the FeNi thin films. The use of the Ta buffer layer leads to a more perfect crystal structure of permalloy films, which, in turn, contributes to a decrease in their coercive force [19,20]. The compositions of the obtained FeNi films were determined by energy dispersive X-ray analysis and in all the cases under consideration it was close to the Fe$_{20}$Ni$_{80}$ permalloy composition with the near-to-zero magnetostriction constant [2]. Magnetic hysteresis loops (both in-plain and out of plane) were measured by a vibrating sample magnetometer. The images of domain structures were obtained by the magneto-optical Kerr effect using an optical microscope (Evico, Dresden, Germany).

The microwave absorption spectra were measured using the equipment of the Krasnoyarsk Regional Center of Research Equipment of the Federal Research Center "Krasnoyarsk Science Center SB RAS" (spectrometer ELEXSYS E580, Bruker, Bremen, Germany). The spectra were acquired at room temperature in the X-band (the resonator pumping frequency was f = 9.43 GHz). The sample was placed into an antinode of the oscillating magnetic field h~ of the cavity resonator, and the external constant magnetic field was applied in the film plane.

3. Results and Discussion

3.1. In-Plane Magnetic Anisotropy

The transition from the thin film state into the "transcritical" state is clearly observed on the hysteresis loops of the single-layer Py/Ta films (Py here and further implies permalloy Fe$_{20}$Ni$_{80}$) with the thicknesses of 50, 100 or 500 nm (Figure 1). This transition results in a sharp increase in the coercive force, change in the loop shape, and almost complete disappearance of the magnetic anisotropy in the film plane. As a consequence of such transition, a perpendicular magnetic anisotropy component appears. Magnetron sputtering of the Py layers is intentionally performed in a field applied in the film plane during the film deposition, which leads to the formation of an in-plane magnetic anisotropy with EA along the direction of the applied field [2]. In Figure 1, this direction corresponds to the angle $\phi = 0°$ between the applied field and the field axis during the deposition.

The formation of EA in the film plane induced under these deposition conditions is confirmed by the shape of the hysteresis loops (Figure 1) for the films with the thicknesses of 50 and 100 nm. The coincidence of the loop shape for the films of 50 and 100 nm means that the film surface does not significantly contribute to the hysteresis. Apparently, it is controlled only by the bulk properties of the film (for example, by the features of the induced magnetic anisotropy). It also means that the hysteresis properties of the Py layer are uniform though the film thickness. The Stoner–Wohlfarth model here well describes the coercivity angular dependence $H_c(\psi)$ in the angular range of $\pm 15°$ on the hard magnetization axis (dashed lines in Figure 1a). At other angles, the $H_c(\phi)$ dependence corresponds qualitatively to the inhomogeneous magnetization associated with the nucleation of reverse magnetization domains and motion of the domain walls. Fitting the $H_c(\phi)$ data in the range of angles $\pm 15°$ from the direction corresponding to the hard magnetization axis by the equation $H_c(\phi) = H_{ip}|\cos(\phi)|$ provides an estimate of the magnetic anisotropy field $H_{ip} = 0.40 \pm 0.08$ kA/m.

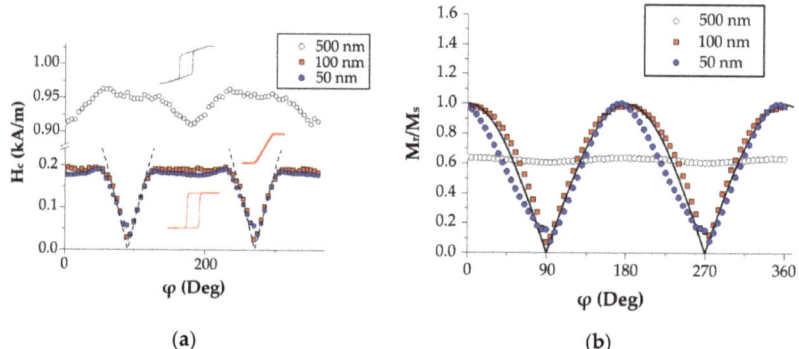

Figure 1. The parameters of the hysteresis loop of permalloy films of various thicknesses: (**a**) coercive force and (**b**) remnant magnetization. The dashed line in (**a**) is the equation $H_c(\phi) = H_{ip}|\cos(\phi)|$; the solid line in (**b**) is $f(\phi) = |\cos(\phi)|$.

The saturation field estimation from the magnetic hysteresis loop measured in the hard magnetization direction is the most common approach for estimating the in-plane magnetic anisotropy field. The magnetization curve of the Py (100 nm) film in the in-plane applied magnetic field perpendicular to in-plane easy magnetization axis (gray symbols in Figure 2a) and its fitting using the formula:

$$M(H) = \begin{cases} M_s \sum f_i \cdot \frac{H}{H_{ip}}, \text{ for } |H| < H_{ip} \\ M_s, \text{ for } |H| > H_{ip} \end{cases} \quad (1)$$

where f_i is the statistic weight of sites with the specific H_{ip} value, which makes it possible to estimate the inhomogeneity of the anisotropy field H_{ip} (inset in Figure 2a). The Kerr image in Figure 2b shows large stripe domains typical for the subcritical (thin-film) state of the film. Below, we show that the film in the transcritical state (thick-film) shows stripe domains that are two orders of magnitude narrower.

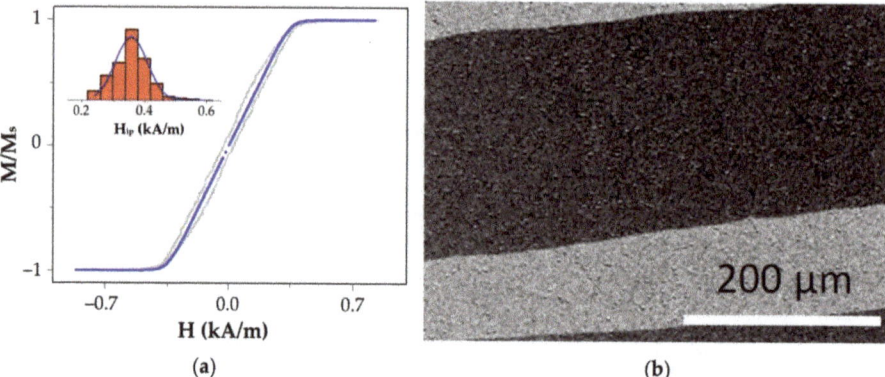

Figure 2. The magnetization curve of the Py (100 nm) film in the in-plane applied magnetic field perpendicular to the in-plane easy magnetization axis (gray symbols) and its fitting using Formula (1) (the inset shows the evaluation result for the film Py (100 nm)) (**a**). The magnetic domain structure in the zero magnetic field; the easy magnetization axis is oriented close to the horizontal direction (**b**).

According to the Stoner–Wohlfarth model, the saturation field in the direction perpendicular to the axis of the easiest magnetization is equal to the magnetic anisotropy field. The loop shape within the framework of the model is nearly linear in the range of $-H_s \cdot (M = M_s)$ to $H_s \cdot (M = M_s)$. Outside this range, the sample is uniformly magnetized to the saturation ($M = \pm M_s$). In the experiment, in a perfectly homogeneous film, a feature near the field $H_s = H_a$ would be observed as a sharp gap on the M(H) curve, or a discontinuity in the susceptibility $\chi(H) = dM/dH$. There can also be a narrow peak in the field dependence of $d\chi/dH$. Since the inhomogeneity of the anisotropy field somewhat blurs the gap in M(H) near H_s, this feature is used to quantify the inhomogeneity of the magnetic anisotropy field [21,22]. An estimate of the magnetic anisotropy field inhomogeneity in the film plane is shown in Figure 2 (inset), with the distribution center $H_{ip} = 357 \pm 5$ A/m, and FWHM (full width at half maximum) = 120 A/m.

Interesting possibilities for estimating the inhomogeneity of magnetic anisotropy are provided by the measurements of the magnetic parameters of local areas in the sample using an automated scanning FMR spectrometer [8,23], in which a miniature micro-strip resonator fabricated on a substrate with a high dielectric constant is employed as a thin film based sensitive element. Near the antinode of the high-frequency magnetic field, a small measuring hole was made in the resonator screen, ensuring the locality of measurements (for the data in Figure 3 the diameter of the measuring hole of the sensor was ~1 mm). The idea to use the microwave techniques based on a conventional homodyne spectrometer, as a microwave microscope was reported in previous publications [24,25]. Bhagat et al. employed a microwave microscope, i.e., a 2 mm diameter hole in a thinned wall of the cavity for the evaluation of the properties of the FeNi film placed outside the cavity in front of the hole [25]. This methodology has the advantages of avoiding extra-large loading and offers a possibility to estimate the homogeneity of the properties by exposing different regions of the sample to microwaves. However, the system had only a manual displacement mode allowing limited number of points. In addition, it was possible to make measurements from two sides of the film deposited onto a glass substrate: from the side of the film and from the side of the substrate. The equipment described in the present work has an advantage of a scanning system and higher resolution of measurements as the scanning hole had a two times lower diameter of the hole.

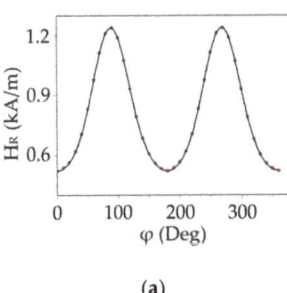
(a)

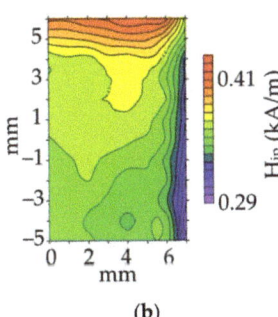
(b)

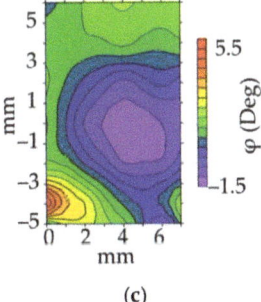
(c)

Figure 3. The angular dependence of the resonant field of a pixel in the center of the single-layer film of Py (100 nm) (a) Inhomogeneity of the uniaxial magnetic anisotropy field in the plane of the single-layer film of Py (100 nm) (b) Orientation inhomogeneity of the in-plane easy axis in the plane of the single-layer film of Py (100 nm) (c) Deviation from the average in-plane easy axis.

In this work, a microwave sensor with a pump frequency f = 1.010 GHz was used for these measurements. In Figure 3, we demonstrate the approach to the quantification of the local magnetic anisotropy (Figure 3a) and the result of studying the inhomogeneity of magnetic anisotropy in the plane of the single-layer film of Py (100 nm) (Figure 3b,c). For the given frequency with the applied field in the film plane, the resonant fields in the single-layer film of Py (100 nm) did not exceed 1.6 kA/m. The angular dependences of the resonant field H_R were recorded in each local area with a step of 1 mm over the entire surface of the film. Then, the main magnetic characteristics of the films were determined from these data on the basis of a phenomenological calculation [23,26]. The obtainment of the anisotropy constant from the angular dependence of the resonant field is based on the dependence of the resonant field on the equilibrium position of the magnetization, which is described by the Stoner–Wohlfarth model (Smit–Beljers approach [26]). When the external magnetic field is applied in the film plane, in this model, the in-plane anisotropy field is the only important parameter. We used the software developed in [23], which has additional options to quantify not only the uniaxial constants, but also higher-order anisotropy contributions. The major contribution to the angular dependence of the resonant field comes from the uniaxial in-plane anisotropy. The maps of the in-plane anisotropy field and orientation of EA (Figure 3) show the local value of the in plane magnetic anisotropy. The inhomogeneity of the anisotropy field along the right and upper edges of the sample (Figure 3) is due to deformations that occur in the process of cutting the sample. A more uniform distribution is observed along the lower and left edges of the sample that are not subjected to the cut process. The uniformity in the anisotropy field is satisfactory and the inhomogeneity in the orientation of the EA over an area of 5×7 mm does not exceed $6 \div 7°$. The average field obtained here $H_{ip} = 340 \pm 20$ A/m is close to the anisotropy field estimated from the hysteresis loops. It is important that direct local measurements of the magnetic anisotropy (Figure 3b,c) provide not only an estimate of the inhomogeneity of magnetic anisotropy, but also visualize the spatial pattern of the distribution of the local magnetic anisotropy.

3.2. Out-of-Plane Magnetic Anisotropy

In the external magnetic field applied in a film plane, the hysteresis loop of the Py film (500 nm) has the shape typical for the "transcritical" state (Figure 4). The evolution of the micromagnetic state corresponding to the descending branch of the loop is a transition from a quasi-uniform state (containing no closing magnetic domains) above H_s to the appearance and development of a stripe structure in the zero external magnetic field [25,27]. The saturation magnetic field H_s is conditioned by the magnetic constants and parameters of the film by the following equation [27,28]:

$$1 - \frac{H_s}{H_{op}} = \frac{1}{2}\sqrt{\frac{2A}{\mu_0 H_{op} M_s}} \cdot L^{-1} \left[1 + \frac{H_{op}}{M_s}\right]^{-1/2} \quad (2)$$

where L is the film thickness, A is the exchange stiffness, M_s is the saturation magnetization and H_{op} is the-out-of plane anisotropy field ($H_{op} = 2K_{op}/M_s$). Using $H_s = 3.18 \pm 0.16$ kA/m determined from the hysteresis loops similar to the one shown in Figure 4 and the constants measured for this film (see the supplement $A = (0.90 \pm 0.05) \cdot 10^{-11}$ J/m, $M_s = 800 \pm 20$ kA/m) and L = 500 nm, one can obtain the value $H_{op} = 8.0 \pm 0.8$ kA/m.

Another approach for the estimation of the H_{op} value is to use the width of the stripe domains in the "transcritical" state. It was shown in [27] that if the domain width is defined as the size of the area of uniform magnetization, where the transverse magnetization component (m_z) contribution is at least 65% of the maximum value, then such a stripe domain width (D_m) can be described by the Murayama equation [27,29]:

$$D_m = \sqrt{L} \cdot \left(A(K_{op} + \frac{\mu_0 M_s^2}{2})/8\mu_0 K_{op} M_s^2\right)^{1/4} \quad (3)$$

Using Kerr image processing (Figure 4b and inset) $D_m = 260 \pm 20$ nm was determined. Solving Equation (2) with the same constants (A and M_s) that were used for solving Equation (2), the estimated D_m and L = 500 nm provides the value $K_{op} = (5.2 \pm 1.6) \cdot 10^3 \frac{J}{m^3}$ or $H_{op} = 10 \pm 3$ kA/m. Note that the confidence interval of this estimate is consistent with the estimate from the field H_s and Equation (2), although the experimental error associated with this method is much higher.

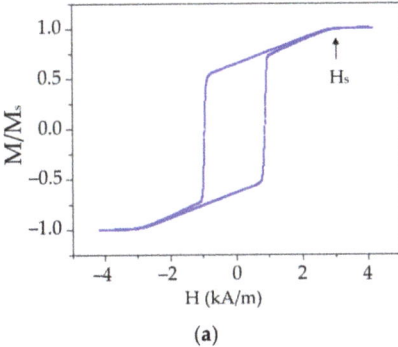
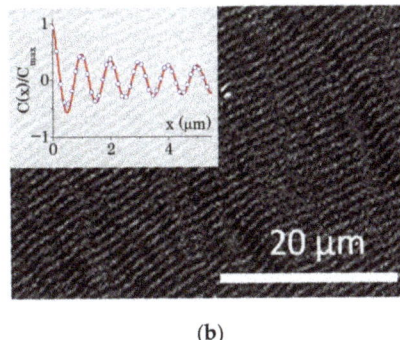

Figure 4. Hysteresis loop of the Py film (500 nm), field applied in the film plane at an angle of 0° to in-plane EA (**a**). The stripe magnetic domain structure in the zero external magnetic field. The inset shows the correlation function in the direction across the stripe structure to the estimate stripe domain width $D_m = 260 \pm 20$ nm (**b**).

When the field is oriented out of the film plane, the ferromagnetic resonance field changes in a wide range due to the high contribution to the magnetic anisotropy constant from the film shape $(-\frac{1}{2}\mu_0 M_s^2)$. In this case, the film magnetization M_s is determined both by the Zeeman energy and the value of the magnetic anisotropy constant $K_{op} - \frac{1}{2}\mu_0 M_s^2$, ($K_{op}$ is the out of plane or perpendicular anisotropy constant). These terms contribute to the angular dependence of the resonant field in a different way making it possible to estimate K_{op} and M_s separately using the Smit–Beljers approach [26]. Thus, both in-plane and out-of-plane anisotropy are present in one film sample. The axis of the out-of-plane rotation was chosen to coincide with the in-plane EA to exclude the influence of the in-plane magnetic anisotropy on the out-of-plane angle dependence of the resonance field. The characteristic angles that are determined by the theoretical expressions for the resonant field and are controlled in the experiment are shown in Figure 5. Thus, for the measurements with the out-of-plane oriented external field, the angle ϕ_H was chosen as $\phi_H = 0°$ to easily study the out of plane magnetic anisotropy.

For the Py film (100 nm), the FMR spectrum (Figure 5a) in the range ϑ of 7° to 90° is well described by a single Lorentzian mode. Furthermore, this mode corresponds to a uniform precession of the magnetization. In the range $(-7° \div 7°)$ no uniform oscillation modes of magnetization are excited and several peaks of spin-wave resonance are observed. Figure 5a shows the resonance fields corresponding to the uniform and the 1st spin wave bulk mode. Fitting the angular dependence of the resonance field for the uniform mode using the Smit–Beljers approach [26] for the Py (100 nm) film, gives the following fitting parameters $M_s = 880 \pm 10$ kA/m, $H_{op} = 8.0 \pm 0.8$ kA/m (see also Table 1). For the Py film (500 nm), the shape of the FMR spectrum is not described by one Lorentzian. However, it can be satisfactorily described by the sum of at least three Lorentzian functions. The multiplicity of the peaks here may be interpreted as the manifestation of inhomogeneity in the thick film, and the peaks in this case are supposed to be associated with some naturally formed layers with the given absorption. It is impossible to specify exactly what these layers are and at what depth each of them is located within the framework of this approach alone. Even so, the separate processing of the data related to these composite peaks can be considered as a useful way of characterizing the film inhomogeneity over the thickness. These layers are apparently the result of the inhomogeneity of elastic deformations over the film thickness. It is hard to observe such layering using transmission electron microscopy. Because of this, we see some value of the proposed approaches using FMR. The issue of this inhomogeneity is apparently related to the film growth mechanism, since it usually determines the inhomogeneity of elastic deformations in deposited films. For example, residual strains in the films can be distributed over the thickness in such a way that compressive stresses are replaced by tensile stresses. A change in the sign of the deformation should lead to a change in the sign of the stress-induced magnetic anisotropy field. The negative anisotropy field in the multilayer sample (Table 1) corresponds to the easy-plane anisotropy. Note that negative H_{op} is observed only in the multilayer sample, where there are more sources of internal deformations due to a larger number of interfaces. The range of angles in which the individual peaks are considered as the result of a uniform precession of the magnetization is ϑ_H of 7° to 90°. The results of fitting of the angular dependence of each peak are given in Table 1. In the multilayered film $(Fe_{20}Ni_{80}/Cu)_5$ with the Py layer thickness of 100 nm, we also observe a non-Lorentzian shaped line, which can also be successfully described as the sum of three Lorentzian peaks. The saturation magnetization and the perpendicular magnetic anisotropy field are determined for each of the three modes and are given in Table 1.

Table 1. The parameters of the single-layered films $Fe_{20}Ni_{80}$ (100 nm and 500 nm) and multilayered film $(Fe_{20}Ni_{80}/Cu)_5$ with the $Fe_{20}Ni_{80}$ layer thickness of 100 nm, determined from the angular dependences of the FMR.

	M_{eff}, kA/m	H_{op}, kA/m
$Fe_{20}Ni_{80}$ single-layer film of 100 nm	880	8.0
$Fe_{20}Ni_{80}$ single-layer film of 500 nm	924	11.9
	894	0
	890	4.0
multilayer $(Fe_{20}Ni_{80}/Cu)_5$ with $Fe_{20}Ni_{80}$ layer thickness of 100 nm	758	0
	740	15.9
	790	−35.8

According to Table 1, the single-layer Py film (100 nm) shows more of the uniform microwave response (only one peak) and a wider range of angles (from 7 to 90 up to 90°) of the uniform precession mode than the Py film of 500 nm (from 25 to 90°). The range of the angles is important as a parameter of the sensor element operating at super high frequencies, because the excitation of inhomogeneous precession modes will inevitably lead to many peaks in the film response. In the multilayered film, this range is closer to that of the single-layered film (from 11 to 90°), which indicates the advantages of the multilayer film sensor design.

The deviations of the peak from the Lorentzian shape can be also viewed as a measure of the inhomogeneity of the microwave response [30–32]. Let us summarize and discuss the observations concerning it. For the single-layered thin film (Py of 100 nm), a single Lorentz peak is observed: the deviation is negligibly small. For the single-layered Py film of 500 nm the deviation is the largest. For the multilayered structure (with the total thickness of 500 nm) this deviation is much lower than the one observed for the single-layered Py film (500 nm), although it somewhat exceeds the deviation observed for the single-layer film (100 nm). The fields H_{op} in the thick film are close to those for the single-layered Py film (100 nm), and the average of the fields over three components coincides with it (Table 1). This is in agreement with the conclusion (see the analysis for Figure 1) that the constant (or field) of the perpendicular magnetic anisotropy of single-layered films is, in general, uniform over the thickness and, therefore, does not change with the thickness. In addition, the field H_{op} for the Py (100 nm) film coincides with the estimate made from the hysteresis loop of the Py film (500 nm) in the "transcritical" state. These observations reveal additional advantages of the multilayered design for GMI sensors, discussed in a number of previous works [33–35].

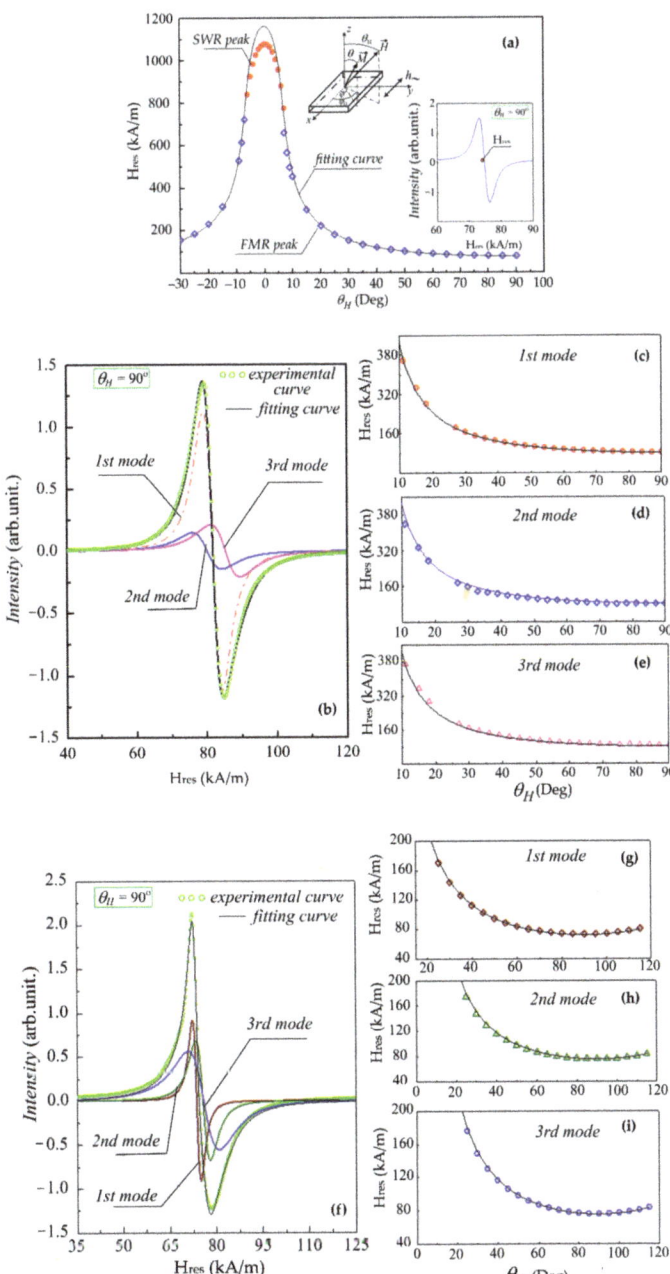

Figure 5. The examples of the microwave spectra of the one-layer films with the thickness 100 nm (inset to Figure (**a**)) and 500 nm (**f**), and the multilayer film (Fe20Ni80/Cu)5 (**b**) measured at $\vartheta_H = 90°$. The solid lines demonstrate the fitting curves of the angular dependences of the resonance fields of the uniform modes fitted according to the Smit–Beljers. The resonance field of the single mode for the 100 nm film (**a**) and the resonance fields of the individual modes for the multilayer film (**c**–**e**) and for the 500 nm film (**g**–**i**) obtained from the fitting of the experimental curve are shown by different symbols.

4. Conclusions

The characterization of magnetic anisotropy using magnetization curves and the ferromagnetic resonance techniques of single-layered and multilayered thin film structures based on a permalloy for magnetic field sensors was performed. It was demonstrated that the proposed approaches allowed for not only the characterization of magnetic anisotropy in the plane and perpendicular to the plane of the thin films, but also the study of their inhomogeneity, which is important in the magnetic film sensor design. It is shown that the quantitative data from the direct integral and local measurements of magnetic anisotropy are consistent with the direct and indirect estimates based on the magnetization curve processing. The example of estimating the perpendicular magnetic anisotropy constant from the width of stripe domains in a film in the supercritical state is provided. The average in-plane magnetic anisotropy of the single-layered Py (50, 100 and 500 nm) films prepared by magnetron sputtering onto a Corning glass is uniform through the thickness of the single-layer film. The inhomogeneity of the perpendicular anisotropy field for the 500 nm film is greater than that for the 100 nm film, and the inhomogeneity of the multilayer film ($Fe_{20}Ni_{80}$ (100 nm)/Cu (3 nm))$_5$ is greater than that for the single-layer of the approximately same thickness.

Author Contributions: Conceptualization, S.V.K., R.S.I. and A.V.S.; methodology, V.N.L. and A.V.S.; software, S.V.K.; validation, V.N.L. and A.V.S.; formal analysis, A.A.P.; investigation, I.G.V., S.A.K., V.N.L., A.V.S and A.A.P.; data curation, V.N.L.; writing—original draft preparation, S.V.K., R.S.I. and A.V.S.; writing—review and editing, S.V.K., R.S.I. and A.V.S.; visualization, A.A.P. and A.V.S.; supervision—all authors have discussed the results and implications and have commented on the manuscript at all stages. All authors have read and agreed to the published version of the manuscript.

Funding: This research was funded by the Russian Science Foundation (RSF), project no. 22-29-00980, https://rscf.ru/en/project/22-29-00980/ (accessed on 20 March 2022).

Institutional Review Board Statement: This work did not involve humans or animals and therefore did not require an Institutional Review Board Statement and approval.

Informed Consent Statement: Not applicable.

Data Availability Statement: Data available from the corresponding author upon reasonable request.

Acknowledgments: This research was funded by the Russian Science Foundation (RSF), project no. 22-29-00980, https://rscf.ru/en/project/22-29-00980/ (accessed on 20 March 2022). We thank A.M. Gor'kovenko for special support.

Conflicts of Interest: The authors declare no conflict of interest.

References

1. Coïsson, M.; Vinai, F.; Tiberto, P.; Celegato, F. Magnetic Properties of FeSiB Thin Films Displaying Stripe Domains. *J. Magn. Magn. Mater.* **2009**, *321*, 806–809. [CrossRef]
2. Kurlyandskaya, G.V.; Elbaile, L.; Alves, F.; Ahamada, B.; Barrué, R.; Svalov, A.V.; Vas'kovskiy, V.O. Domain Structure and Magnetization Process of a Giant Magnetoimpedance Geometry FeNi/Cu/FeNi(Cu)FeNi/Cu/FeNi Sensitive Element. *J. Phys. Condens. Matter* **2004**, *16*, 6561–6568. [CrossRef]
3. Komogortsev, S.V.; Varnakov, S.N.; Satsuk, S.A.; Yakovlev, I.A.; Ovchinnikov, S.G. Magnetic Anisotropy in Fe Films Deposited on SiO2/Si(001) and Si(001) Substrates. *J. Magn. Magn. Mater.* **2014**, *351*, 104–108. [CrossRef]
4. Solovev, P.N.; Izotov, A.V.; Belyaev, B.A. Numerical Study of Structural and Magnetic Properties of Thin Films Obliquely Deposited on Rippled Substrates. *J. Phys. Condens. Matter* **2021**, *33*, 495802. [CrossRef] [PubMed]
5. Gubbiotti, G.; Sadovnikov, A.; Beginin, E.; Sheshukova, S.; Nikitov, S.; Talmelli, G.; Asseberghs, I.; Radu, I.P.; Adelmann, C.; Ciubotaru, F. Magnonic Band Structure in CoFeB/Ta/NiFe Meander-Shaped Magnetic Bilayers. *Appl. Phys. Lett.* **2021**, *118*, 162405. [CrossRef]
6. Noginova, N.; Gubanov, V.; Shahabuddin, M.; Gubanova, Y.; Nesbit, S.; Demidov, V.V.; Atsarkin, V.A.; Beginin, E.N.; Sadovnikov, A.V. Ferromagnetic Resonance in Permalloy Metasurfaces. *Appl. Magn. Reson.* **2021**, *52*, 749–758. [CrossRef]
7. Skomski, R. Nanomagnetics. *J. Phys. Condens. Matter* **2003**, *15*, R841–R896. [CrossRef]
8. Belyaev, B.A.; Izotov, A.V. Ferromagnetic Resonance Study of the Effect of Elastic Stresses on the Anisotropy of Magnetic Films. *Phys. Solid State* **2007**, *49*, 1731–1739. [CrossRef]

9. Saito, N.; Fujiwara, H.; Sugita, Y. A New Type of Magnetic Domain Structure in Negative Magnetostriction Ni-Fe Films. *J. Phys. Soc. Jpn.* **1964**, *19*, 1116–1125. [CrossRef]
10. Svalov, A.V.; Kurlyandskaya, G.V.; Hammer, H.; Savin, P.A.; Tutynina, O.I. Modification of the "Transcritical" State in Ni75Fe16Cu5Mo4 Films Produced by RF Sputtering. *Tech. Phys.* **2004**, *49*, 868–871. [CrossRef]
11. Gardner, D.S.; Schrom, G.; Paillet, F.; Jamieson, B.; Karnik, T.; Borkar, S. Review of On-Chip Inductor Structures with Magnetic Films. *IEEE Trans. Magn.* **2009**, *45*, 4760–4766. [CrossRef]
12. Hunter, D.; Osborn, W.; Wang, K.; Kazantseva, N.; Hattrick-Simpers, J.; Suchoski, R.; Takahashi, R.; Young, M.L.; Mehta, A.; Bendersky, L.A.; et al. Giant Magnetostriction in Annealed Co1−xFex Thin-Films. *Nat. Commun.* **2011**, *2*, 518. [CrossRef] [PubMed]
13. Schmalz, J.; Kittmann, A.; Durdaut, P.; Spetzler, B.; Faupel, F.; Höft, M.; Quandt, E.; Gerken, M. Multi-Mode Love-Wave SAW Magnetic-Field Sensors. *Sensors* **2020**, *20*, 3421. [CrossRef] [PubMed]
14. Kurlyandskaya, G.V.; Fernández, E.; Svalov, A.; Burgoa Beitia, A.; García-Arribas, A.; Larrañaga, A. Flexible Thin Film Magnetoimpedance Sensors. *J. Magn. Magn. Mater.* **2016**, *415*, 91–96. [CrossRef]
15. Agra, K.; Mori, T.J.A.; Dorneles, L.S.; Escobar, V.M.; Silva, U.C.; Chesman, C.; Bohn, F.; Corrêa, M.A. Dynamic Magnetic Behavior in Non-Magnetostrictive Multilayered Films Grown on Glass and Flexible Substrates. *J. Magn. Magn. Mater.* **2014**, *355*, 136–141. [CrossRef]
16. Melnikov, G.Y.; Lepalovskij, V.N.; Svalov, A.V.; Safronov, A.P.; Kurlyandskaya, G.V. Magnetoimpedance Thin Film Sensor for Detecting of Stray Fields of Magnetic Particles in Blood Vessel. *Sensors* **2021**, *21*, 3621. [CrossRef]
17. Amos, N.; Fernandez, R.; Ikkawi, R.; Lee, B.; Lavrenov, A.; Krichevsky, A.; Litvinov, D.; Khizroev, S. Magnetic Force Microscopy Study of Magnetic Stripe Domains in Sputter Deposited Permalloy Thin Films. *J. Appl. Phys.* **2008**, *103*, 07E732. [CrossRef]
18. Svalov, A.V.; Aseguinolaza, I.R.; Garcia-Arribas, A.; Orue, I.; Barandiaran, J.M.; Alonso, J.; FernÁndez-Gubieda, M.L.; Kurlyandskaya, G.V. Structure and Magnetic Properties of Thin Permalloy Films Near the "Transcritical" State. *IEEE Trans. Magn.* **2010**, *46*, 333–336. [CrossRef]
19. Mao, M.; Leng, Q.; Huai, Y.; Johnson, P.; Miller, M.; Tong, H.-C.; Miloslavsky, L.; Qian, C.; Wang, J.; Hegde, H. Characterization of ion beam and magnetron sputtered thin Ta/NiFe films. *J. Appl. Phys.* **1999**, *85*, 5780–5782. [CrossRef]
20. Saravanan, P.; Hsu, J.-H.; Tsai, C.-L.; Singh, A.K.; Alagarsamy, P. Effect of Ta underlayer on thickness-depending magnetic properties of Ni-Fe films. *IEEE Trans. Magn.* **2015**, *51*, 2006604. [CrossRef]
21. Barandiaran, J.M.; Vazquez, M.; Hernando, A.; Gonzalez, J.; Rivero, G. Distribution of the Magnetic Anisotropy in Amorphous Alloys Ribbons. *IEEE Trans. Magn.* **1989**, *25*, 3330–3332. [CrossRef]
22. Yoo, E.; Samardak, A.Y.; Jeon, Y.S.; Samardak, A.S.; Ognev, A.V.; Komogortsev, S.V.; Kim, Y.K. Composition-Driven Crystal Structure Transformation and Magnetic Properties of Electrodeposited Co–W Alloy Nanowires. *J. Alloys Compd.* **2020**, *843*, 155902. [CrossRef]
23. Belyaev, B.A.; Izotov, A.V.; Leksikov, A.A. Magnetic Imaging in Thin Magnetic Films by Local Spectrometer of Ferromagnetic Resonance. *IEEE Sens. J.* **2005**, *5*, 260–267. [CrossRef]
24. Hubert, A.; Schäfer, R. *Magnetic Domains: The Analysis of Magnetic Microstructures*; Springer: Berlin/Heidelberg, Germany, 1998.
25. Bhagat, S.M. Ferromagnetic Resonance. In *Metals Handbook*; ASM International-Materials Park: Novelty, OH, USA, 1986; Volume 10.
26. Smit, J.; Beljers, H.G. Ferromagnetic Resonance Absorption in BaFe12O12, a Highly Anisotropic Crystal. *Philips Res. Rep.* **1955**, *10*, 113–130.
27. Solovev, P.N.; Izotov, A.V.; Belyaev, B.A.; Boev, N.M. Micromagnetic Simulation of Domain Structure in Thin Permalloy Films with In-Plane and Perpendicular Anisotropy. *Phys. B Condens. Matter* **2021**, *604*, 412699. [CrossRef]
28. Sugita, Y.; Fujiwara, H.; Sato, T. Critical thickness and perpendicular anisotropy of evaporated permalloy films with stripedomains. *Appl. Phys. Lett.* **1967**, *10*, 229–231. [CrossRef]
29. Murayama, Y. Micromechanics on Stripe Domain Films. I. Critical Cases. *J. Phys. Soc. Jpn.* **1966**, *21*, 2253–2266. [CrossRef]
30. Meshcheryakov, V.F.; Vasil'ev, A.G.; Timonin, K.V.; Khorin, I.A. Structural Inhomogeneities and Magnetic Properties of Co/Cu Multilayer Films. *Crystallogr. Rep.* **2002**, *47*, 1063–1071. [CrossRef]
31. Ustinov, V.V.; Rinkevich, A.B.; Vazhenina, I.G.; Milyaev, M.A. Microwave Giant Magnetoresistance and Ferromagnetic and Spin-Wave Resonances in (CoFe)/Cu Nanostructures. *J. Exp. Theor. Phys.* **2020**, *131*, 139–148. [CrossRef]
32. Vazhenina, I.G.; Iskhakov, R.S.; Milyaev, M.A.; Naumova, L.I.; Rautskii, M.V. Spin Wave Resonance in the [(Co0.88Fe0.12)/Cu]N Synthetic Antiferromagnet. *Tech. Phys. Lett.* **2020**, *46*, 1076–1079. [CrossRef]
33. Buznikov, N.A.; Safronov, A.P.; Orue, I.; Golubeva, E.V.; Lepalovskij, V.N.; Svalov, A.V.; Chlenova, A.A.; Kurlyandskaya, G.V. Modelling of Magnetoimpedance Response of Thin Film Sensitive Element in the Presence of Ferrogel: Next Step toward Development of Biosensor for in-Tissue Embedded Magnetic Nanoparticles Detection. *Biosens. Bioelectron.* **2018**, *117*, 366–372. [CrossRef]
34. García-Arribas, A. The Performance of the Magneto-Impedance Effect for the Detection of Superparamagnetic Particles. *Sensors* **2020**, *20*, 1961. [CrossRef]
35. Kurlyandskaya, G.V.; Blyakhman, F.A.; Makarova, E.B.; Buznikov, N.A.; Safronov, A.P.; Fadeyev, F.A.; Shcherbinin, S.V.; Chlenova, A.A. Functional Magnetic Ferrogels: From Biosensors to Regenerative Medicine. *AIP Adv.* **2020**, *10*, 125128. [CrossRef]

Article

A Uniform Magnetic Field Generator Combined with a Thin-Film Magneto-Impedance Sensor Capable of Human Body Scans

Tomoo Nakai

Industrial Technology Institute, Miyagi Prefectural Government, Sendai 981-3206, Japan; nakai-to693@pref.miyagi.lg.jp

Abstract: A detection system for magnetic inclusions of large bulk, such as that of a whole human body, is proposed in this paper. The system consists of both a uniform magnetic field generating apparatus capable of the insertion of a whole human body and also of a high-sensitivity magnetic sensor array installed in the strong magnetic field. The system can detect the magnetic inclusion simultaneously through its magnetization, which is advantageous for detecting low-remanence magnetic materials, such as a cluster of nanoparticles. The thin-film magneto-impedance sensor was reported to be capable of tolerating strong magnetic fields of more than 3000 Gauss (0.3 T) in the substrate's normal direction and can retain its sensitivity even in strong fields. Through a combination of both uniformity of strength and the placement of its directionally aligned, static magnetic field in a particular measurement area and its array of single-dimensional thin-film magneto-impedance sensors, it was reported that it can estimate a magnetic sample's 3D position by using a simple equation. The aim of the system developed in this study is to nondestructively detect a cluster of magnetic nanoparticles in a human body and also to detect the position and the concentration of the clustered magnetic particles. In this paper, a prototype system consisting of a magnetic field generator with an area of W500 mm × L400 mm × H210 mm and a uniform magnetic field of 370 Gauss (37 mT) is reported. It also reported that the thin-film magneto-impedance sensor installed in the system verified the detection of 2 mm × 1 mm small ellipsoidal magnetic chips at a distance of 27 mm from the sensor element.

Keywords: magnetic thin film; magneto-impedance sensor; nondestructive inspection; human body

Citation: Nakai, T. A Uniform Magnetic Field Generator Combined with a Thin-Film Magneto-Impedance Sensor Capable of Human Body Scans. *Sensors* **2022**, *22*, 3120. https://doi.org/10.3390/s22093120

Academic Editor: Galina V. Kurlyandskaya

Received: 29 March 2022
Accepted: 16 April 2022
Published: 19 April 2022

Publisher's Note: MDPI stays neutral with regard to jurisdictional claims in published maps and institutional affiliations.

Copyright: © 2022 by the author. Licensee MDPI, Basel, Switzerland. This article is an open access article distributed under the terms and conditions of the Creative Commons Attribution (CC BY) license (https:// creativecommons.org/licenses/by/ 4.0/).

1. Introduction

A thin-film magneto-impedance sensor [1–6] is useful for detecting magnetic materials in industrial and medical products [7–15]. The magneto-impedance sensor has the ability to detect a cluster of magnetic nanoparticles from a particular distance [13,16–18]. A system which can detect soft, small magnetic pieces using a thin-film magneto-impedance sensor within a strong static field is proposed, and it can detect submillimeter-sized magnetic particles nondestructively [19–23]. Such particles are detected by the sensor simultaneously with magnetization using a particular magnetic structure installed with sensors. The thin-film sensor is installed in a strong magnetic field and retained its sensitivity [21]. The reason the sensor is able to detect a small particle in the strong magnetic field is the tolerance against the surface normal field of the thin-film sensor, which is due to the demagnetizing field along the thickness direction of the sensor. A system which can estimate the position of a magnetic contaminant is proposed using a combination of a uniform vertical magnetic field and an array of in-plane, single-directional magnetic field sensors [23]. It can estimate both the 3D position and also the value of the magnetization. In the proposed system, the sensing direction of the sensor, which is the in-plane direction of the thin-film sensor, is set in the perpendicular direction to the applied strong magnetic field. Position and size estimation are carried out using the measured signal waveforms obtained by the nearer two sensors while feeding the sample linearly to the measurement area.

In this study, a uniform and strong magnetic field generator for a particular large volume applicable to a human body scan is proposed, in which both the position and the volume of the magnetic concentration are possible to estimate. Recent progress in medical technology utilizes magnetic nanoparticles that can be introduced into the human body. Resovist®, a kind of super paramagnetic iron oxide (SPIO) nanoparticle, is well known for its use in contrast-enhancement for magnetic resonance imaging (MRI). It consists of a mixture of Fe_2O_3 and Fe_3O_4 nanoparticles having paramagnetic properties. It has a bio-compatibility with the human body, thus many trials of medical applications using iron oxide magnetic nanoparticles, such as those for magnetic separation [24], magnetic drug delivery [25], magnetic targeting [26], and hyperthermia for cancer treatment [27], have been undertaken recently. Detection of locally concentrated magnetic nanoparticles, such as those present in a cancer or a lymph node, is becoming important for the purpose of preoperative detection of cancer metastasis. A high-sensitivity magnetic field sensor combined with such a field generator is a candidate for use in such a medical diagnostic test.

2. Concept of the Measurement System

In this section, a magnetic field generator which is capable of applying a whole human body scan with a combination of a thin-film magneto-impedance sensor for detecting the magnetic leakage field from a cluster of magnetic nanoparticles is proposed, and the concept of this system is introduced.

Figure 1 shows a schematic image of the proposed inspection system. It consists of a soft, rectangular magnetic core and magnets with opposite magnetic poles facing each other from each inner side of the core. The magnetic core works as a closed-loop magnetic field in order to make the inner magnetic field greater. On the inner surface of the magnets there are magnetic flux homogenization plates that consist of soft magnetic material of high-saturation magnetic flux density. Due to the effect of this homogenization plate, it is possible to compose the magnet on one side of several separated magnets that make a uniform magnetic field. After this, the proposed magnetic structure is able to make a uniform field in strength and of a constant directional field in the space between the upper and lower magnets. This measurement system was developed for the application of a small-sized work piece [22], and we have also proposed a method for estimating the position of a magnetic particle [23].

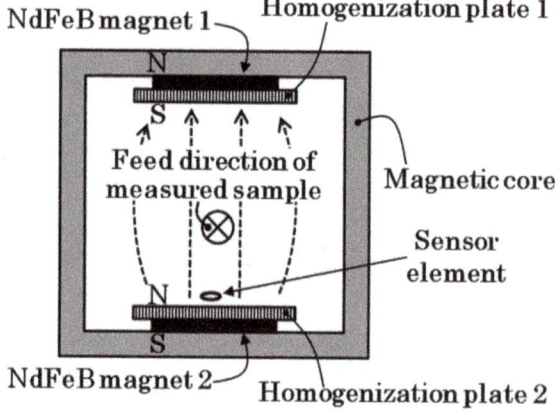

Figure 1. Schematic image of proposed inspection system. (The dotted arrow explains a generated magnetic field).

Figure 2 is an image of the measurement of a human body using the proposed inspection system. A cluster of magnetic particles is magnetized by the generated uniform and constant directional field, and an array of magnetic sensors which are set inside the magnetic field is able to detect the field coming from the magnetized particle cluster. This system offers the advantage of detecting low-remanence magnetic particles due to the fact that the sensor measures the magnetic particles simultaneously with the magnetization inside the strong magnetic field. In this system, the magnetic sensor is a thin-film sensor in which the sensing direction is along the direction of the film plane, and which sets it in the uniform and constant directional strong magnetic field with the sensing direction set perpendicular to the magnetic field. The sensor works by retaining high sensitivity even in such a strong magnetic field due to the property of tolerance against the surface of the normal magnetic field. Based on our previous study, the position estimation of the small magnetic piece was realized using a measurement system which possesses the same structure shown in Figure 2.

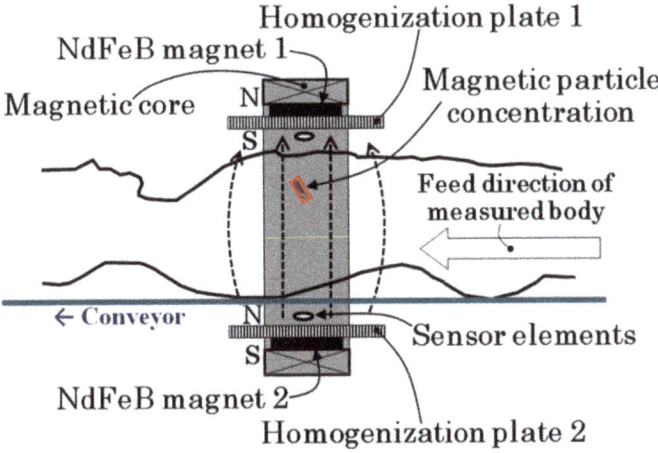

Figure 2. Image of the measurement of a human body using the proposed inspection system.

The study induced an analytical equation for estimating 3D position based on the equation for a magnetic field generated by a magnetic dipole. The resultant estimation shows that the height of the small piece, h, is calculated by the equation $h = 2 \times XSP$ when the chips are run through just above the sensor, where XSP is a position of extreme value of the measured magnetic waveform obtained by the sensor. Figure 3 explains the relationship between the feeding height of the magnetic dipole and its measured waveform of magnetic flux density obtained by the sensor, which is shown in Ref. [23]. The waveform indicates variation as a function of the feeding position of the dipole at a particular height. It shows that the relationship between the feeding height h and the position of extreme point XSP is $h = 2 \times XSP$. The position of extreme point XSP is an important parameter for position estimation [23]. Then, the longitudinal dimensions of the measurement area must have a larger value, which is the value of $h'/2$, where h' is the height of the expected maximum measurement area of a magnetic piece. In this study, the height of the measurement area was assumed to be 225 mm, which is available from the human body. The average value of the chest depth of a Japanese male is reported to be 212 mm [28]. Owing to the height of the measurement area, the longitudinal dimension is assumed to be 112.5 mm as a minimum value, which is based on the demands of the height estimation. The dimensions of the prototype system were designed as 200 mm on each side, taking into consideration the effect of the magnetic field expanding in the fringe area. Because this dimension is needed for both the + and − directions, the actual longitudinal dimensions of the system were set at 400 mm and were in the feeding direction. In this study, the measurement area, which has a uniform vertical magnetic field, was designed as W700 mm × L400 mm × H225 mm. This longitudinal dimensions can be decreased when the sensor arrays are set on both sides of the measurement area, especially when they are set on the surface of the lower and upper sides of the magnetic homogenization plate.

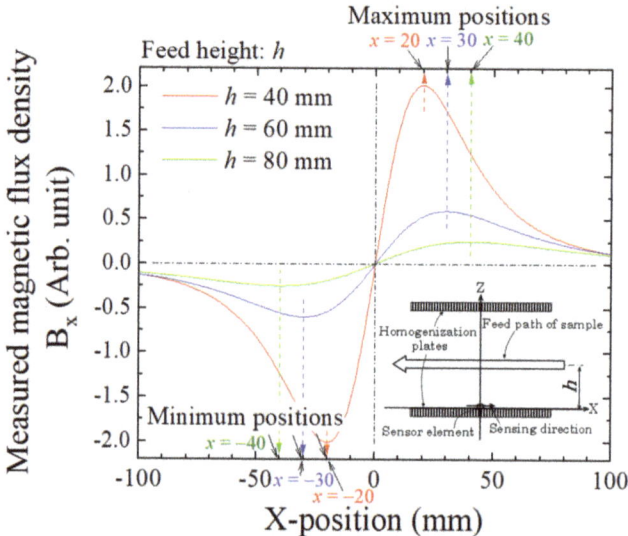

Figure 3. Relationship between the feeding height of a magnetic dipole and the position of an extreme point of measured magnetic flux density [23].

In our proposed system, the magnetic sensor is a thin-film magneto-impedance sensor driven by 400 MHz high-frequency electric circuit [20]. The merit of this circuit was introduced by our previous article. It has the ability to realize a high sensitivity for driving the thin-film MI sensor, and also makes it possible to align the sensors in a suitable bias field position in the strong magnetic field of the measurement area. The MI sensor needs a particular value of the magnetic bias field, so it must be set by controlling the position and azimuthal angle of the sensor element within the strong field of the system. In our prototype system, a particular slight distribution of the vertical field, which has a partial vector along the sensor's sensing direction, is generated and makes it possible to control the sensor bias point. The driver circuit was suitably designed for this purpose. An electrically adjustable attenuator is set in our circuit and can switch the operation mode from differential amplification to single-ended amplification by setting the attenuation value to the maximum. In this case, the signal output of this circuit indicates the value of the sensor impedance. This circuit is effectively used for adjusting the sensor position and direction by monitoring the sensor impedance value for the purpose of adjusting the bias point of the sensor operation.

Figure 4 shows a schematic explanation of the bias field adjustment. The sensor's driver circuit, in which the attenuator is set as the maximum attenuation, outputs a signal that is negatively proportional to the sensor impedance. In this case, if the profile of the magnetic field distribution is known in advance, then it is possible to set the sensor in a suitable position. It is desirable to design the partial vector of the magnetic field to have a slight variation along the sensor's sensing direction. In this study, the direction is matched to the feeding direction of the system. In this case, it is possible to adjust the sensor bias point by controlling the sensor position mechanically in the feeding direction. It is well known that high-frequency electric circuits which drive more than 100 MHz have difficulty setting an electrical switch in the circuit due to the demand of impedance matching. The proposed driver circuit has the advantage of this function in addition to realizing a high-sensitivity measurement using the MI sensor. The combination properties of an MI sensor and a magnetic field generator are shown in the latter part of this article.

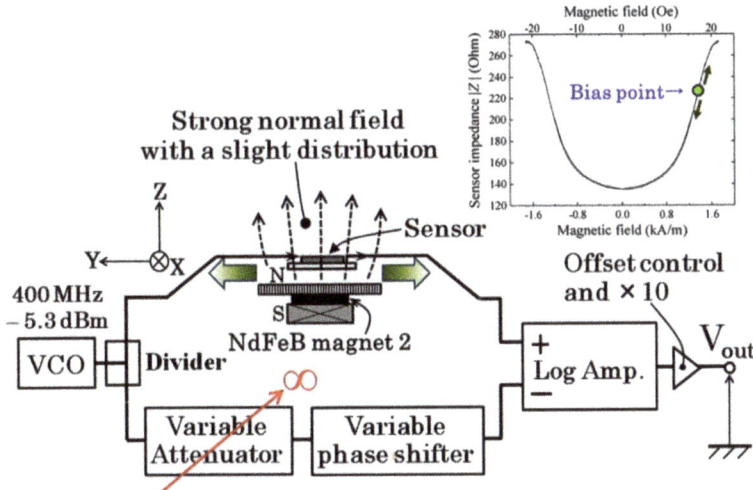

Figure 4. Schematic explanation of the bias field adjustment using a field distribution.

3. Experimental Results

In this section, an actual fabrication of the proposed system is shown. Verifications of the structure of the generated magnetic field and also verification of the sensor installation set in the strong magnetic field with the sensing direction perpendicular to the strong magnetic field are shown here.

3.1. Experimental Apparatus

The structure of the magnetic field generating apparatus which was designed and fabricated in this study is explained.

Figure 5 indicates the design layout of our proposed field generation apparatus. This is a magnetic field generator which is made of square sections of NdFeB magnets and magnetic homogenization plates. The left figure is of the side view and the right one is the plan view of the apparatus. It consists of a pair of units having the upper and lower ones with the opposite magnetic poles facing each other. One of the units has 18 magnets, with the dimensions of each magnet being W100 mm × L100 mm × H20 mm, to form a magnet array of 3 × 6 matrix layout. The magnet is a product that is being marketed. The magnetic field homogenization plate has dimensions of W740 mm × L450 mm × H12 mm. It was made of S45C soft magnetic steel. The distance between magnets was set as 20 mm width X and 45 mm length Y. The middle law of the magnets in the three laws were directly fixed on the homogenization plate, which is shown in the left side of Figure 5. The magnets laws on both sides had a particular distance from the plate for the purpose of making the in-plane directional bias field around the upper side of the directly fixed middle-law magnets. The position of the sensor array was actually a particular distance from the central position in order to control the sensor bias point at a suitable magnetic field value by controlling the sensor longitudinal position. The magnets of each law were fixed on a base plate made of S45C steel. This base plate has dimensions of W900 mm × L120 mm × H12 mm and was fixed on magnets on another side of the homogenization plate. The base plates were designed to be connected with the side steel walls of the apparatus to form a closed magnetic structure, which is the same structure of the magnetic core in Figure 1. In this study, the distance between the two homogenization plates was designed as 225 mm, and a uniform magnetic field with a slight distribution is generated between them.

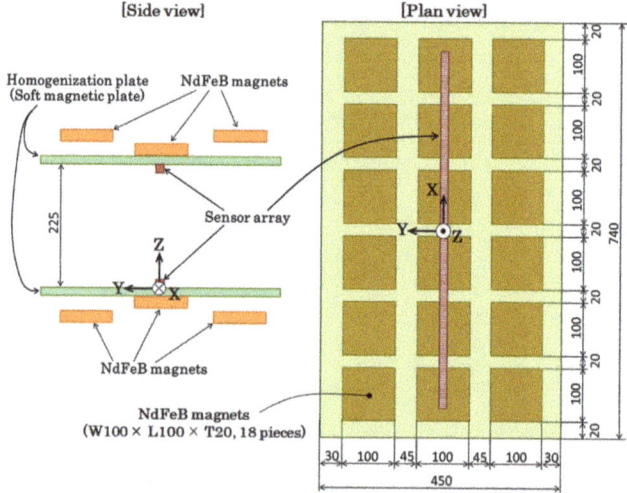

Figure 5. Design layout of the proposed field generation apparatus. (Unit: mm).

Figure 6 shows a perspective photograph of the fabricated measurement system. This system has a moving table for carrying the measured work piece, such as a human body or industrial products. The materials used in this system were selected as non-magnetic materials, such as SUS304, aluminum and thermosetting resins. An electrical actuator and sliding rail units for feeding the carrying table were set just above the floor to minimize the magnetic effect on the measurement sensors. In this photo, there is a gaussmeter on the carrying table for verifying the generated magnetic field. A digital oscilloscope for measuring and memorizing the distribution of the magnetic field was set beside the apparatus. The measurement of the magnetic field profile was carried out using the automatic feeding table by positioning the hall probe in various width and height positions.

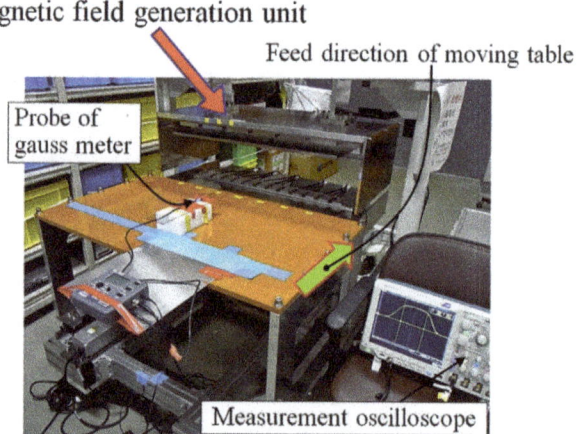

Figure 6. Perspective view of the fabricated measurement system.

3.2. Magnetic Field Distribution

An experimental confirmation of the fabricated magnetic field distribution was carried out and reported here. It confirms the wide area uniformity of the magnetic field and also that the magnetic field is at the sensor position which is suitable for sensor operation.

Figure 7 shows a schematic of the measurement procedure of the magnetic field profile. In this measurement, magnetic flux density in Z-direction B_z was measured by the feeding Hall probe at a constant speed of 100 mm/s. The original point of the measurement coordinates was set in the middle of the surface of the lower homogenization plate. The measurement area ranges from −300 mm to +300 mm in the feeding Y-direction, from −500 mm to +500 mm in the width X-direction and from +38 mm to +174 mm in the vertical Z-direction. The position on the carrying table in which the Hall probe was set just above the surface was with z = 38 mm.

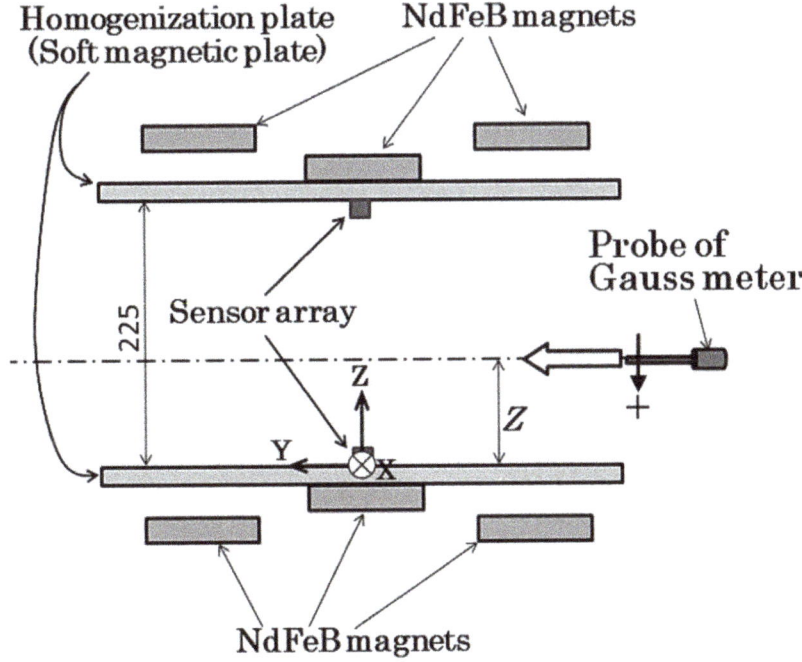

Figure 7. Schematic of the measurement procedure of the magnetic field profile. (Unit: mm).

From here, the measurement results of the generated magnetic field by our proposed system are shown.

Figure 8 shows the measured magnetic field when the vertical position was z = 38 mm, which is just above the upper surface of the carrying table. The B_z suddenly rose up almost above the edge of the homogenization plate, which was y = ±225 mm, and there was an almost flat B_z area above the plate. The value at the center of the plate was 370 G (37 mT). The waving variation existed around the top of the table in which the variation range was lower than 2%. In this case, the B_z was apparently constant as a function of transverse position x.

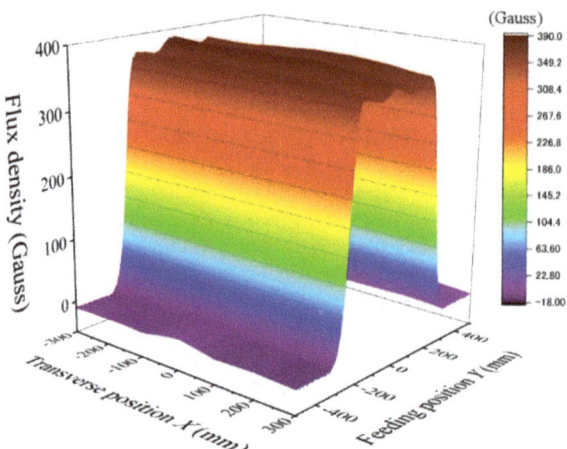

Figure 8. Measured magnetic field B_z at the vertical position $z = 38$ mm.

Figure 9 shows the case of $z = 93$ mm, which is 20 mm lower from the vertical middle height. In this case, the value of the center position was the same value, $B_z = 370$ G (37 mT), as the previous result. The field value gradually decreased as it got closer to the edge of the homogenization plate. At the edge of the transverse range, $x = \pm 300$ mm, the value was 330 G (33 mT), which was 11% decreased from the center value. This was expected by the preliminarily carried out magnetic field simulation and was caused by the effect of magnetic flux line expansion outward from the strong area.

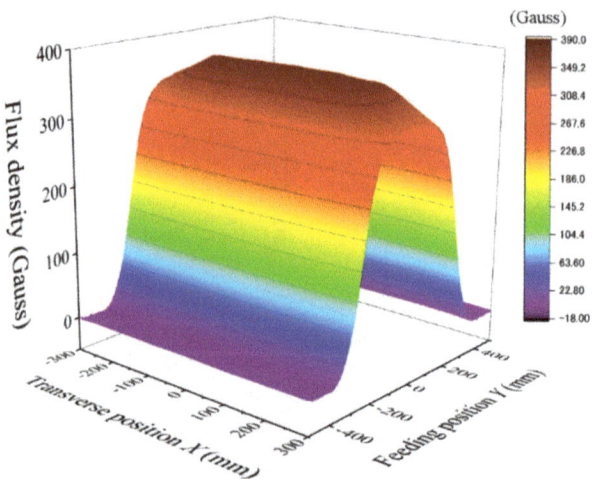

Figure 9. Measured magnetic field B_z at the vertical position $z = 93$ mm.

Figure 10 is the overlapping expression of the field variation as a parameter of vertical height z. The horizontal axis represents the feeding position Y, and the vertical axis represents the flux density B_z. The transverse position was fixed at $x = 0$, which was on the center line. The parameter z ranged from $z = 38$ mm to $z = 174$ mm, which was within the distance between the homogenization plates, 225 mm. The result show that the vertical flux density was almost the same value independent of the height position. Of special note, the value of the central position was a constant 370 G (37 mT). A slight difference existed at around the fringe area of the homogenization plate. This result shows a verification that our proposed structure is able to generate an almost constant vertical magnetic field of 370 G (37 mT) within a W500 mm × L400 mm × H210 mm area, with degradation in the fringe area of less than 10% compared with the center value.

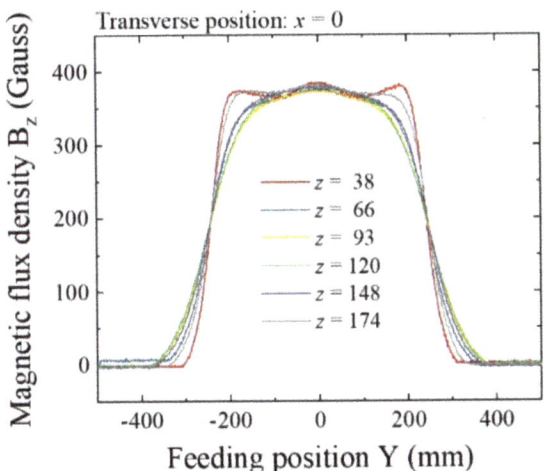

Figure 10. Overlapping expression of the field variation B_z on feeding position Y as a parameter of vertical height Z. (Unit of z: mm).

From here the magnetic structure designed for controlling the bias field of the MI sensor is explained. As shown in the previous explanation of this paper, the MI sensor needs a particular bias field for operation. In the case of our sensor, which is made of $Co_{85}Nb_{12}Zr_3$ amorphous thin film, the bias field is roughly 17 Oe (1.35 kA/m). In order to utilize a high-sensitivity magnetic sensor having a tolerance against the strong field directed in the substrate's surface normal direction of the thin-film MI sensor, the bias field is indispensable. In this section, the concept of the formation of the bias field structure in this system, which is suitable for a sensor array configuration having multiple sensors in a line at intervals, is introduced.

The method of bias field adjustment was introduced in the previous section in this paper, which was carried out using the sensor position controlling mechanism along the tangential direction of the surface of the homogenization plate. The sensor driver circuit in our proposal made it possible to adjust the sensor position mechanically in the strong magnetic field, which is more than 20 times larger compared with the bias field. The magnetic field generating apparatus was designed for this purpose. It has a distributed vertical field just above the middle law of the magnet array. The distribution has a partial vector along the Y-direction which changes as a function of the Y-position. If this distribution has a particular range of magnetic field, the bias field control is realized.

For the purpose of verifying the bias field generation in this system, magnetic field simulation was carried out due to the difficulty of the actual measurement of it inside the vertical strong field.

Figure 11 shows the compared results of the measurement and simulation. This is the profile of magnetic flux density B_z as a function of the feeding directional Y-position. The measurement is the same result as is shown in Figure 10. It shows the cases with the vertical positions z = 38 mm and z = 93 mm. The simulated results are shown by dashed lines and the measurements are shown by solid lines. The simulation was carried out using Maxwell 3D (ANSYS, Inc., Canonsburg, PA, USA). The results are in good agreement.

Therefore, the simulation is a reliable estimate of the field distribution in the vicinity of the homogenization plate. The sensor element is designed to be set at $z = 5$ mm in this system due to a mechanical restriction, which is a 5 mm altitude from the surface of the lower homogenization plate. A difficulty in the precise measurement of magnetic flux density comes from the difficulty in accessing the Hall probe to the surface of the homogenization plate in a strong magnetic field with a strict position arrangement against the coordinate axis.

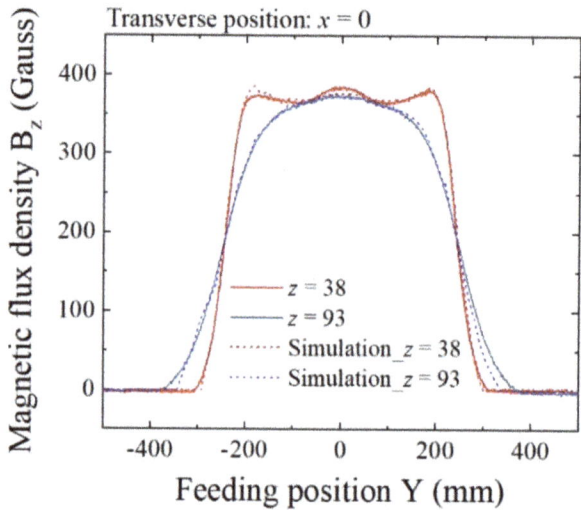

Figure 11. Comparison of measurement and simulation of B_z at different heights Z. (Unit of z: mm).

Figure 12 shows the simulated results of magnetic flux density at the sensor height of $z = 5$ mm. The horizontal axis represents the Y-position, which is the feeding direction. The original point is the central position of the homogenization plate. In this system, the sensor is able to move mechanically, in parallel with the Y-axis, by keeping the sensing direction in the Y-direction using a sliding plate onto which is mounted a sensor element. Figure 12 shows that the magnetic flux in the Y-direction linearly changes between $y = -40$ mm and $y = +40$ mm, and the variation range corresponds to the bias field from $B_y = -20$ G (-2 mT) to +20 G (+2 mT). It also shows that this controlling field is kept within the transversal position range, with $x = \pm 300$ mm. Based on Figure 5, the parameter $x = 0$ corresponds to the middle position between the plate magnets, the $x = 150$ mm corresponds to the position 30 mm from the center of a plate magnet, and the $x = 300$ mm corresponds to the position at the center of a plate magnet. This variation range for B_y is suitable for applying the sensor bias field to our sensor element, having the bias point at about 17 Oe (17 G (1.7 mT) in air circumstances). In this case, it is possible to set the bias field at a suitable value when the position of the sensor element is set at about $y = 30$ mm. The actual fabricated magnetic structure has the possibility to include material non-uniformity and also to include dimensional and attachment mechanical error. Even in this case, our proposed system can adjust the bias point by controlling the sensor position by monitoring the sensor output signal that indicates sensor impedance.

Figure 13 shows the variations in flux density in the transverse X-direction with B_x, as a function of the feeding Y-position in the range of the bias controlling line. The B_x field is in the transverse direction to the sensing direction. If it is the in-plane transverse direction of the thin-film element, then a particular weakness exists against the magnetic field in this direction. Our sensor element is capable of tolerating transverse fields of more than 50 Oe

(50 G (5mT) in air conditions). This simulated result shows that the sensor element is able to set and work within the range $x = \pm 300$ mm.

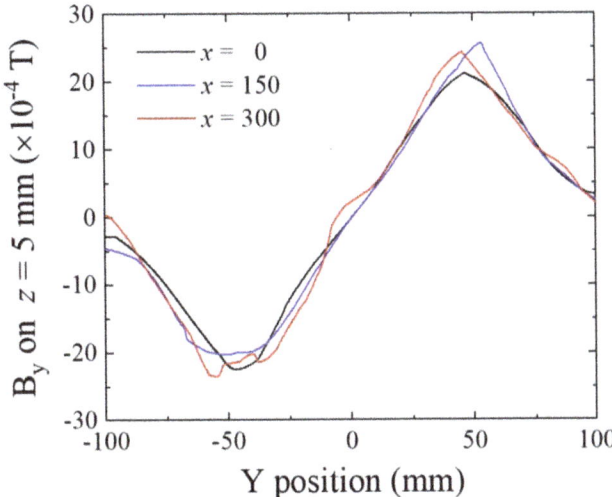

Figure 12. Simulated variations in magnetic flux density B_y on Y-position at the sensor height $z = 5$ mm as a parameter of transverse position X. (Unit of x: mm).

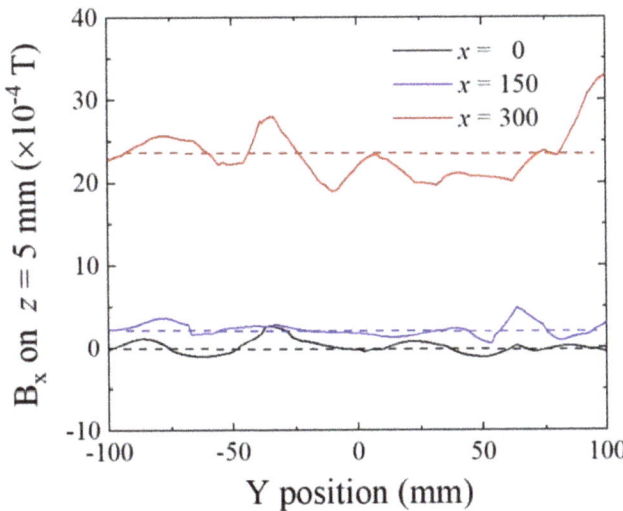

Figure 13. Simulated variations in transverse magnetic flux density B_x on Y-position at the sensor height $z = 5$ mm as a parameter of transverse position X. (Unit of x: mm).

Based on the results in this section, the proposed magnetic field generation apparatus is capable of applying a suitable bias field to the area of the sensor array installation. The bias point can be set by controlling the Y-position of each element aiming at a particular, suitable magnetic bias field by monitoring the sensor output. The range of the controlling field is designed as a suitably larger value of the bias field in order to include the uncertainty of the material's non-uniformity and also dimensional and attachment errors.

3.3. Verification of Sensor Performance

In this section, an experimental verification of the measurement system in this paper is reported. A single thin-film magneto-impedance sensor was installed in the magnetic field generating structure reported in this paper, and it confirmed the validity and operational performance of the system.

The thin-film magneto-impedance sensor and driver circuit which were used in this study were the same ones which were reported previously by us [22]. A single sensor element was installed for the purpose of confirming the functional performance, although the magnetic field generation system was designed for installing multiple-array sensors on a particular bias field transversal line. We used this system to attempt to detect a millimeter-sized steel chip mounted on the surface of the complex-shaped cast aluminum work piece. Based on recent progress in medical technology, it is possible to now include conductive metal materials in the human body which are used for replacing the functions of the human body. Titanium has been used for this purpose due to its biocompatibility. Examples of such applications are as follows: artificial joints and bone, surgical bone plates and bolts and stents and implants. Although titanium possesses very slight paramagnetic properties, it is regarded as a conductive non-magnetic material. A conventional metal detector is well known to be strongly affected by conductive material. An alternating magnetic field is used for this conventional method, and a small magnetization in the vicinity of the non-magnetic conductive material causes difficulties due to the eddy currents in the conductive material. The eddy current generates strong noise in the detection signal, although the system proposed in this study can detect a magnetized piece in the vicinity of such conductive material. In this study, in order to simulate a cluster of magnetic nanoparticles in the vicinity of a complexly shaped piece artificial bone, we tried to detect a small magnetic chip placed on the surface of an aluminum casting. This is an initial study of our proposed detection system. Further study using an actual bio-medical sample will be carried out as the next steps.

Figure 14 is a photo of the measured small steel chip. It has a 2 mm length and 1 mm width and an ellipsoidal shape resembling the shape and size of a sesame seed. It was mounted on the end-flange surface of a cast aluminum work piece, as is shown in Figure 15. The work piece was fed through the measurement area using the feeding table of this system. The distance between the sensor element and the steel chip was 27 mm in the vertical direction, which was a just above nearest position between them during feeding.

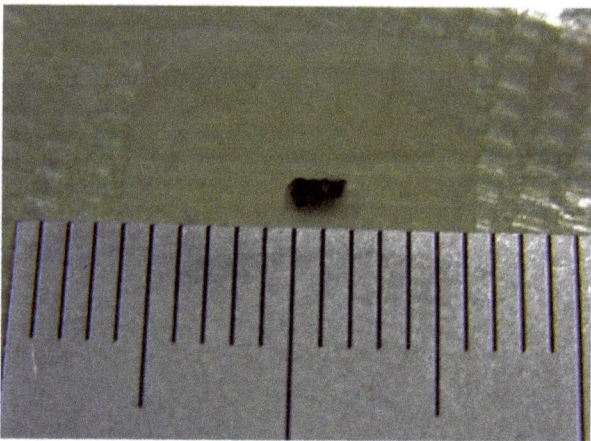

Figure 14. Photograph of the measured small steel chip.

Figure 15. Photograph of the small steel chip mounted on the end-flange of an aluminum casting with a complex shape.

Figure 16 shows the photo of the measurement. The flange surface mounting the steel chip was set as it was placed at the bottom of the work piece, and it was mounted on the feeding table. The work piece was positioned such that the chip was ran through just above the sensor element. The dimensions of the aluminum work piece were a height of 200 mm and a width of 380 mm.

Figure 16. Photograph of the measurement using the fabricated unit.

Figure 17 is a typical measurement result without a magnetic chip. The aluminum casting without the chip was mounted on the feeding table. The table started moving at the time $t = 0$, where a step-up signal was detected, and it stopped at time $t = 105$ s, where a step-down signal was detected. The feeding speed in this study was 10 mm/s. The sensor can detect 1 mG in the feeding direction as 0.1 V output. The measured result in this figure shows an increasing tendency as a function of time, and then as a function of the table position. The reason for this increasing tendency was the existence of a steel plate at the bottom of the feeding table that was used to fix it mechanically to the feeding actuator. The variation range was almost 2.3 V of sensor output, corresponding to a 23 mG variation in static magnetic field in the sensor's sensing direction.

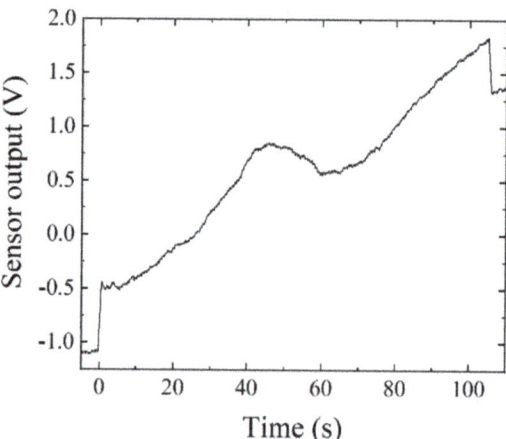

Figure 17. Typical measurement result without magnetic chip. (Feeding speed: 10 mm/s).

Figure 18 shows the result when the small steel chip was set on the work piece. It clearly detected the existence of the millimeter-sized small chip at a distance of 27 mm from the sensor element using the proposed system in the study. The system can detect it even in the vicinity of non-magnetic conductive bulk metal.

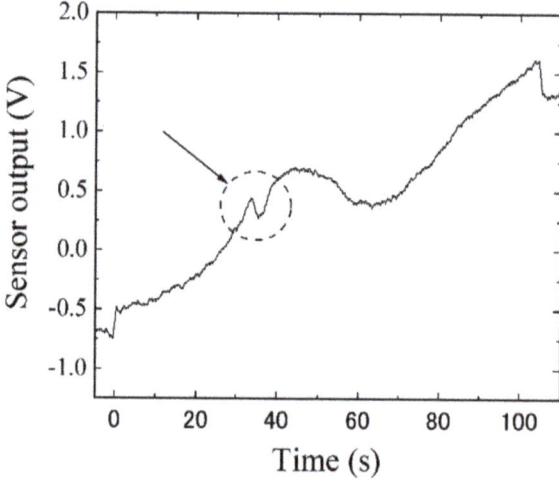

Figure 18. Measurement result when the small steel chip was set on the work piece, as shown in Figure 15. The dotted circle indicates the signal of magnetic chip. (Feeding speed: 10 mm/s).

Figure 19 is a display hard-copy of the measured digital oscilloscope. Figure 18 is a direct plot of its digitized and memorized sampling data. It is understood that the detection was clearly carried out, even though there was no application of any post-filtering procedures for eliminating noise from the measured signal. A consideration of the cause and level of the noise in this measurement and the detection limit are explained in the following discussion section.

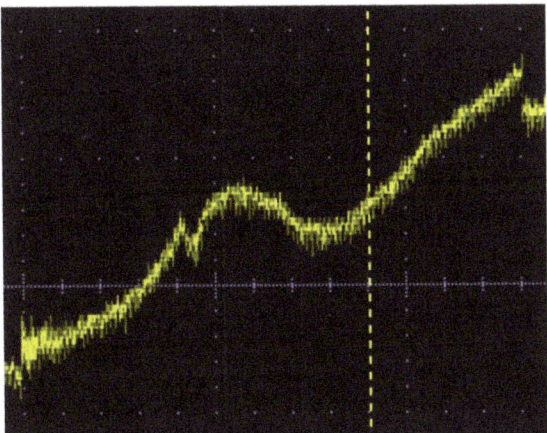

Figure 19. Display hard-copy of the measured digital oscilloscope, during the measurement of Figure 18.

4. Discussion

In this section, the target specifications of the system developed for medical applications is discussed.

From the view point of sensor sensitivity, a prior study utilized a Hall probe which had a sensitivity of 1 µT for detecting magnetic nanoparticle concentrations [29]. It achieved the detection of 140 µg of iron which was contained in 5µL of a fluid of "superparamagnetic iron oxide nanoparticles" at a distance of 10 mm. It was detected by applying a 50 mT static magnetic field to magnetize the nanoparticles. It utilized a special magnetic structure which could install the Hall probe at the magnetic null point. Based on this prior study, the proposed system in this paper has an advantage in sensitivity. The applied static field in this paper is 37 mT, about 3/4 of the prior study's. In the measurement system which detects magnetic particles using a magnetic sensor simultaneously with magnetization in a static magnetic field, there is a relationship between the magnetization strength and the detection sensitivity of the particle. Considering this effect, an improvement in the sensitivity of the sensor is needed for our sensor system to put into practice the position and size estimation. In order to improve the sensor's sensitivity, noise reduction in the sensor's driver circuit is needed.

Regarding the size of the sensor element, the smaller element size has higher precision in position and size estimation, especially when the detected particle is becoming nearer to the sensor element. The reason is the spatial distribution of the detected magnetic field coming from the magnetized particle. The 1 mm sensor element in this paper has a disadvantage from this view point. A newly developed, small size magneto-impedance sensor would be applicable [30]. Additionally, the development of an estimation method for a precise magnetic field using a large sensor element would be considerable. The magneto-impedance sensor has the ability to detect magnetic particles tens of microns in size [19], which are feeding along in the vicinity of the element. There is a particular possibility for developing a method which can estimate a magnetic field as if it were detected by a point sensor using a magneto-impedance strip-shaped element.

From the view point of the realization of the sensor array using multiple sensors, the development of a small unit of the sensor's driver circuit having a low noise and high-sensitivity is indispensable. A fundamental concept of the driver circuit was already proposed, as shown in this paper, and thus an effort towards miniaturization is the point of future development.

The sensor's driver circuit which was used in this experiment was the same one as in the previous report [20–22]. The uncertainty of the measurement was discussed in the previous paper, in which the circuit noise waveform was shown in Figure 16 in Section 3. Discussion of Ref. [20]. A 0.2 V periodically repeated dipping noise was present with the time interval of 0.67 s. The same figure is reposted in Figure 20 in this paper. This noise is assumed to come from the 400 MHz oscillator, and would be erased by a more sophisticated circuit design. The measurement noise, which is shown in the actual display of the oscilloscope, as shown in Figure 19, was the same level as that in Figure 20. Therefore, by refining our driver circuit, the detection level of the magnetic small piece is expected to be well improved. It is the subject of a future study.

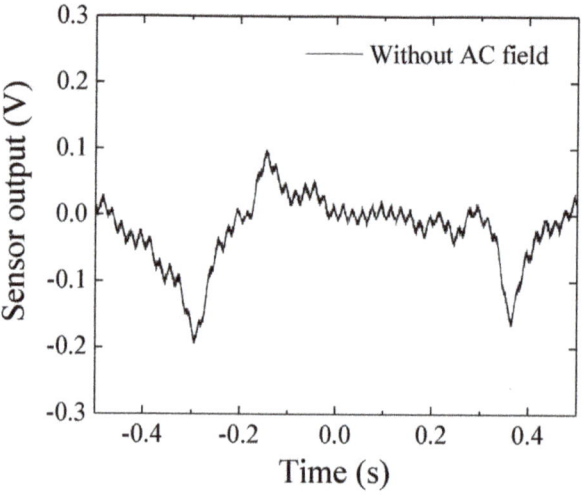

Figure 20. The 0.2 V periodical dipping noise of the sensor's driver circuit [20].

The detection limit of this measurement system is discussed as the final aspect of this paper. It is easily understood that the limit is determined by the sensor sensitivity and also the strength of the magnetic dipole, with the latter having the same meaning as the magnetic moment of the measured sample. The magnetic moment is derived from a multiplication of the magnetization and the sample volume. It is also affected by the distance between the magnetic moment and the sensor. With increasing distance, the strength of the magnetic field coming from the magnetic moment at the sensor position decreases. In order consider the system sensitivity, we have to consider three parameters: sensor sensitivity, the magnitude of the magnetic moment and the detection distance. The fundamental equation for this consideration is a simple one, which was already discussed in our previous work [19,23]. Figure 21 is one of the reposts of this work [19]. It shows a variation in magnetic flux density at the sensor element position as a function of feeding height. The magnetization was assumed as 1 T, and the diameter of the sphere particle was a parameter. Based on this consideration, a ϕ 20 μm magnetic sphere can be detected by a sensor having a 10^{-8} T sensitivity at a distance of 3 mm. In order to expand the consideration for the case of a cluster of nanoparticles, we have to consider the dispersed magnetic particles in a particularly shaped volume. Dispersion of the particle concentration inside the volume also needs to be considered, and these are expected as subjects of future study.

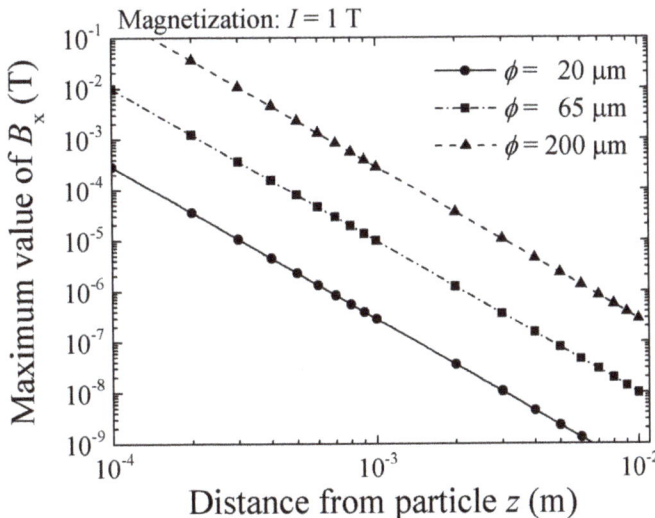

Figure 21. Variations in magnetic flux density at sensor element position as a function of feeding height z [19]. (The magnetization of particle: 1 T. The diameter of the sphere particle ϕ: parameter. The configuration of measurement system is the same as this paper).

5. Summary

A uniform and strong magnetic field generator was developed which is applicable to a human body scan. A sensor system in combination with a uniform strong magnetic field and also an array of thin-film magneto-impedance sensors within the strong magnetic field was designed, fabricated, and experimentally evaluated. The magnetic field in the measurement area was uniform in strength and was directionally aligned with a structure having adequate volume to insert a human body. The thin-film sensor was set in the vicinity of the magnet with its sensing direction perpendicular to the strong magnetic field. This system has the ability to detect a stray field of magnetic particles and also to estimate their 3D position and concentration. As a result, a prototype of a magnetic field generator with an area of W500 mm × L400 mm × H210 mm and with a uniform magnetic field of 370 Gauss was developed, and a thin-film magneto-impedance sensor was installed in it. Experimental verification showed that the 2 mm × 1 mm ellipsoidal small magnetic chip, which generated about 1 mG (10^{-7} T) at the sensor, was detected at a distance 27 mm from the sensor.

Funding: This research was funded by the budget of Miyagi prefecture for the research and development of industrial technology.

Acknowledgments: The author gratefully acknowledges the Katsuyuki Sawada, Hikichi Seiko Co., Ltd., for his support of the trial production of the magnetic generator.

Conflicts of Interest: The author declares no conflict of interest.

References

1. Mohri, K.; Bushida, K.; Noda, M.; Yoshida, H.; Panina, L.V.; Uchiyama, T. Magneto-impedance element. *IEEE Trans. Magn.* **1995**, *31*, 2455–2460. [CrossRef]
2. Uchiyama, T.; Mohri, K.; Panina, L.; Furuno, K. Magneto-impedance in sputtered amorphous films for micro magnetic sensor. *IEEE Trans. Magn.* **1995**, *31*, 3182–3184. [CrossRef]
3. Mohri, K.; Uchiyama, T.; Shen, L.P.; Cai, C.M.; Honkura, Y.; Aoyama, H. Amorphous wire and CMOS IC based sensitive micro-magnetic sensors utilizing magneto-impedance (MI) and stress-impedance (SI) effects and applications. In Proceedings of the 2001 International Symposium on Micromechatronics and Human Science, Nagoya, Japan, 9–12 September 2001. [CrossRef]

4. Raposo, V.; Vázquez, M.; Flores, A.G.; Zazo, M.; Iñiguez, J.I. Giant magnetoimpedance effect enhancement by circuit matching. *Sens. Actuators A Phys.* **2003**, *106*, 329–332. [CrossRef]
5. Yabukami, S.; Suzuki, T.; Ajiro, N.; Kikuchi, H.; Yamaguchi, M.; Arai, K. A high frequency carrier-type magnetic field sensor using carrier suppressing circuit. *IEEE Trans. Magn.* **2001**, *37*, 2019–2021. [CrossRef]
6. Zhao, W.; Bu, X.; Yu, G.; Xiang, C. Feedback-type giant magneto-impedance sensor based on longitudinal excitation. *J. Magn. Magn. Mater.* **2012**, *324*, 3073–3077. [CrossRef]
7. Kurlyandskaya, G.V.; De Cos, D.; Volchkov, S.O. Magnetosensitive transducers for nondestructive testing operating on the basis of the giant magnetoimpedance effect: A review. *Russ. J. Nondestruct. Test.* **2009**, *45*, 377–398. [CrossRef]
8. Vacher, F.; Alves, F.; Gilles-Pascaud, C. Eddy current nondestructive testing with giant magneto-impedance sensor. *NDT E Int.* **2007**, *40*, 439–442. [CrossRef]
9. Tehranchi, M.; Ranjbaran, M.; Eftekhari, H. Double core giant magneto-impedance sensors for the inspection of magnetic flux leakage from metal surface cracks. *Sens. Actuators A Phys.* **2011**, *170*, 55–61. [CrossRef]
10. Chiriac, H.; Tibu, M.; Moga, A.E.; Herea, D.D. Magnetic GMI sensor for detection of biomolecules. *J. Magn. Magn. Mater.* **2005**, *293*, 671–676. [CrossRef]
11. Kumar, A.; Mohapatra, S.; Fal-Miyar, V.; Cerdeira, A.; García, J.A.; Srikanth, H.; Gass, J.J.; Kurlyandskaya, G.V. Magnetoimpedance biosensor for nanoparticle intracellular uptake evaluation. *Appl. Phys. Lett.* **2007**, *91*, 143902. [CrossRef]
12. Kim, D.; Kim, H.; Park, S.; Lee, W.; Jeung, W.Y. Operating Field Optimization of Giant Magneto Impedance (GMI) Devices in Micro Scale for Magnetic Bead Detection. *IEEE Trans. Magn.* **2008**, *44*, 3985–3988. [CrossRef]
13. Blanc-Béguin, F.; Nabily, S.; Gieraltowski, J.; Turzo, A.; Querellou, S.; Salaun, P.Y. Cytotoxicity and GMI bio-sensor detection of maghemite nanoparticles internalized into cells. *J. Magn. Magn. Mater.* **2009**, *321*, 192–197. [CrossRef]
14. Kurlyandskaya, G.V. Giant magnetoimpedance for biosensing: Advantages and shortcomings. *J. Magn. Magn. Mater.* **2009**, *321*, 659–662. [CrossRef]
15. Chiriac, H.; Herea, D.D.; Corodeanu, S. Microwire array for giant magneto-impedance detection of magnetic particles for biosensor prototype. *J. Magn. Magn. Mater.* **2007**, *311*, 425–428. [CrossRef]
16. Llandro, J.; Palfreyman, J.J.; Ionescu, A.; Barnes, C.H.W. Magnetic biosensor technologies for medical applications: A review. *Med. Biol. Eng. Comput.* **2010**, *48*, 977–998. Available online: https://link.springer.com/article/10.1007/s11517-010-0649-3 (accessed on 25 March 2022). [CrossRef]
17. Fodil, K.; Denoual, M.; Dolabdjian, C.; Treizebre, A.; Senez, V. In-flow detection of ultra-small magnetic particles by an integrated giant magnetic impedance sensor. *Appl. Phys. Lett.* **2016**, *108*, 173701. [CrossRef]
18. Kurlyandskaya, G.V.; Portnov, D.S.; Beketov, I.V.; Larrañaga, A.; Safronov, A.P.; Orue, I.; Medvedev, A.I.; Chlenova, A.A.; Sanchez-Ilarduya, M.B.; Martinez-Amesti, A.; et al. Nanostructured materials for magnetic biosensing. *Biochim. Biophys. Acta BBA-Gen. Subj.* **2017**, *1861*, 1494–1506. [CrossRef]
19. Nakai, T. Study on Detection of Small Particle Using High-Frequency Carrier-Type Thin Film Magnetic Field Sensor with Subjecting to Strong Normal Field. The Papers of Technical Meeting on Physical Sensor, IEE Japan. 2016. PHS-16-15. pp. 11–20. (In Japanese). Available online: http://id.nii.ac.jp/1031/00093680/ (accessed on 30 April 2019).
20. Nakai, T. Magneto-impedance sensor driven by 400 MHz logarithmic amplifier. *Micromachines* **2019**, *10*, 355. [CrossRef]
21. Nakai, T. Sensitivity of thin film magnetoimpedance sensor in 0.3 T surface normal magnetic field. *IEEJ Trans. Electr. Electron. Eng.* **2020**, *15*, 1230–1235. [CrossRef]
22. Nakai, T. Nondestructive detection of magnetic contaminant in aluminum casting using thin film magnetic sensor. *Sensors* **2021**, *21*, 4063. [CrossRef]
23. Nakai, T. Estimation of Position and Size of a Contaminant in Aluminum Casting Using a Thin-Film Magnetic Sensor. *Micromachines* **2022**, *13*, 127. [CrossRef] [PubMed]
24. Olsvik, O.; Popovic, T.; Skjerve, E.; Cudjoe, K.S.; Hornes, E.; Ugelstad, J.; Uhlén, M. Magnetic separation techniques in diagnostic microbiology. *Clin. Microbiol. Rev.* **1994**, *7*, 43–54. [CrossRef] [PubMed]
25. Langer, R. New Methods of Drug Delivery. *Science* **1990**, *249*, 1527–1533. [CrossRef]
26. Polyak, B.; Friedman, G. Magnetic targeting for site-specific drug delivery: Applications and clinical potential. *Expert Opin. Drug Deliv.* **2009**, *6*, 53–70. [CrossRef]
27. Pankhurst, Q.A.; Connolly, J.; Jones, S.K.; Dobson, J. Applications of magnetic nanoparticles in biomedicine. *J. Phys. D Appl. Phys.* **2003**, *36*, 13. [CrossRef]
28. AIST Human Body Data Base. Available online: https://www.airc.aist.go.jp/dhrt/91-92/data/list.html (accessed on 25 March 2022).
29. Sekino, M.; Kuwahata, A.; Ookubo, T.; Shiozawa, M.; Ohashi, K.; Kaneko, M.; Saito, I.; Inoue, Y.; Ohsaki, H.; Takei, H.; et al. Handheld magnetic probe with permanent magnet and Hall sensor for identifying sentinel lymph nodes in breast cancer patients. *Sci. Rep.* **2018**, *8*, 1195. [CrossRef]
30. Kikuchi, H.; Umezaki, T.; Shima, T.; Sumida, C.; Oe, S. Impedance Change Ratio and Sensitivity of Micromachined Single-Layer Thin Film Magneto-Impedance Sensor. *IEEE Magn. Lett.* **2019**, *10*, 8107205. [CrossRef]

Article

A Model for the Magnetoimpedance Effect in Non-Symmetric Nanostructured Multilayered Films with Ferrogel Coverings

Nikita A. Buznikov [1] and Galina V. Kurlyandskaya [2,3,*]

[1] Scientific and Research Institute of Natural Gases and Gas Technologies–Gazprom VNIIGAZ, Vidnoye, Razvilka, 142717 Moscow, Russia; n_buznikov@mail.ru
[2] Department of Electricity and Electronics, Basque Country University UPV/EHU, 48940 Leioa, Spain
[3] Department of Magnetism and Magnetic Nanomaterials, Institute of Natural Sciences and Mathematics, Ural Federal University, 620002 Ekaterinburg, Russia
* Correspondence: kurlyandskaya.gv@ehu.eus; Tel.: +34-9460-13237; Fax: +34-9460-13071

Abstract: Magnetoimpedance (MI) biosensors for the detection of in-tissue incorporated magnetic nanoparticles are a subject of special interest. The possibility of the detection of the ferrogel samples mimicking the natural tissues with nanoparticles was proven previously for symmetric MI thin-film multilayers. In this work, in order to describe the MI effect in non-symmetric multilayered elements covered by ferrogel layer we propose an electromagnetic model based on a solution of the 4Maxwell equations. The approach is based on the previous calculations of the distribution of electromagnetic fields in the non-symmetric multilayers further developed for the case of the ferrogel covering. The role of the asymmetry of the film on the MI response of the multilayer–ferrogel structure is analyzed in the details. The MI field and frequency dependences, the concentration dependences of the MI for fixed frequencies and the frequency dependence of the concentration sensitivities are obtained for the detection process by both symmetric and non-symmetric MI structures.

Keywords: magnetic multilayers; magnetoimpedance; modeling; magnetic sensors; magnetic biosensors; magnetizable nanoparticles; ferrogels

1. Introduction

Demands for fast development of small biomedical devices have increased the interest in magnetic biosensors. They are compact analytical devices with magnetic physico-chemical transducers for the evaluation of the concentration of components of interest [1,2]. In general, magnetic sensors can be divided into two groups: magnetic sensors for label-free detection processes [3,4] and magnetic devices for detection of magnetic labels [1,2,5,6]. Although label-free magnetic detector prototypes were of special interest recently [7,8], a major part of the studies is related to the possibility to detect magnetizable nanoparticle concentrations [9,10]. There are different solutions for the detection: in "in vitro" experimental models [1,10–12], nanoparticles inside living cells [13,14], in continuous flow in medical devices or in blood flow [15,16], as a part of the implants or embedded into a natural tissue or artificial composites mimicking a natural tissue [5,17–19]. Many types of magnetic field sensors were tested in a simple "laboratory" device configuration just to ensure the proof of the concept. A recent overview might be useful for advanced reading on this subject [20]. The improved levels of sensitivity now available with respect to the applied magnetic field allow one to propose very new applications of magnetic field sensors which would be unthinkable at lower levels of sensitivity [21,22].

Magnetoimpedance (MI) is one of the effects ensuring very high sensitivity with respect to magnetic field and detection of fields of the order of up to 10^{-8} Oe. The MI phenomenon consists in the change of the total impedance of a ferromagnetic conductor in the presence of an external magnetic field [23]. The interest in MI is related to the design of small magnetic-field detectors for different application areas–from non-destructive testing and vehicle control to the magnetic detection of the signals closely related to the living system functionality and magnetic marker detection [9,19,24]. The magnetoimpedance phenomenon was previously observed and studied in different soft ferromagnets obtained by rapid quenching, electrodeposition, magnetron sputtering and other methods [25]. However, from the point of view of the compatibility with existing semiconductor electronics and in view of specially requested packaging of sensor array and miniaturization, multilayered film-based MI elements with the total thickness of magnetic layers are one of the most suitable magnetic materials for magnetoimpedance applications [26]. The MI effect was predicted to reach the highest value, theoretically described and experimentally tested for three-layered structures in MI "sandwich" configuration (Figure 1a) for which two ferromagnetic layers (F) of the equal thickness are separated by the conductive non-ferromagnetic central layer (C) of the same thickness as the ferromagnetic layers [27,28]. In this case, when the thickness of the top (FT) and the bottom (FB) ferromagnetic layers is the same, the structure is called "symmetric" [29]. As the theory predicted the highest MI value for symmetric structures, the non-symmetric structures (with FT $\neq$ FB, Figure 1b) were not experimentally studied.

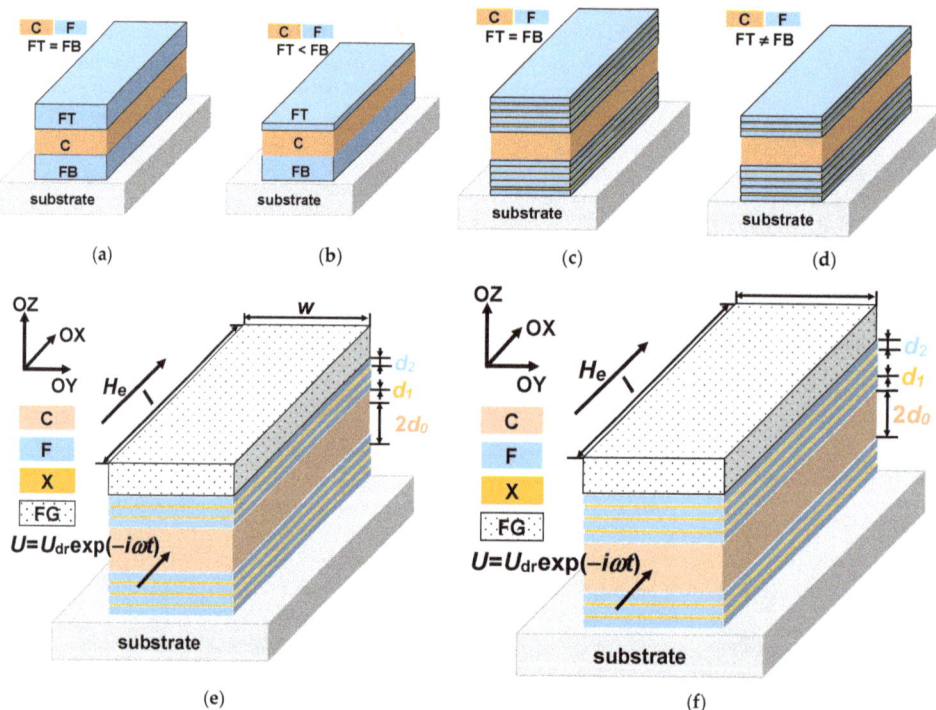

Figure 1. Schematic representation of the MI multilayered structures. Symmetric structures of "sandwich" type with equal thicknesses of ferromagnetic and conductive layers (**a**). Non-symmetric structure of "sandwich" type with different thicknesses of ferromagnetic top and bottom layers (**b**). Symmetric MI structure with equal thicknesses of top and bottom identical multilayers consisting of ferromagnetic soft magnetic sub-layers separated by non-ferromagnetic spacers and conductive central layer (**c**). Non-symmetric MI structures with different thicknesses of top and bottom multilayers consisting of ferromagnetic soft magnetic sub-layers separated by non-ferromagnetic spacers and conductive central layer (**d**). Studied non-symmetric multilayered films: the multilayer with ferrogel placed onto the surface of thin layer (**e**) and onto the surface of thick layer (**f**).

In the case of permalloy thin films, the need of nanostructuring, i.e., substitution of thick ferromagnetic FeNi layers by the multilayered structures of [F/X]$_m$ (Figure 1c) was rather carefully discussed in the literature [30,31]. This step is necessary due to the existence of the transition into a "transcritical" state [31,32]. Non-symmetric multilayered structures (Figure 1d) were studied experimentally and theoretically [26,29]. Non-symmetric MI structures with open magnetic flux can be obtained by sputtering deposition of top and bottom ferromagnetic layers with different thickness [5,26]. However, there are no theoretical studies for the case of magnetic nanoparticles (MNPs) detection by a magnetic biosensor with non-symmetric MI elements (Figure 1d).

In the course of the development of MI thin-film-based prototypes for the detection of MNPs embedded in natural tissue an intermediate solution was proposed. Synthetic composites consisting of a polymer matrix with embedded MNPs–ferrogels substituted for the biological tissues at the first stage of the development of the biosensor [5,17]. MNPs in ferrogel are dispersed swollen in water forming an elastic cross-linked polymeric network. The mobility of MNPs in a ferrogel depends on two parameters: the diameter of the MNPs and the distance between adjacent crosslinks of the gel network. Ferrogels' properties in many senses can mimic the main properties of biological samples, being similar to the cytoskeleton. Their structure in the simplest way can be described as a polymer network with electric charges localized on macromolecule filaments of the network and free counterions. The last are dispersed inside the polymeric network in the liquid phase [17,33]. Although ferrogels-covering detection by the MI symmetric structures was demonstrated experimentally and a satisfactory model was proposed [5,34], the non-symmetric structures for such detection were still not analyzed.

In this work, we propose a new model in order to describe the MI effect in non-symmetric multilayered elements covered by a ferrogel layer. The theoretical approach is based on the previous calculations of the distribution of electromagnetic fields in non-symmetric multilayered films further developed for the case of the ferrogel covering. The influence of the asymmetry of the film on the MI response of the multilayer–ferrogel structure is analyzed.

2. Model
2.1. Field Disrtibution and MI in Multilayer–Ferrogel Structure

The studied $[F/X]_m/F/C/[F/X]_n/F$ multilayered structure is a rectangular MI element with transverse effective magnetic anisotropy. The induced magnetic anisotropy is formed due to application of the technological magnetic field during multilayer deposition. The MI element consists of a highly conductive non-ferromagnetic central layer C of a thickness $2d_0$ and two external multilayers: ferromagnetic top and bottom parts. The external multilayers contain soft magnetic sub-layers F of thickness d_2 separated by non-ferromagnetic spacers X of thickness d_1. The corresponding conductivities of the materials C, X and F are σ_0, σ_1 and σ_2. The multilayered element length and width are l and w, respectively. The film structure is non-symmetric, that is, the top external multilayer is either thicker than the bottom one, $m > n$, either thinner than the bottom multilayer, $m < n$. The layer of ferrogel with the thickness of d_3 is placed on the top surface of the film structure (Figure 1e,f). The driving voltage is used to feed the element: $U = U_{dr}\exp(-i\omega t)$ (t is the time ω is the angular frequency of the electromagnetic filed and i is the imaginary unit).

The driving voltage is applied to the MI element (Figure 1e) in the geometry of longitudinal MI. the external magnetic field H_e is parallel to the long side of the element in the flowing current direction. We assume the dependence of the electromagnetic fields on the coordinate perpendicular to the film plane (x-coordinate) only. Such an assumption is possible due to the fact that the length of the multilayer and its width are much higher than the thickness of the multi-layered structure. That is it the one-dimensional approximation is used. The Maxwell equations can be solved and the amplitudes of the longitudinal electric e_j and the transverse magnetic h_j fields in the sub-layers can be written as follows [29]:

$$e_j = (cp_k/4\pi\sigma_k)[A_j\cosh(p_kx) + B_j\sinh(p_kx)], \qquad (1)$$

$$h_j = A_j\sinh(p_kx) + B_j\cosh(p_kx). \qquad (2)$$

Here $j = 1, \ldots 2(m + n + 1) + 1$ is the sub-layer number; subscript $k = 0, 1$ and 2 corresponds to the central conductive layer, spacer and magnetic sub-layer, respectively. A_j and B_j are the constants; $p_k = (1 - i)/\delta_k$; $\delta_k = c/(2\pi\omega\sigma_k\mu_k)^{1/2}$; c is the speed of light in vacuum; σ_k and μ_k are the conductivity and the transverse permeability values for the material k. For the non-ferromagmagnetic central layer and spacers we assume $\mu_0 = \mu_1 = 1$.

The distribution of the fields within ferrogel layer can be expressed as follows [35]:

$$e_g = [A_g \cosh(p_3 x) + B_g \sinh(p_3 x)]/\varepsilon^{1/2}, \quad (3)$$

$$h_g = A_g \sinh(p_3 x) + B_g \cosh(p_3 x). \quad (4)$$

Here e_g and h_g are the amplitudes of the electric and magnetic field in the ferrogel; A_g and B_g are the constants; ε is the permittivity of the ferrogel and $p_3 = -i\omega\varepsilon^{1/2}/c$.

In order to describe the distribution of the field outside the structure consisting of a MI multilayered element and a ferrogel the approximate solution for the vector potential in the previously obtained general form was used [36,37]. The corresponding field amplitudes in the particular geometry for the external regions can be be expressed as follows:

$$e_{ext,q} = C_q \frac{i\omega l}{2cw} \left[\frac{1}{2w} \ln\left(\frac{R+w}{R-w}\right) - \frac{4x}{l} \arctan\left(\frac{wl}{2Rx}\right) + \frac{w}{l} \ln\left(\frac{R+l}{R-l}\right) \right], \quad (5)$$

$$h_{ext,q} = C_q (2/w) \arctan(wl/2Rx). \quad (6)$$

Here the subscripts $q = 1$ and $q = 2$ correspond to the bottom and top external region, respectively; C_q are the constants and $R = (l^2 + w^2 + 4x^2)^{1/2}$.

To describe completely the distribution of the electromagnetic fields within the studied structure, the constants in Equations (1)–(6) should be found. The continuity conditions for the amplitudes of the electric and magnetic fields at the interfaces between different sub-layers of the film, $j < 2(m + n + 1)$, can be presented in the form:

$$e_j = e_{j+1}, \quad h_j = h_{j+1}. \quad (7)$$

Additional restrictions for the values of the field amplitudes are obtained from the excitation condition of the multilayered film. We should take into account that the driving voltage is applied to the multilayer region $-t_1 < x < t_2$, where $t_1 = d_0 + nd_1 + (n+1)d_2$ and $t_2 = d_0 + md_1 + (m+1)d_2$. Then, the boundary conditions are given by the following expressions:

$$\begin{aligned} e_{ext,1}(-t_1) + U_{dr}/l &= e_1(-t_1), \\ h_{ext,1}(-t_1) &= h_1(-t_1 - d_3), \\ e_{2(m+n+1)+1}(t_2) &= e_g(t_2) + U_{dr}/l, \\ h_{2(m+n+1)+1}(t_2) &= h_g(t_2), \\ e_g(t_2 + d_3) &= e_{ext,2}(t_2 + d_3), \\ h_g(t_2 + d_3) &= h_{ext,2}(t_2 + d_3). \end{aligned} \quad (8)$$

Equations (7) and (8) allow one to find all constants in Equations (1)–(6). After that, the impedance Z of the multilayered structure can be obtained as a ratio of the applied voltage to the total current flowing through the multilayer [33,34]:

$$Z = U_{dr} \left[w \int_{-t_1}^{t_2} \sigma(x) e(x) dx \right]^{-1} = \frac{4\pi}{cw} \times \frac{U_{dr}}{h_{2(m+n+1)+1}(t_2) - h_1(-t_1)}. \quad (9)$$

2.2. Static Magnetization Distribution and Transverese Permeability

The field and frequency dependences of the MI response in the multilayered structure are determined by the transverse permeability μ_2 in the soft magnetic sub-layers. At relatively high frequencies, the permeability μ_2 is governed by the magnetization rotation [25]. It is assumed that all magnetic sub-layers have uniaxial in-plane magnetic anisotropy, and the angle ψ of deviation of the anisotropy axis from the transverse direction is relatively small. The deviation of the anisotropy axis from the transverse direction in real multilayers is related to the influence of different factors, in particular, the shape anisotropy and local non-uniformities.

The influence of the ferrogel layer on the MI response is related to stray fields induced by MNPs. The stray fields change the magnetization distribution in the soft magnetic sub-layers and affect the transverse permeability μ_2. To describe the influence of the stray fields on the MI effect it is assumed that the ferrogel layer generates a spatially uniform effective field H_p in the multilayer [34,35]. The value of H_p is proportional to the concentration of MNPs in the ferrogel, since it was found that the ferrogel saturation magnetization M_{sat} increases linearly with the concentration of MNPs of iron oxide which are biocompatible materials widely used in biomedical applications [34,36,38].

Following the approach developed previously [34], the dependence of the ferrogel magnetization M_g on the external magnetic field is approximated by the following linear function:

$$M_g = M_{sat}(H_e - H_c)/(H_1 - H_c). \tag{10}$$

where H_c is the coercive force of the ferrogel and H_1 is the value of the external magnetic field close saturation, i.e., $M_g \approx M_{sat}$. By the assumption, the effective field H_p is oriented in the opposite direction with respect to the magnetization vector in the layer of ferrogel. To simplify calculations, we assume that the effective stray field H_p is proportional to the ferrogel saturation magnetization M_{sat}. In addition, we neglect the spatial distribution of the field H_p over the multilayer thickness. Although the approach simplifies the real distribution of the stray fields, it allows one to describe qualitatively the effect of the MNPs on the MI response in the multilayer–ferrogel structure.

The magnetization distribution in the magnetic sub-layers can be calculated by the minimization of the free energy. Because of the above described procedure [34] the following equation for the equilibrium magnetization angle θ can be obtained:

$$H_a \sin(\theta - \psi)\cos(\theta - \psi) - H_p \sin(\theta - \varphi) - H_e \cos\theta = 0. \tag{11}$$

Here H_a is the anisotropy field in the magnetic sub-layers and $\varphi = \arcsin(M_g/M_{sat})$. Note that at $H_e \ll H_1$, Equation (11) can be simplified since $\varphi \approx 0$.

The transverse permeability μ_2 in the soft magnetic sub-layers can be found by means of solution of the linearized Landau–Lifshitz equation [25], which results in the following expression [34]:

$$\mu_2 = 1 + \frac{\omega_m[\omega_m + \omega_1 - i\kappa\omega]\sin^2\theta}{[\omega_m + \omega_1 - i\kappa\omega][\omega_1 - \gamma H_a \sin^2(\theta - \psi) - i\kappa\omega] - \omega^2}. \tag{12}$$

Here $\omega_m = 4\pi\gamma M$, M is the saturation magnetization of the magnetic sub-layers, γ is the gyromagnetic constant, κ is the Gilbert damping parameter and:

$$\omega_1 = \gamma[H_a \cos^2(\theta - \psi) - H_p\cos(\theta - \varphi) + H_e\sin\theta]. \tag{13}$$

3. Results

The proposed model allows one to analyze the MI effect of non symmetric multilayered elements with a ferrogel covering. The simulations were carried out for multilayered films with a copper central layer, permalloy $Fe_{20}Ni_{80}$ magnetic sub-layers with close to zero magnetostriction and titanium spacers.

Further, we use the following parameters of the permalloy sub-layers: the saturation magnetization $M = 750$ G, the conductivity $\sigma_2 = 3\cdot 10^{16}$ s^{-1}, the anisotropy field $H_a = 6$ Oe, the anisotropy axis deviation angle $\psi = -0.1\pi$ and the Gilbert damping parameter $\kappa = 0.02$. It is assumed that the thickness of the copper central layer $2d_0 = 500$ nm and the conductivity $\sigma_0 = 5\cdot 10^{17}$ s^{-1}. For the titanium spacers, we take $d_1 = 3$ nm and $\sigma_1 = 5\cdot 10^{16}$ s^{-1}.

The length of the multilayers is 1 cm and their width is 0.2 mm. The ferrogel layer thickness is 1 mm, and the permittivity of the ferrogel $\varepsilon = 80$.

To describe a relative variation of the impedance we introduce the MI ratio $\Delta Z/Z$, which is given by the following relation: $\Delta Z/Z = [Z(H_e) - Z(H_0)]/Z(H_0)$, where $H_0 = 100$ Oe is the external field sufficient for magnetic saturation of the multilayered structure. It was found that for multilayers with the parameters mentioned above, the maximum values of the MI ratio are achieved within the frequency range from 50 to 100 MHz [29]. Figure 2 shows the field dependence of the MI ratio $\Delta Z/Z$ calculated at the frequency $f = \omega/2\pi = 100$ MHz for the non-symmetric multilayer without ferrogel and multilayer with ferrogel for different values of the effective stray field H_p. The results are presented only for the range of the positive fields, since the MI ratio is symmetric with respect to the sign of the external field. The pure gel is a hydrogel without MNPs, for which stray fields are equal to zero ($H_p = 0$) [34,35].

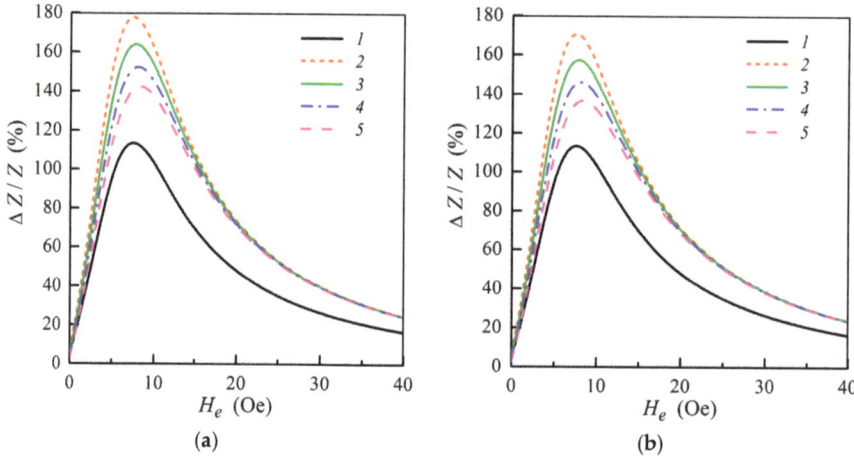

Figure 2. MI ratio $\Delta Z/Z$ as a function of the external field H_e at $f = \omega/2\pi = 100$ MHz for the multilayered element with ferrogel placed onto the surface of thin layer ($m = 3$, $n = 4$) (**a**) and onto the surface of thick layer ($m = 4$, $n = 3$) (**b**) (see also Figure 1e,f). Curves 1, multilayered film without gel; curves 2, film with pure gel (hydrogel) ($H_p = 0$); curves 3, $H_p = 0.25$ Oe; curves 4, $H_p = 0.5$ Oe; curves 5, $H_p = 0.75$ Oe. Parameters used for calculations are $l = 1$ cm, $w = 0.2$ mm, $2d_0 = 500$ nm, $d_1 = 3$ nm, $d_2 = 100$ nm, $M = 750$ G, $H_a = 6$ Oe, $\psi = -0.1\pi$, $\sigma_0 = 5 \times 10^{17}$ s^{-1}, $\sigma_1 = 5 \times 10^{16}$ s^{-1}, $\sigma_2 = 3 \times 10^{16}$ s^{-1}, $\kappa = 0.02$, $d_3 = 1$ mm, $H_c = 6.5$ Oe, $H_1 = 750$ Oe and $\varepsilon = 80$.

When the gel sample is placed onto the surface of the magnetoimpedance element (Figure 1e,f) MI ratio increases due to high permittivity of the gel as it was observed previously for the symmetric structures [33,34]. The effective stray field increases with the concentration of MNPs in the ferrogel due to a growth of the ferrogel saturation magnetization. As a result, the MI ratio decreases with an increase of the stray field H_p. However, observed behavior (in the case of non-symmetric MI elements) is differ from the results obtained for corresponding symmetric structures [34,35].

It follows from Figure 2 that the MI ratio $\Delta Z/Z$ is higher when the ferrogel layer is placed onto the surface of the thin layer. The dependence of the MI ratio on the position of the ferrogel layer is related to changes in the distribution of the electromagnetic fields within the multilayer in the presence of the ferrogel. It was demonstrated that the MI ratio is maximal for the symmetric multilayer without ferrogel at not too high frequencies [29]. Thus, the symmetric distribution of the electromagnetic fields within the multilayer is preferable in order to achieve highest values of the MI ratio. In the case when the ferrogel is placed onto the surface of the thin layer, the field distribution within the multilayer becomes more symmetric. On the contrary, when the ferrogel is placed onto the surface of the thick layer, the asymmetry in the field distribution is enhanced, which results in the decrease of the MI ratio (Figure 2).

To analyze the effect of the ferrogel layer positioning with respect to the magnetic layers of the non-symmetric structure on the MI response, we use the maximum MI ratio $(\Delta Z/Z)_{max}$, which corresponds to the maximum value of $\Delta Z/Z$ at a fixed frequency. Figure 3 shows the frequency dependence of $(\Delta Z/Z)_{max}$ calculated for the non-symmetric multilayer with different positions of the pure gel (hydrogel) without MNPs. Within the low frequency range, the application of the gel onto the surface of thin layer (Figure 1e) allows one to obtain higher values of $(\Delta Z/Z)_{max}$.

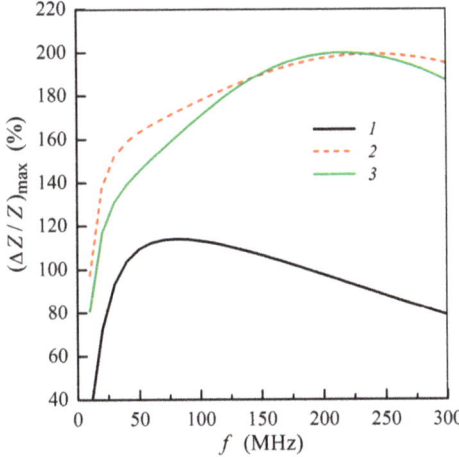

Figure 3. Frequency dependence of maximum MI ratio $(\Delta Z/Z)_{max}$: curve 1, multilayered film without gel; curve 2, multilayer with pure gel (hydrogel) placed onto the surface of thin layer ($m = 3$, $n = 4$); curve 3, multilayer with pure gel placed onto the surface of thick layer ($m = 4$, $n = 3$) (see also Figure 1e,f). Other parameters used for calculations are the same as in Figure 2.

There is a frequency range approximately from 25 to 50 MHz where the difference between MI ratio for two positions under consideration exceeds 10% of $(\Delta Z/Z)_{max}$, which is sizeable difference. At sufficiently high frequencies (above 150 MHz), the values of the maximum MI ratio depend slightly on the gel position: whether it is placed onto the thin or thick multilayer part (Figure 3).

The influence of the stray fields induced by MNPs on the frequency dependence of the maximum MI ratio is illustrated in Figure 4. The presence of the stray fields results in a decrease of $(\Delta Z/Z)_{max}$ value with an increase of the concentration of MNPs in the ferrogel [7,34]. This fact is due to a decrease of the transverse permeability in the magnetic sub-layers under action of the stray fields. The dependences shown in Figure 4 describes qualitatively experimental results obtained previously [5,7,20,34]. The calculated frequency dependences of $(\Delta Z/Z)_{max}$ are similar for both the positions of the ferrogel layer, however, the maximum MI ratio is higher when the ferrogel is placed onto the surface of thin multilayer (Figure 4).

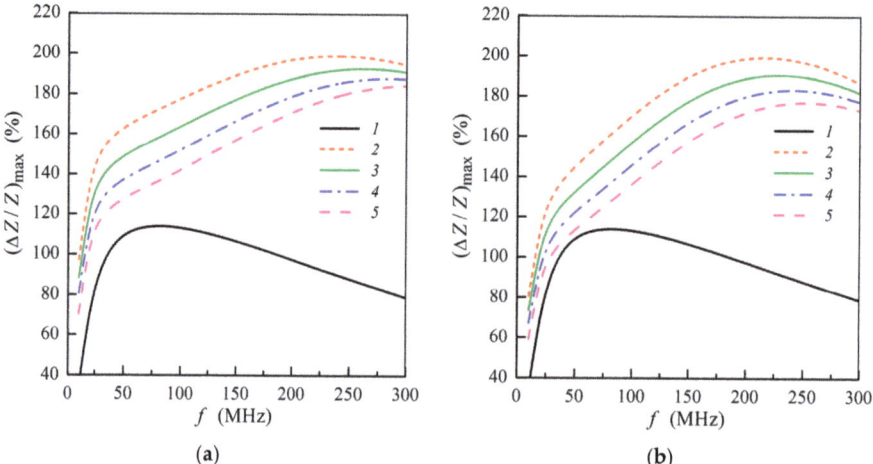

Figure 4. Frequency dependence of maximum MI ratio $(\Delta Z/Z)_{max}$ for the multilayered sensitive element with ferrogel placed onto the surface of thin layer ($m = 3$, $n = 4$) (**a**) or onto the surface of thick layer ($m = 4$, $n = 3$) (**b**) (see also Figure 1e,f). Curves 1 correspond to the multilayered film without gel; curves 2, film with pure gel (hydrogel) ($H_p = 0$); curves 3 $H_p = 0.25$ Oe; curves 4, $H_p = 0.5$ Oe; curves 5, $H_p = 0.75$ Oe. Other parameters used for calculations are the same as in Figure 2.

The MI response is sensitive to the stray field H_p created by the MNPs, and hence the studied multilayer–ferrogel structure could be used to determine the concentration of the MNPs in the ferrogel in a simple and reliable way by MI measurements. In order to simulate the nanoparticle concentration dependence of the MI, the maximum MI ratio $(\Delta Z/Z)_{max}$ as a function of the stray fields intensity H_p is obtained. Figure 5 shows the dependence of $(\Delta Z/Z)_{max}$ on H_p calculated at different frequencies in a low and intermediate frequency range. For both ferrogel layer possible positions (at the surface of the thin or thick layer, see also Figure 1e,f), the dependence has a nearly-linear behavior for all frequencies, what can be useful for practical purposes of the rapid definition of the concentration of MNPs.

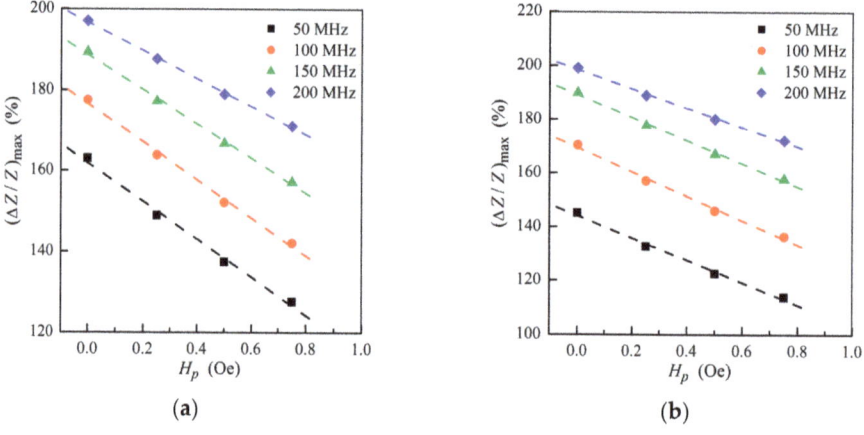

Figure 5. Maximum MI ratio $(\Delta Z/Z)_{max}$ as a function of the field H_p at different frequencies for the multilayer with ferrogel placed onto the surface of thin layer ($m = 3$, $n = 4$) (**a**) and onto the surface of thick layer ($m = 4$, $n = 3$) (**b**) (see also Figure 1e,f). Other parameters used for calculations are the same as in Figure 2.

Let us introduce the concentration sensitivity of the MI response S_c, which is defined as follows:

$$S_c = -\frac{\partial (\Delta Z/Z)_{max}}{\partial H_p}. \tag{14}$$

Figure 6 shows the frequency dependence of the concentration sensitivity calculated for different multilayer–ferrogel structures at selected frequencies. In a low-frequency range, maximum concentration sensitivity is achieved for symmetric multilayer. For the non-symmetric multilayers, the sensitivity S_c is higher when the ferrogel is placed onto the surface of the thin magnetic layer of the MI element. For three studied multilayered films, the values of S_c become the same at the frequency of 150 MHz. At higher frequencies, the concentration sensitivity drops and has higher values for the non-symmetric multilayers (Figure 6).

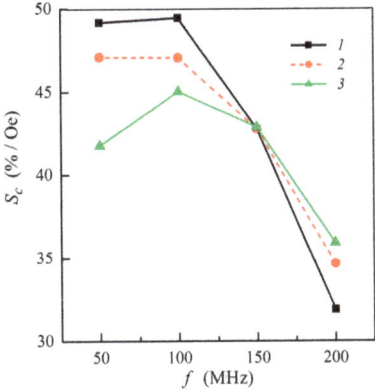

Figure 6. Frequency dependence of the concentration sensitivity S_c: curve 1, symmetric multilayered film ($m = n = 4$); curve 2, non-symmetric multilayer with ferrogel placed onto the surface of thin layer ($m = 3, n = 4$); curve 3, non-symmetric multilayer with ferrogel placed onto the surface of thick layer ($m = 4, n = 3$). Other parameters used for calculations are the same as in Figure 2.

4. Discussion

Magnetic biosensors for MNPs detection in natural tissues are a recent imaging area. Although the possibility of the MNP detection inside living cells was demonstrated almost 15 years ago [13–15], the detection of the MNPs inside cells (typical sizes are 10–100 µm) is quite similar to an "in vitro" detection of biomolecular labels [1,11,39]. The principle of the magnetic label detection is an evaluation of the sum of the stray fields of all magnetizable particles. The MI effect showed its capability for sensing stray fields of MNPs at relatively large distances due to the very high sensitivity with respect to the applied magnetic field. The low sensitivity was a reason why for a long time the development of magnetic biosensors was no possible [39,40].

Recently we have reported w promising way for the detection of magnetic particles in blood vessels in the course of model experiments with multilayered [FeNi (100 nm)/Cu (3 nm)]$_5$/Cu (500 nm)/[Cu (3 nm)/[FeNi (100 nm)]$_5$ magnetoimpedance sensitive elements [41].

In addition, we discussed a hypothetical procedure in which the definition of the local concentration of magnetic carriers could be also used for correct definition of the start time for magnetic hyperthermia. Such an approach, i.e., combination of a number of functions assigned to the magnetic nanoparticles is a very hot topic, indicating some new theranostic directions [42].

Two asymmetric geometries for the detection of a flat ferrogel covering seem to be more useful for regenerative medicine cases. However, one can think about controlled drug release, biocompatible valves and actuators controlled by development of a magnetic field, or monitoring of the matrices for growing cells and tissues including growing conditions in a magnetic field as it was previously discussed for symmetric multilayered MI-sensitive elements [43]. In the case of controlled drug release it can be combined with magnetic hyperthermia [42,44] and control of the local concentration of MNPs by a magnetic field sensor.

As the next step, based on the model developed here, a module for commercial multi-physics software could be built. Our previous experience with symmetric MI structures in simpler cases was positive and fruitful [45]. Then, after verification, technology transfer to the high-tech industry should become possible.

The distance between the sensor surface and the magnetic label is a critically important parameter for the correct evaluation of the change in the output signal of a magnetic sensor. Ideal magnetic labels for biomedical detection are identical spherical superparamagnetic biocompatible MNPs, i.e., they all carry the same magnetic moments in an applied magnetic field of a certain strength. In our previous work, we proposed treating the assay of identical superparamagnetic labels as an additional layer of the multilayered structure [29]. Here, because of the theoretical comparison of symmetric and non-symmetric multilayered structures covered by a ferrogel layer we do observe the difference. The results obtained can be useful for improving the magnetoimpedance biodetector sensitivity.

In the described model, we were based on the results of our previous experiments, in which it was demonstrated that the magnetization of a polyacrylamide ferrogel with iron oxide nanoparticles increases linearly with increasing particle concentration [5,34]. Accordingly, the model assumes that the stray fields generated by the ferrogel are also proportional to the particle concentration. Moreover, the proportionality coefficient in the linear dependence of the field H_p on the concentration has not been discussed anywhere, since it cannot be determined from general considerations.

In principle, one can imagine a ferrogel in which the particles behave differently (for example, some sort of ordering of the antiferromagnetic type can occur). In this case, the dependence of the ferrogel magnetization on the concentration will indeed become nonlinear, but in any case, the magnetization will increase with increasing concentration. Qualitatively, the results of the model will not change for more complex magnetization behaviour, it is only necessary to change the linear dependence for another increasing function. However, the analysis of more complex dependences of the field H_p on the particle concentration is beyond the scope of this work.

The number of layers in MI structure analyzed here was proposed based on previous experimental and theoretical studies of FeNi-based multilayers [17,26,28]. High MI in a frequency range below 100 MHz is observed for $[F/X]_m/F/C/[F/X]_n/F$ with m ≈ n and with the thickness of the $[F/X]_m$ multilayered structures of about 0.5 μm. This roughly gives 20 < n, m < 3 for reasonable number of the sub-layers. A large m value results in the increase of the contribution of the interfaces, lower m causes transition into a "transcritical" state [30,31,46]–both causing the decay of MI effect value and sensitivity.

Special efforts for the development of MI biosensors is made both theoretical and experimental development of new designs of low field sensors with enhanced sensitivity, the possibility of magnetic noise reduction, simple design allowing the point-of-care usage, devises, adapted to microfluidic technologies, lab-on-a-chip packaging and sensor array design [47–49]. Magnetic biosensors are in their way of step-by-step replacement of many current diagnostic systems in laboratories and medical care points.

5. Conclusions

In this work, a theoretical description of the MI effect in non-symmetric multilayered elements covered by a ferrogel layer is developed. A model based on a solution of the Maxwell equations in the one-dimensional approximation allows one to find the distribution of electromagnetic fields in non-symmetric multilayers with ferrogel coverings. The symmetric distribution of the electromagnetic fields within the multilayer insures the highest values of the MI ratio. In the case of non-symmetric MI multilayered structures when the ferrogel is placed on the surface of the thin layer, the field distribution within the multilayer becomes more symmetric. When the ferrogel is placed onto the thick layer, the asymmetry in the field distribution is enhanced, resulting in the decrease of the MI ratio.

Within the low frequency range, placing the ferrogel onto the surface of thin layer increases the maximum MI ratio $(\Delta Z/Z)_{max}$. At frequencies above 150 MHz, the values of $(\Delta Z/Z)_{max}$ depend only slightly on the ferrogel position (onto thin or thick multilayer part). The presence of the stray fields results in a decrease of $(\Delta Z/Z)_{max}$ value with an increase of the concentration of MNPs in the ferrogel due to a decrease of the transverse permeability in the magnetic sub-layers under action of the stray fields. The calculated frequency dependences of $(\Delta Z/Z)_{max}$ are similar for both the positions of the ferrogel layer. However, the maximum MI ratio is higher when the ferrogel is placed onto the surface of a thin layer. The studied multilayer–ferrogel structure could be used to determine the concentration of the MNPs in the ferrogel by MI measurements, since the MI response is sensitive to the stray field H_p. For both ferrogel layer positions (onto the surface of thin or thick layer), the concentration dependence $(\Delta Z/Z)_{max}(H_p)$ has a nearly linear behavior for all frequencies. The sensitivity of the MI response for non-symmetric multilayers calculated for different multilayer–ferrogel structures at selected frequencies is higher when the ferrogel is placed onto the surface of a thin magnetic layer.

Author Contributions: Conceptualization, N.A.B. and G.V.K.; Funding acquisition, G.V.K.; Investigation, N.A.B. and G.V.K.; Methodology, N.A.B. and G.V.K.; Writing—original draft, N.A.B. and G.V.K. Authors are contributed equally to the research and manuscript preparation. All authors have read and agreed to the published version of the manuscript.

Funding: This research was funded by the University Basque Country UPV/EHU, Research Groups Funding (IT1245-19).

Institutional Review Board Statement: This work did not involve humans or animals and therefore it did not require the Institutional Review Board Statement and approval.

Informed Consent Statement: Not applicable.

Data Availability Statement: Data available from the corresponding author upon reasonable request.

Acknowledgments: Technical and human support provided by SGIker (Universidad del País Vasco UPV/EHU) is gratefully acknowledged. Authors would like to thank Andrey V. Svalov for stimulating discussion.

Conflicts of Interest: The authors declare no conflict of interest.

References

1. Baselt, D.R.; Lee, G.U.; Natesan, M.; Metzger, S.W.; Sheehan, P.E.; Colton, R. A biosensor based on magnetoresistance technology. *Biosens. Bioelectron.* **1998**, *13*, 731–739. [CrossRef]
2. Cardoso, S.; Leitao, D.C.; Dias, T.M.; Valadeiro, J.; Silva, M.D.; Chicharo, A.; Silverio, V.; Gaspar, J.; Freitas, P.P. Challenges and trends in magnetic sensor integration with microfluidics for biomedical applications. *J. Phys. D Appl. Phys.* **2017**, *50*, 213001. [CrossRef]
3. Grimes, C.A.; Mungle, C.S.; Zeng, K.; Jain, M.K.; Dreschel, W.R.; Paulose, M.; Ong, G.K. Wireless magnetoelastic resonance sensors: A critical review. *Sensors* **2002**, *2*, 294–313. [CrossRef]
4. Marín, P.; Marcos, M.; Hernando, A. High magnetomechanical coupling on magnetic microwire for sensors with biological applications. *Appl. Phys. Lett.* **2010**, *96*, 262512. [CrossRef]
5. Kurlyandskaya, G.V.; Portnov, D.S.; Beketov, I.V.; Larrañaga, A.; Safronov, A.P.; Orue, I.; Medvedev, A.I.; Chlenova, A.A.; Sanchez-Ilarduya, M.B.; Martinez-Amesti, A.; et al. Nanostructured materials for magnetic biosensing. *Biochim. Biophys. Acta (BBA) Gen. Subj.* **2017**, *1861*, 1494–1506. [CrossRef]
6. Beato-López, J.J.; Pérez-Landazábal, J.I.; Gómez-Polo, C. Magnetic nanoparticle detection method employing non-linear magnetoimpedance effects. *J. Appl. Phys.* **2017**, *121*, 163901. [CrossRef]
7. Schmalz, J.; Kittmann, A.; Durdaut, P.; Spetzler, B.; Faupel, F.; Hoft, M.; Quandt, E.; Gerken, M. Multi-mode love-wave SAW magnetic-field sensors. *Sensors* **2020**, *20*, 3421. [CrossRef] [PubMed]
8. Li, M.; Matyushov, A.; Dong, C.; Chen, H.; Lin, H.; Nan, T.; Qian, Z.; Rinaldi, M.; Lin, Y.; Sun, N.X. Ultra-sensitive NEMS magnetoelectric sensor for picotesla DC magnetic field detection. *Appl. Phys. Lett.* **2017**, *110*, 143510. [CrossRef]

9. Kurlyandskaya, G.V.; Levit, V.I. Advanced materials for drug delivery and biosensors based on magnetic label detection. *Mater. Sci. Eng. C* **2007**, *27*, 495–503. [CrossRef]
10. Wang, T.; Guo, L.; Lei, C.; Zhou, Y. Ultrasensitive determination of carcinoembryonic antigens using a magnetoimpedance immunosensor. *RSC Adv.* **2015**, *5*, 51330–51336. [CrossRef]
11. Yang, Z.; Wang, H.H.; Dong, X.W.; Yan, H.L.; Lei, C.; Luo, Y.S. Giant magnetoimpedance based immunoassay for cardiac biomarker myoglobin. *Anal. Methods* **2017**, *9*, 3636–3642. [CrossRef]
12. Chiriac, H.; Herea, D.D.; Corodeanu, S. Microwire array for giant magnetoimpedance detection of magnetic particles for biosensor prototype. *J. Magn. Magn. Mater.* **2007**, *311*, 425–428. [CrossRef]
13. Kumar, A.; Mohapatra, S.; Fal-Miyar, V.; Cerdeira, A.; Garcia, J.A.; Srikanth, H.; Gass, J.; Kurlyandskaya, G.V. Magnetoimpedance biosensor for Fe_3O_4 nanoparticle intracellular uptake evaluation. *Appl. Phys. Lett.* **2007**, *91*, 143902. [CrossRef]
14. Devkota, J.; Howell, P.; Mukherjee, P.; Srikanth, H.; Mohapatra, S.; Phan, M.H. Magneto-reactance based detection of MnO nanoparticle-embedded Lewis lung carcinoma cells. *J. Appl. Phys.* **2015**, *117*, 17D123. [CrossRef]
15. Blanc-Béguin, F.; Nabily, S.; Gieraltowski, J.; Turzo, A.; Querellou, S.; Salaun, P.Y. Cytotoxicity and GMI bio-sensor detection of maghemite nanoparticles internalized into cells. *J. Magn. Magn. Mater.* **2009**, *321*, 192–197. [CrossRef]
16. García-Arribas, A.; Martínez, F.; Fernández, E.; Ozaeta, I.; Kurlyandskaya, G.V.; Svalov, A.V.; Berganzo, J.; Barandiaran, J.M. GMI detection of magnetic-particle concentration in continuous flow. *Sens. Actuators A* **2011**, *172*, 103–108. [CrossRef]
17. Kurlyandskaya, G.V.; Fernandez, E.; Safronov, A.P.; Svalov, A.V.; Beketov, I.V.; Burgoa Beitia, A.; Garcıa-Arribas, A.; Blyakhman, F.A. Giant magnetoimpedance biosensor for ferrogel detection: Model system to evaluate properties of natural tissue. *Appl. Phys. Lett.* **2015**, *106*, 193702. [CrossRef]
18. Safronov, A.P.; Mikhnevich, E.A.; Lotfollahi, Z.; Blyakhman, F.A.; Sklyar, T.F.; Larrañaga Varga, A.; Medvedev, A.I.; Fernández Armas, S.; Kurlyandskaya, G.V. Polyacrylamide ferrogels with magnetite or strontium hexaferrite: Next step in the development of soft biomimetic matter for biosensor applications. *Sensors* **2018**, *18*, 257. [CrossRef] [PubMed]
19. Uchiyama, T.; Mohri, K.; Honkura, Y.; Panina, L.V. Recent advances of pico-Tesla resolution magnetoimpedance sensor based on amorphous wire CMOS IC MI Sensor. *IEEE Trans. Magn.* **2012**, *48*, 3833–3839. [CrossRef]
20. Sandhu, A.; Southern, P.; de Freitas, S.C.; Knudde, S.; Cardoso, F.A.; Freitas, P.P.; Kurlyandskaya, G.V. *Sensing Magnetic Nanoparticles 172–227 in Magnetic Nanoparticles in Biosensing and Medicine Ed*; Nicholas, J.D., Adrian, I., Justin, L., Eds.; Cambridge University Press: Cambridge, UK, 2019.
21. Gloag, L.; Mehdipour, M.; Chen, D.; Tilley, R.D.; Gooding, J.J. Advances in the Application of Magnetic Nanoparticles for Sensing. *Adv. Mater.* **2019**, *31*, 1904385. [CrossRef]
22. Fujiwara, K.; Oogane, M.; Kanno, A.; Imada, M.; Jono, J.; Terauchi, T.; Okuno, T.; Aritomi, Y.; Morikawa, M.; Tsuchida, M. Magnetocardiography and magnetoencephalography measurements at room temperature using tunnel magneto-resistance sensors. *Appl. Phys. Express* **2018**, *11*, 023001. [CrossRef]
23. Dolabdjian, C.; Ménard, D. Giant magneto-impedance (GMI) magnetometers. In *High Sensitivity Magnetometers*; Grosz, A., Haji-Sheikh, M.J., Mukhopadhyay, S.C., Eds.; Springer: Berlin/Heidelberg, Germany, 2017; pp. 103–126.
24. Wang, T.; Zhou, Y.; Lei, C.; Luo, J.; Xie, S.; Pu, H. Magnetic impedance biosensor: A review. *Biosens. Bioelectron.* **2017**, *90*, 418–435. [CrossRef] [PubMed]
25. Knobel, M.; Vázquez, M.; Kraus, L. Giant magnetoimpedance. In *Handbook of Magnetic Materials*; Buschow, K.H.J., Ed.; Elsevier: Amsterdam, The Netherlands, 2003; Volume 15, pp. 497–563.
26. Kurlyandskaya, G.V.; Chlenova, A.A.; Fernández, E.; Lodewijk, K.J. FeNi-based flat magnetoimpedance nanostructures with open magnetic flux: New topological approaches. *J. Magn. Magn. Mater.* **2015**, *383*, 220–225. [CrossRef]
27. Morikawa, T.; Nishibe, Y.; Yamadera, H.; Nonomura, Y.; Takeuchi, M.; Taga, Y. Giant magneto-impedance effect in layered thin films. *IEEE Trans. Magn.* **1997**, *33*, 4367–4372. [CrossRef]
28. Panina, L.V.; Mohri, K. Magneto-impedance in multilayer films. *Sens. Actuators A* **2000**, *81*, 71–77. [CrossRef]
29. Buznikov, N.A.; Kurlyandskaya, G.V. Magnetoimpedance in symmetric and non-symmetric nanostructured multilayers: A theoretical study. *Sensors* **2019**, *19*, 1761. [CrossRef] [PubMed]
30. Sugita, Y.; Fujiwara, H.; Sato, T. Critical thickness and perpendicular anisotropy of evaporated permalloy films with stripe domains. *Appl. Phys. Lett.* **1967**, *10*, 229–231. [CrossRef]
31. Kurlyandskaya, G.V.; Elbaile, L.; Alves, F.; Ahamada, B.; Barrué, R.; Svalov, A.V.; Vas'kovskiy, V.O. Domain structure and magnetization process of a giant magnetoimpedance geometry FeNi/Cu/FeNi(Cu)FeNi/Cu/FeNi sensitive element. *J. Phys. Condens. Matter* **2004**, *16*, 6561–6568. [CrossRef]
32. Correa, M.A.; Viegas, A.D.C.; da Silva, R.B.; de Andrade, A.M.H.; Sommer, R.L. GMI in FeCuNbSiB\Cu multilayers. *Phys. B* **2006**, *384*, 162–164. [CrossRef]
33. Kennedy, S.; Roco, C.; Délérisa, A.; Spoerria, P.; Cezara, C.; Weavera, J.; Vandenburghd, H.; Mooney, D. Improved magnetic regulation of delivery profiles from ferrogels. *Biomaterials* **2018**, *161*, 179–189. [CrossRef]
34. Buznikov, N.A.; Safronov, A.P.; Golubeva, E.V.; Lepalovskij, V.N.; Orue, I.; Svalov, A.V.; Chlenova, A.A.; Kurlyandskaya, G.V. Modelling of magnetoimpedance response of thin film sensitive element in the presence of ferrogel: Next step toward development of biosensor for in-tissue embedded magnetic nanoparticles detection. *Biosens. Bioelectron.* **2018**, *117*, 366–372. [CrossRef]

35. Blyakhman, F.A.; Buznikov, N.A.; Sklyar, T.F.; Safronov, A.P.; Golubeva, E.V.; Svalov, A.V.; Sokolov, S.Y.; Melnikov, G.Y.; Orue, I.; Kurlyandskaya, G.V. Mechanical, electrical and magnetic properties of ferrogels with embedded iron oxide nanoparticles obtained by laser target evaporation: Focus on multifunctional biosensor applications. *Sensors* **2018**, *18*, 872. [CrossRef]
36. Sukstanskii, A.; Korenivski, V.; Gromov, A. Impedance of a ferromagnetic sandwich strip. *J. Appl. Phys.* **2001**, *89*, 775–782. [CrossRef]
37. Gromov, A.; Korenivski, V.; Haviland, D.; van Dover, R.B. Analysis of current distribution in magnetic film inductors. *J. Appl. Phys.* **1999**, *85*, 5202–5204. [CrossRef]
38. Kurlyandskaya, G.V.; Fernández, E.; Svalov, A.; BurgoaBeitia, A.; García-Arribas, A.; Larrañaga, A. Flexible thin film magnetoimpedance sensors. *J. Magn. Magn. Mater.* **2016**, *415*, 91–96. [CrossRef]
39. Llandro, J.; Palfreyman, J.J.; Ionescu, A.; Barnes, C.H.W. Magnetic biosensor technologies for medical applications: A review. *Med. Biol. Eng. Comput.* **2010**, *48*, 977–998. [CrossRef]
40. Roychoudhury, A. Magnetic-based sensing. In *Nanotechnology in Cancer Management: Precise Diagnostics toward Personalized Health Care*; Khondakar, K.R., Kaushik, A.K., Eds.; Elsevier: Amsterdam, The Netherlands, 2021; Volume 15, pp. 149–184.
41. Melnikov, G.Y.; Lepalovskij, V.N.; Svalov, A.V.; Safronov, A.P.; Kurlyandskaya, G.V. Thin film sensor for detecting of stray fields of magnetic particles in blood vessel. *Sensors* **2021**, *21*, 3621. [CrossRef]
42. Tishin, A.M.; Shtil, A.A.; Pyatakov, A.P.; Zverev, V.I. Developing antitumor magnetic hyperthermia: Principles, materials and devices. *Recent Pat. Anti Cancer Drug Discov.* **2016**, *11*, 360–375. [CrossRef]
43. Kurlyandskaya, G.V.; Blyakhman, F.A.; Makarova, E.B.; Buznikov, N.A.; Safronov, A.P.; Fadeyev, F.A.; Shcherbinin, S.V.; Chlenova, A.A. Functional magnetic ferrogels: From biosensors to regenerative medicine. *AIP Adv.* **2020**, *10*, 125128. [CrossRef]
44. Regmi, R.; Bhattarai, S.R.; Sudakar, C.; Wani, A.S.; Cunningham, R.; Prem, P.; Vaishnava, P.P.; Naik, R.; Oupicky, D.; Lawes, G. Hyperthermia controlled rapid drug release from thermosensitive magnetic microgels. *J. Mater. Chem.* **2010**, *20*, 6158–6163. [CrossRef]
45. Kozlov, N.V.; Chlenova, A.A.; Volchkov, S.O.; Kurlyandskaya, G.V. The study of magnetic permeability and magnetoimpedance: Effect of ferromagnetic alloy characteristics. *AIP Conf. Proc.* **2020**, *2313*, 030050.
46. Kurlyandskaya, G.V.; Svalov, A.V.; Fernandez, E.; Garcia-Arribas, A.; Barandiaran, J.M. FeNi-based magnetic layered nanostructures: Magnetic properties and giant magnetoimpedance. *J. Appl. Phys.* **2010**, *107*, 09C502. [CrossRef]
47. Fodil, K.; Denoual, M.; Dolabdijan, C.; Treizebre, A.; Senez, V. In-flow detection of ultra-small magnetic particles by an integrated giant magnetic impedance sensor. *Appl. Phys. Lett.* **2016**, *108*, 173701. [CrossRef]
48. Beato, J.; Pérez-Landazábal, J.; Gómez-Polo, C. Enhanced magnetic nanoparticle detection sensitivity in non-linear magnetoimpedance-based sensor. *IEEE Sens. J.* **2018**, *18*, 8701–8708. [CrossRef]
49. Garcia-Arribas, A. The performance of the magneto-impedance effect for the detection of superparamagnetic particles. *Sensors* **2020**, *20*, 1961. [CrossRef] [PubMed]

Article

Magnetoimpedance Thin Film Sensor for Detecting of Stray Fields of Magnetic Particles in Blood Vessel

Grigory Yu. Melnikov [1], Vladimir N. Lepalovskij [1], Andrey V. Svalov [1], Alexander P. Safronov [1] and Galina V. Kurlyandskaya [1,2,*]

[1] Institute of Natural Sciences and Mathematics, Ural Federal University, Ekaterinburg 620002, Russia; grigory.melnikov@urfu.ru (G.Y.M.); vladimir.lepalovsky@urfu.ru (V.N.L.); safronov@iep.uran.ru (A.V.S.); Andrey.svalov@urfu.ru (A.P.S.)

[2] Departamento de Electricidad y Electrónica, Universidad del País Vasco UPV/EHU, 48080 Bilbao, Spain

* Correspondence: kurlyandskaya.gv@ehu.eus

Abstract: Multilayered [FeNi (100 nm)/Cu (3 nm)]$_5$/Cu (500 nm)/[Cu (3 nm)/[FeNi (100 nm)]$_5$ structures were used as sensitive elements of the magnetoimpedance (MI) sensor prototype for model experiments of the detection of magnetic particles in blood vessel. Non-ferromagnetic cylindrical polymer rod with a small magnetic inclusion was used as a sample mimicking thrombus in a blood vessel. The polymer rod was made of epoxy resin with an inclusion of an epoxy composite containing 30% weight fraction of commercial magnetite microparticles. The position of the magnetic inclusion mimicking thrombus in the blood vessel was detected by the measurements of the stray magnetic fields of microparticles using MI element. Changes of the MI ratio in the presence of composite can be characterized by the shift and the decrease of the maximum value of the MI. We were able to detect the position of the magnetic composite sample mimicking thrombus in blood vessels. Comsol modeling was successfully used for the analysis of the obtained experimental results and the understanding of the origin the MI sensitivity in proposed configuration. We describe possible applications of studied configuration of MI detection for biomedical applications in the field of thrombus state evaluation and therapy.

Keywords: magnetic nanomaterials; magnetic multilayers; magnetic particles; magnetic composites; magnetoimpedance; stray fields; biomedical applications; COMSOL modeling

1. Introduction

Nanostructured magnetic materials have become a subject of interest in the last decades. In many senses, their popularity is based on the development of new fabrication techniques and significant improvements of existing synthesis methods [1–3]. They attract special attention from both the point of view fundamental physical and chemical properties and variety of practical applications [3–5]. Among other systems magnetic nanoparticles were in focus for their investigation in technical devices [6–8] and medical conditions [9,10]. Medical applications require the most strict protocol for nanomaterials and nanodrugs processing [10]. The key point is evaluation of the same parameter by different techniques and thorough characterization of nanodrugs at all stages of production, transportation and diagnostics or therapy. A very special need in fabrication techniques with well controlled conditions of elaboration of large batches become very clear [10,11]. The majority of the methods of fabrication of magnetic nanoparticles (MNPs) are chemical techniques [12]. However, the electrophysical techniques of the electric explosion of wire or laser target evaporation provide the largest single batches and green synthesis processing with low amount of required solvents [11].

The second important type of magnetic nanostructures is thin films and multilayered structures. They are most often used as sensitive parts of magnetic field sensors and biosensors [13,14]. One of the reasons of special success of magnetic films for sensor applications is their compatibility with existing semiconductor electronics and high potential of packaging [15]. There are different magnetic effects observed in thin film structures capable of supporting the development of magnetic field sensors with sensitivity sufficient to detect either biomagnetic fields or stray fields of magnetic nanoparticles used as biomolecular labels [16]. Their efficiency and choice depend on particular conditions of the detection and many parameters can contribute to the decision and device design. It might be a corrosion stability in biological fluids, power consumption, optimum response in the temperature range under consideration and many others. Importantly, in the case of small magnetic field detectors the sensitivity with respect to applied magnetic field in the field range of a few oersted is a key parameter [17,18]. Among others, magnetoimpedance effect (MI) observed in different magnetic materials [19] was shown to be effective for magnetic biosensing both biomagnetic fields and stray fields of magnetic particles [20].

One of important features of present day nanomaterials applications is their operation in conditions when at least two magnetic components are involved in the technological or biomedical process. The search for optimum conditions of the application (optimum functional properties) becomes quite complex because the best conditions for one of the materials might be far from being optimum for another one. It this case the whole system optimization goes in a direction of the search for most favorable work interval in which each one of the materials might be functioning not with the best but acceptable performance. The simplest example of such system is a magnetic particle creating stray fields in the external magnetic field and magnetic field sensor. Application of external magnetic field increases the magnetic moment and the value of the stray fields created in the presence of magnetic particles: there is an extended interval in which the rule "the higher the external field the higher the stray fields" is correct. However, external magnetic field detector may have the highest sensitivity with respect to the applied magnetic field in the field range where the stray fields of the particle are still too small to be detected. In this case magnetic label detection become impossible. Summarizing, in each particular application involving two or more magnetic materials there is a goal to search for synergetic combinations in order to reach the best performance.

Nowadays magnetic field sensors attract attention for a variety of applications related to detection of biocomposites containing magnetic nano- or microparticles. The possibility of the detection of superparamagnetic nanoparticles of iron oxide by MI multilayered sensitive element was shown in the case of such biomimetic materials as ferrogels [18,21]. Those measurements were made in close proximity mode: the ferrogel sample was placed directly onto the surface of MI multilayered sensitive element. However, the sensitivity of MI detectors should be sufficient for magnetic particles (MPs) detection at the distances of the order of 100 microns including detection of the presence of aggregates of a complex geometry.

One of such applications is a detection of MPs agglomerates in the blood vessel. A pathological blood clot that forms during the life of a patient in the lumen arteries or veins is denominated as a thrombus. Diseases caused by thrombosis are from the list of the leading causes of death. Thrombosis is currently being treated by surgery, which consists in performing a complex operation with a high-risk complication. Another way is the therapy with the use of thrombolytics medication, having many side effects. One of the ways to reduce side effects is targeted drug delivery to a blood clot using MPs [22]. For example, biocompatible MPs of magnetite can be employed [2,21]. During the therapy it is critically important to determine the local concentration of magnetic particles in the thrombus region. Here we propose the next step of the development of this interesting idea. The task of the local concentration definition can be solved using a MI sensor, detecting stray fields of magnetic particles forming agglomerates in the thrombus area.

The aim of this work is to study the effect of stray magnetic fields of microparticles of iron oxide in a model geometry imitating a blood clot in a blood vessel. It is done by experimental study of magnetoimpedance responses of multilayered sensitive elements based on permalloy at different positions magnetic sample made of composite material (epoxy resin and magnetite particles). In addition, Comsol modeling of stray fields is used for the analysis of the obtained experimental results.

2. Materials and Methods

2.1. Materials

For magnetoimpedance sensitive element, we selected configuration with open magnetic flux (Figure 1) [21]. It consisted of a number of layers of the same width and length deposited onto the glass substrate one on the top of the other (Figure 1). Metallic masks were used during the magnetron sputtering deposition of the layers. The MI element was deposited through the mask with the shape of the stripe (0.5 × 10 mm) by dc magnetron sputtering onto a glass substrate. The deposition was prepared with a background pressure of 3×10^{-7} mbar and a working Ar pressure of 3.8×10^{-3} mbar. During the deposition process an in-plane constant magnetic field H_d = 100 Oe was applied along the short side of the MI elements for inducing a transverse uniaxial in-plane magnetic anisotropy. Therefore, the direction parallel to the short side of the rectangular stripe along which the technological magnetic field was oriented during film deposition was expected to be an easy magnetization axis (EMA). The direction along the long side of the rectangular MI element was expected to be the hard magnetization axis (HMA). The deposition rates for each part of the multilayered structure (copper and permalloy) were calculated in previous calibrations. As it was proposed in previous works [21,23–25] the magnetic layers of permalloy before and after the Cu-lead were nanostructured by Cu-spacers in order to avoid the transition into a "transcritical" state [26–28].

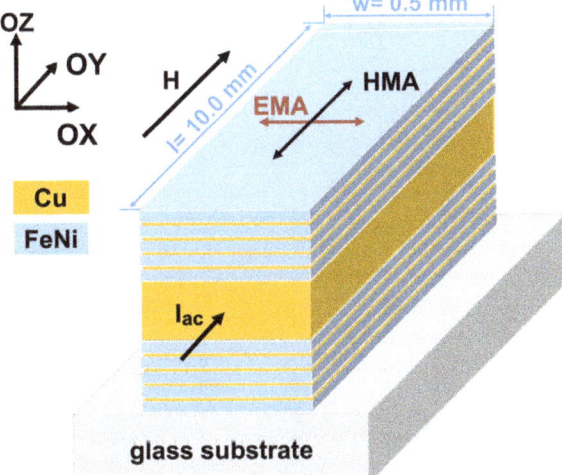

Figure 1. Scheme of multilayered [FeNi (100 nm)/Cu (3 nm)]$_5$/Cu (500 nm)/[Cu (3 nm)/[FeNi (100 nm)]$_5$ MI sensitive element. EMA—easy magnetization axis; HMA—hard magnetization axis; I$_{ac}$—high frequency alternating current flowing along the rectangular MI element; H—external magnetic field applied during magnetic measurements. Technological magnetic field during deposition was applied along the short side of the MI element.

Figure 1 shows the general arrangement of the layers of magnetoimpedance [Fe$_{21}$Ni$_{79}$ (100 nm)/Cu (3 nm)]$_5$/Cu (500 nm)/[Cu (3 nm)/[Fe$_{21}$Ni$_{79}$ (100 nm)]$_5$ elements with a copper conductive central lead. One can see that the used configuration is a symmetric MI structure.

In addition, despite the complexity of the structure it consists of the magnetic layers of two compositions: Cu and FeNi. For their deposition pure Cu and alloyed FeNi targets were used. Although in our previous works discussing both experimental and theoretical results for FeNi/Ti and FeNi/Cu based systems the higher MI variations were found for FeNi/Ti based multilayered element [21,29], FeNi/Cu based elements are much easier to fabricate.

Cylindrical polymer rod with the diameter of about 5.1 mm and approximately 54 mm length was used for model experiments. The sample mimicking thrombus in a blood vessel was made of epoxy polymer, and it was combined of three sections along its length. Both end sections (25 mm each) were non-ferromagnetic epoxy polymer, the central section (4 mm) was a ferromagnetic inclusion. It was made of a filled epoxy composite containing 30% weight fraction of commercial (Alfa Aesar, Ward Hill, MA, USA) magnetite submicron sized particles (Fe_3O_4 phase—94 weight%; Fe_2O_3 phase—1 weight%; FeO(OH) phase—5 weight%). According to SEM microphotographs the particles were quasi-spherical and their caliper diameter fall within 0.05–0.50 micron with a median value 0.30 μm. Saturation magnetization of magnetite particles was 84 emu/g, remnant magnetization was 6.6 emu/g, and coercivity was 78 Oe.

The model sample was prepared with the epoxy resin KDA (Chimex Ltd., St. Petersburg, RF) as a polymer matrix. Epoxy resin was mixed with a tri(ethyl)-tetra(amine) hardener (Epital, Moscow, Russia) in a 6:1 weight ratio. Magnetite particles were vigorously mixed with liquid epoxy resin at 25 °C for 10 min in order to obtain a homogeneous mixture. The curing of the combined model sample was performed in a polyethylene tube in a stepwise manner. At the first step the blank epoxy composition was cured in the tube, which was vertically positioned, then the filled epoxy composition was cured on top of the blank epoxy polymer, and finally the blank epoxy composition was placed and cured on top of the magnetic central section. Each curing step was done for 2 h at 70 °C. The control sample (control) was also prepared by curing the blank epoxy resin cylinder without magnetic particles (0.0% concentration composite).

2.2. Experimental Methods

The average grain size of the microparticles and their shapes were evaluated by scanning electron microscopy, SEM (JEOL JSM-7000 F, Japan) for the set of the particles spread onto carbon conductive tape. The magnetic properties of the multilayered sensitive elements were studied using magneto-optical Kerr-microscopy (MOKE). Both magnetic domains and magnetic hysteresis loops were studied using Kerr-microscope (Evico Magnetics GmbH, Dresden, Germany). The magnetic hysteresis loops of epoxy resin and magnetic filled composite were measured at room temperature by a vibration sample magnetometer 7407 VSM (Lake Shore Cryotronics, Westerville, OH, USA).

The longitudinal MI effect was measured, i.e., external applied magnetic field (H) and the high frequency alternating flowing current (AC) were parallel to each other (Figure 2). MI measurements were made by the automatic system based on Agilent HP E 4991 An impedance analyzer (Agilent, Santa Clara, CA, USA) at room temperature in an external magnetic field created by Helmholtz coils. The in-plane magnetic field in the range of ±150 Oe was applied along the long side of the MI element in the direction of the alternating current flow for the currents in the frequency range of 1–400 MHz. This frequency range was selected taking into account both the length of the MI element and its expected sensitivity with respect to an applied field [21]. As before, the system was carefully calibrated for the extraction of the intrinsic impedance values [15,21].

The experimental detection of the magnetic insert mimicking thrombus in blood vessels was carried out as a model experiment as follows. The MI element was incorporated into a "microstripe" line using silver paint. The impedance of the MI element was measured at various positions of the center of the magnetic inset with respect to a center of the MI element. The magnetic insert inside the non-ferromagnetic tube was placed at a distance of 1.1 ± 0.2 (mm) above the MI element surface (Figure 2). The tube was parallel to the surface of the MI element and its axis was perpendicular to the long side of the element. The tube and the magnetic insert can be displaced along the axis perpendicular to the long side of the MI element (along the OX axis).

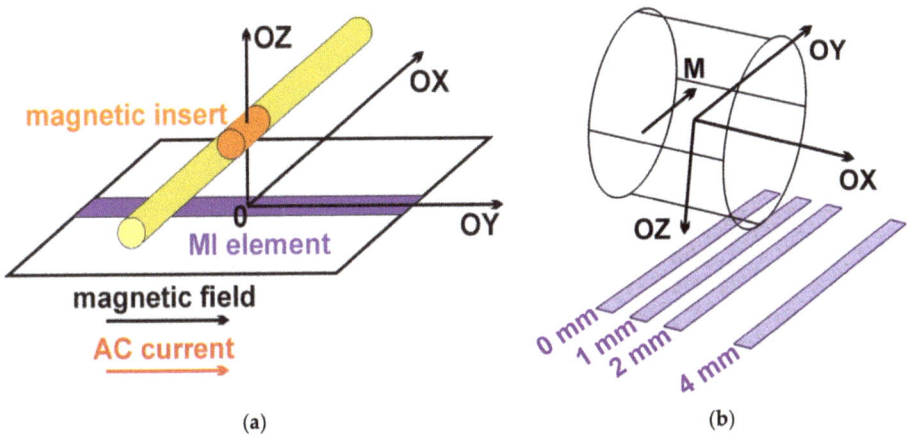

Figure 2. Scheme of the model experiment for detection of magnetic insert mimicking thrombus in the blood vessel by a magnetoimpedance multilayered sensitive element (**a**). Scheme of Comsol modeling of the stray fields of the cylindrical sample of magnetic composite. Positions at the distances of 0, 1, 2, and 4 mm along the OX direction are shown as different rectangular elements for obtaining a visual effect. Magnetization value M = 2.5 G (**b**).

Magnetic composite inset was previously magnetized in a high external magnetic field up to 10 kOe, resulting in the remnant magnetization of the sample due to the magnetic hysteresis [30]. During the experiment, the remnant magnetization (M_r = 2.5 G) of the inset was directed along the OY axis and remained constant in the external field range. The zero position (an origin) was considered to be the center of the MI element (Figure 2).

For the MI response description, we used the magnetoimpedance ratio (MI ratio) (1), the MI ratio sensitivity (2) and the MI response (3) which were calculated as follows:

$$\Delta Z/Z = 100 \times \frac{Z(H) - Z(H_{max})}{Z(H_{max})} \quad (1)$$

$$S(\Delta Z/Z) = \frac{\delta(\Delta Z/Z)}{\delta H} \quad (2)$$

$$\Delta(\Delta Z/Z) = \Delta Z/Z_{control} - \Delta Z/Z_{position} \quad (3)$$

where H_{max} = 100 Oe and $\delta(\Delta Z/Z)$ is the change MI ratio for the total impedance at the change of the magnetic field δH = 0.1 Oe, $\Delta Z/Z_{control}$ is the total impedance MI ratio with the control; $\Delta Z/Z_{position}$ is the total impedance MI ratio with the magnetic insert located in the certain position. For MI frequency dependence analysis, the change of the maximum value of the total impedance $\Delta Z/Z_{max}$ was provided. The experimental error in the impedance determination was within 1%. To estimate the random error in the setting of the magnetic insert position, the experiment was carried out three times on different days. Random error was estimated as follows:

$$\dot{\Delta}(\Delta Z/Z) = \sigma \times t \quad (4)$$

where σ—standard deviation of $\Delta Z/Z$, t = 4.3—Student coefficient of t-distribution for n = 3 [31].

2.3. Magnetic Stray Field Distribution Modeling

Nowadays computer modeling becomes more and more requested in the cases when magnetic filled composites are involved in the detection process. Among other reasons of usefulness of them we would like to mention the fact that the existing techniques of the structural evaluation of magnetic materials based on magnetic nanoparticles in many senses are still under development [32,33]. The magnetic stray field distribution of the magnetic insert on the surface of the MI multilayered sensitive element was modeled in Comsol MultiPhysics (AC/DC Module) (Comsol LLC, Switzerland, License № 17074991). In the Comsol model, y—the component of the magnetic field Hy along the long side of the MI element and x—the component of the magnetic field Hx along the short side of the MI element in XY plane were calculated. The geometries of MI element (0.5 mm × 10.0 mm) and the magnetic insert diameter 5.0 mm were taken to be the same as in the experimental MI measurements. The origin was set in a center of the magnetic insert and coordinates of the MI element center defined the position of the MI element. The magnetic distribution was calculated at different MI element positions along the OX axis (0.0, 1.0, 2.0 and 4.0 mm—these positions are shown as different rectangular elements for obtaining a visual effect), the OY coordinate 0.0 mm, the OZ coordinate 1.1 mm (Figure 2b).

The magnetization of the magnetic insert (M) was directed along the OY axis and according to the experimental data for hysteresis loop measurements M(H) it was about 2.5 G in the zero field (remnant magnetization). The detailed discussion of the magnetic properties of the filled composite sample will be given in Section 3.

3. Results

3.1. Structure and Magnetic Properties

Figure 3 shows the hysteresis loops and selected examples of magnetic domain structure of multilayered [FeNi (100 nm)/Cu (3 nm)]$_5$/Cu (500 nm)/[Cu (3 nm)/[FeNi (100 nm)]$_5$ MI sensitive element measured along EMA and HMA. It is clearly seen that deposition in an external magnetic field resulted in formation of well-defined uniaxial induced magnetic anisotropy along the short side of the rectangular stripe. The shapes of the hysteresis loops are in accordance with magnetic domain structures and special features of magnetization processes for magnetization along EMA and HMA. Magnetic domains in zero applied field (after saturation) are typical for soft ferromagnet bar domains with in-plain orientation of magnetization corresponding to opposite directions in "white" and "black" patterns. Closure domains near the borders of the MI element were shown to have complex structure [34] but this point is not under consideration in the present work. Magnetization process along the short side was characterized by domain wall displacements, coercive force defined for the top layer was of the order of 9 Oe (Figure 4b).

It is worth mentioning that magnetooptical signal passes to the permalloy for at most 40 nm. Taking into account direct and return pass, we obtained 20 nm information length [30]. For HMA the magnetization process occurs by the pure rotation for a magnetic field applied along the long side of the MI element. Thus, multilayered nanostructures have induced in-plane uniaxial magnetic anisotropy along the short side of the element and the anisotropy field of about 5 Oe. This ensures the work interval of the magnetic field sensor in the range below 5 Oe being quite convenient for applications.

Figure 4a shows SEM images of iron oxide nanoparticles that were used for fabrication of epoxy composites. Despite the fact that the shape of the microparticles varies, in a most general description we can say that they tend to be quasi-spherical and their average caliper size is about 300 nm. As a result of the analysis of the magnetic hysteresis loops the magnetic properties of the epoxy composite can be described as follows: coercivity Hc = 85 Oe, specific saturation magnetic moment m_s = 23 emu/g, residual specific magnetic moment mr = 1.8 emu/g (Figure 3a), saturation magnetization Ms = 29 G, remnant magnetization Mr = 2.5 G (Figure 4b,c).

The specific magnetic moment of the control composite (without magnetic particles) can be described as weak diamagnetic response, which was four orders of magnitude smaller than a magnetic response of the magnetic composite itself. The specific magnetic moment of saturation of magnetic microparticles, calculated from magnetic measurements of the composite and in accordance with the direct measurements of the NPs is of the order of 65 emu/g, which is in accordance with the values for magnetic particles of magnetite of such a size [35].

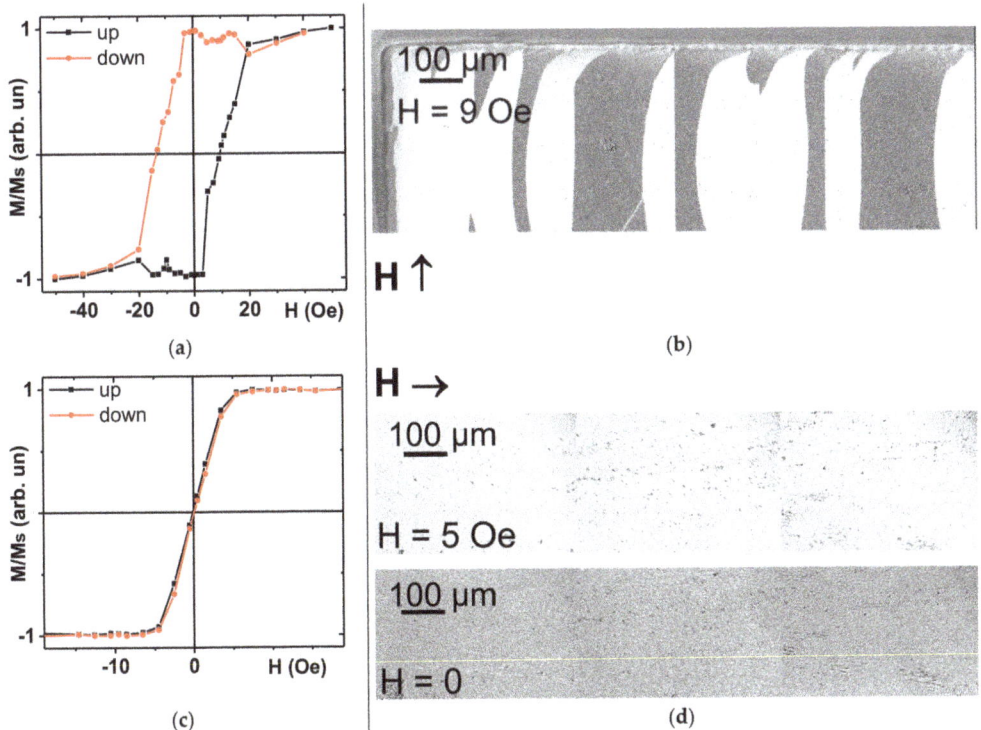

Figure 3. Magnetic hysteresis loops and selected examples of magnetic domain structures of MI element measured by Kerr–microscopy. (**a**,**b**) Along the short side of the MI element (EMA) or (**c**,**d**) long side of MI element (EMA). Surface magnetic domain images for the MI element: magnetic field of 9 Oe (coercive force) applied along the EMA (**b**); magnetic field of 5 Oe (approaching the anisotropy field) or close to zero applied along the HMA.

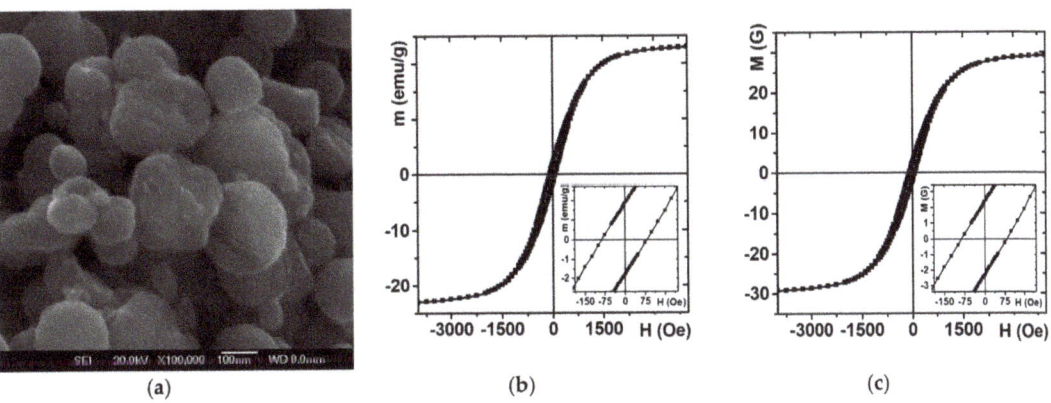

Figure 4. General view of magnetic particles used as a filler for fabrication of the epoxy composites—SEM (**a**). Magnetic properties of the magnetic composite: (**b**) specific magnetic moment; (**c**) magnetization.

3.2. Detection by MI Sensitive Element

Figure 5 shows MI responses (maximum MI ratio) of multilayered sensitive element as a function of the frequency of the driving current. For the whole magnetization cycle of the MI element there are "up" (from −150 Oe to 150 Oe) and "down" (from 150 Oe to −150 Oe) branches of the MI curves. Inset for Figure 5a shows the field dependences of the MI ratios for the total impedance. One can see that the field dependence of the MI ratio "up" and "down" curves are "mirror-symmetric" with respect to each other. However, each one of them is not symmetric with respect to the H = 0 axis. Such a behavior was previously observed many times and for each particular MI material and sensitive element the reason or a number of reasons causing such a behavior can be different. We will briefly mention some of the reasons—high order type of effective magnetic anisotropy, longitudinal static magnetic hysteresis, insufficient value of the applied field, etc. [36–38] but their discussion is also not considered for the present work.

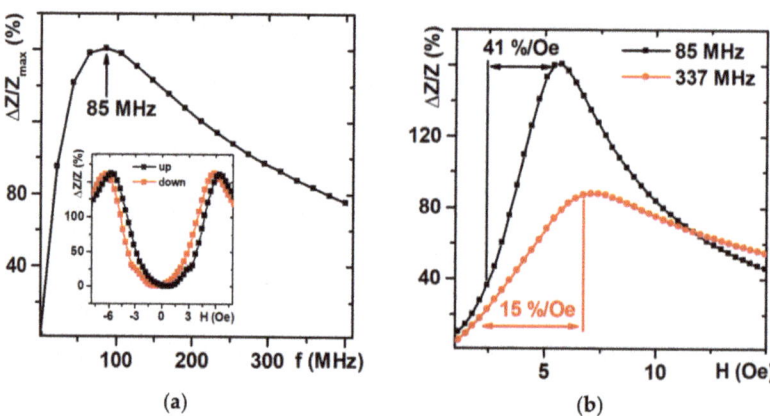

Figure 5. (a) Frequency dependence of the maximum value of the MI ratio. In the insert the field dependence of the MI ratio at 85 MHz high-frequency current. (b) "Down" curves of field dependence of the MI ratio at 85 and 337 MHz of the driving current frequency.

The maximum MI ratio (close to 160%) observed at the frequency of 85 MHz also corresponds to a maximum of the magnetic field sensitivity (41%/Oe). We define the working interval of magnetic sensor as the field range corresponding to maximum sensitivity: here it is 3–5 Oe. After 87 MHz as the driving current frequency increases, the MI ratio decreases, and the sensitivity and the maximum value of the MI ratio also are decreasing (Figure 5b). Magnetoimpedance is a spectroscopic technique—the measurements can be done and evaluated for a number of the frequencies of the interest. We therefore selected two values of the driving current frequency for the analysis: the frequency for the highest sensitivity with respect to the applied field and high frequency for which dielectric contributions are high but the magnetic field sensitivity of the multilayered element is still sufficient for the detection process.

Earlier we have investigated the MI responses of the multilayered film sensitive elements in the presence of micro- and nanoparticles on their surface or using polyacrylamide gel coverings [15,18]. Gels and ferrogels contain a large amount of water in their structures, contributing to MI signals due to high dielectric constant, which is especially strong at high frequencies. This circumstance defines strict conditions for the quality of the gel or ferrogel samples: their shape and weight must be controlled thoroughly. We therefore decided at the first stage of the development of the model experiments for thrombus detection to use composites with low dielectric response of the matrix. Epoxy resin dielectric constant (ε = 3.49 at 25 °C) is low in comparison with dielectric constant of liquid water being around 78.4 for low frequency ranges below 100 kHz. In a MHz range such a difference becomes even higher. With this selection of the matrix MI measurements become less demanding to the deviation of the weight and the shape of the samples of the composites, the dielectric effect can be neglected as making very small contribution.

Further, for simplicity we consider the "down" MI ratio curves only. As the magnetic field composite sample approaches the MI element, the stray magnetic fields of the magnetic particles affecting the surface of the MI element increase. It leads to the MI ratio decreases and MI curves peaks become shifted along the H axis (Figure 6). According to the existing theoretical approaches [21,37,38] such a behavior in the longitudinal MI geometry can be explained as follows. The shift of the MI ratio curves is due to the appearance of the parallel component of the effective magnetic field directed along the long side of the MI element. The decrease in the maximum MI ratio value is due to the perpendicular component of the magnetic field directed perpendicular to the long side of the MI element. The shift of the MI curves without the strong change of the maximum MI ratio (−4 and 4 mm) is similar to the case of an additional shift field opposite to the external magnetic field along the long side of the element. The largest decrease in the maximum value of the MI ratio corresponds to the position when the center of the magnetic filled composite sample is situated just above the center of the MI element.

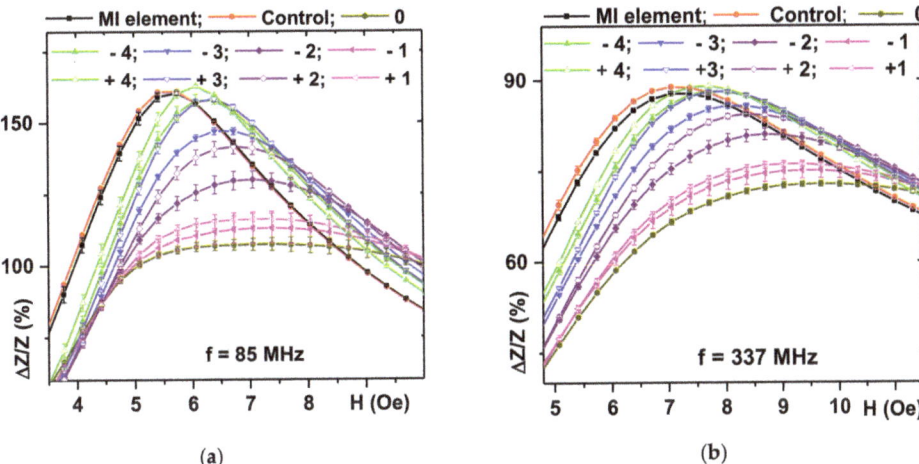

Figure 6. Field dependences of the MI ratio at the driving current frequencies of 85 (**a**) and 337 MHz (**b**) for different positions of epoxy composite samples (see also Figure 2a). MI responses of multilayered [FeNi (100 nm)/Cu (3 nm)]$_5$/Cu (500 nm)/[Cu (3 nm)/[FeNi (100 nm)]$_5$ MI sensitive element itself and epoxy resin sample without magnetic particles are also given. Numbers indicate displacements in mm of the center of the cylindrical composite from zero position corresponding to the point just above the center of the MI element being elevated by 1.2 mm from the free side of the element.

Maximum variations of the MI ratio in course of the change of the position of the filled composite were observed at the driving current frequency of 85 MHz corresponding to the maximum MI sensitivity with respect to external magnetic field (Figure 6a). In some cases, MI curves related to the symmetric positions of the magnetic composite sample (for example, −3 and 3 mm) do not match to each other. One of the reasons is possible magnetic heterogeneity of the composite as for the filling rather large particles were used. Another reason could be an experimental error in the displacement definition. Despite this uncertainty, the results are very clear. The position of the magnetic filled composite sample was well detected based on the analysis of the maximum value of the MI ratio.

For the analysis of the maximum MI value, the maximum response of the MI element was observed at the position when the center of the composite was above the center of the MI element: the MI response is almost zero at the magnetic composite position (−4 and 4 mm) (Figure 7a). However, the ΔZ/Z(H) curve maximum is quite wide. For practical reasons, the detection of the maximum is not always the optimum procedure (it requires some iteration measurements for the definition of peak position).

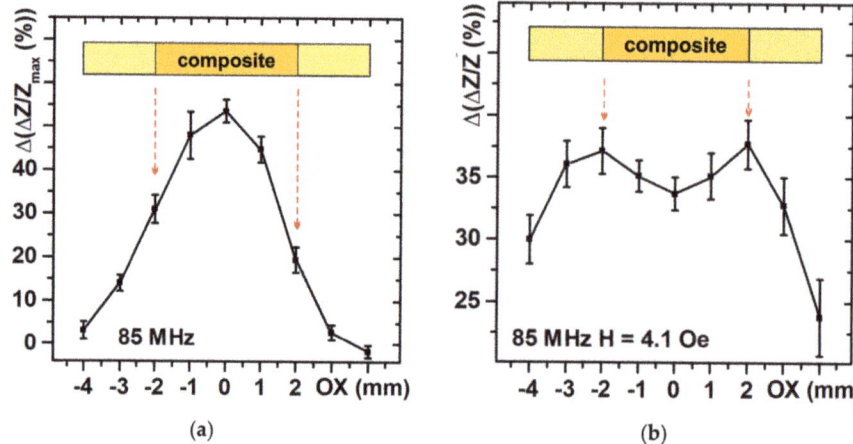

Figure 7. GMI response at the different positions of the magnetic insert: maximum MI ratio (a), in the sensitive range of the magnetic field (b). Arrows point to the points that correspond to the magnetic insert boundaries.

Let us select the sensitive range of the external magnetic field and analyze MI responses for H = 4.1 Oe. At the magnetic composite position (−4 mm, 4 mm) ΔZ/Z(H = 4.1 Oe) ratio value was about 30%. It indicates that, although, magnetic composite is situated far from the center of MI element, even the magnetic stray fields can be easily detected. However, the MI response at the central position of the magnetic insert (0 mm) corresponds to a local minimum and ΔZ/Z(H = 4.1 Oe) was weaker than in the cases for ΔZ/Z_{max} (Figure 7b). The reason can be that the stray magnetic field shifts effective external field point (4.1 Oe) from the sensitive range as the effective field is a superposition of the external field and stray field strength corresponding to each particular space point. Moreover, the perpendicular to the long side of the MI element magnetic stray field component can reduce the working interval of magnetic sensor as well. Therefore, the next step in understanding of the results of the composite detection was made in the direction of the calculation of the geometry of the stray fields of the cylindrical composite.

3.3. Magnetic Modeling

According to the Comsol-based modeling, the stray magnetic field distribution is uniform throughout the thickness of thin film element, thus we consider stray magnetic field on the surface of the MI element. As the position of the center of magnetic composite sample approaches the position just above the center of the MI element, the H_y stray magnetic field component increases. At the position of the magnetic composite just above the center of the MI element (OX = 0 mm), the stray magnetic field value was as high as about 3 Oe. It was concentrated in the central region of the MI element (±2 mm along OY axis) and it had the opposite sign with respect to the Y coordinate. At the edges of the MI element, stray magnetic fields had the same direction as the OY axis being about 1 Oe. When the magnetic composite position changes away from the center of the MI element, magnetic field concentration region is expanded and the stray magnetic field strength decreases (Figure 8a).

The H_x stray magnetic field component is almost zero at the position OX = 0 mm. At the other positions (OX: 1, 2 and 4 mm) stray magnetic field strength was about 1 Oe. It was oriented in opposite directions with respect to the OX axis in symmetric regions around the OY = 0 mm position (Figure 8b). In the presence of the stray magnetic field along the long side of the MI element (along the OY axis) without component along the short side (along the OX axis), the MI curves just shift along the field axis and the maximum MI ratio does not change. In the model magnetic fields distribution at the magnetic composite position just above the MI element H_x value is almost zero. However, the MI measurements show the largest decrease in the MI ratio (Figure 6). The reason for this is that the magnetic composite consists of relatively large magnetic particles each of which has an individual magnetic anisotropy axis. Thus, an effective magnetic anisotropy axis for the whole magnetic composite sample is affected by the distribution of the individual axes of the particles. Additionally, as the magnetic composite has a cylindrical form, demagnetization factors contribute to the position of the effective magnetic anisotropy axis as well. Unfortunately, in such circumstances we cannot calculate exact parameters for the direction of the magnetization in the magnetic composite. However, we can suggest some deviation of the cylindrical magnetic composite sample magnetization from the OY axis in the XY plane and describe the result qualitatively.

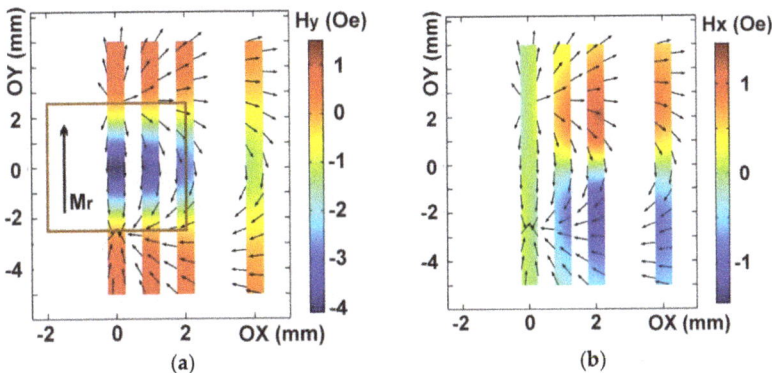

Figure 8. Stray magnetic fields distribution on the surface of the MI multilayered element for different positions of the magnetic composite sample placed at a distance of 1.2 mm from the surface of the MI element. Magnetic composite magnetization is directed along the OY axis: Hy stray magnetic field component along the long side of the MI element (**a**); Hx stray magnetic field component along the short side of the GMI element (**b**). Mr—the magnetization direction of the magnetic composite in XY plane.

Let us consider that the magnetic composite sample magnetization is directed at an angle of 15° with respect to the OY axis. The Hy stray magnetic fields component distribution is almost the same as in the case without magnetization deviation (Figure 9a). However, the Hx stray magnetic fields component distribution is significantly different. The Hx stray magnetic fields component is about -1 Oe and oriented in the opposite direction with respect to the OX axis throughout the MI element length at the position of the magnetic insert is above the GMI element (OX = 0 mm). At the other positions (OX: 1 mm, 2 mm, 4 mm), magnetic field distribution can be described by selection of two characteristic regions. The magnetic stray field in these regions is oriented in opposite directions along the OX axis (as in the case without magnetization deviation), but these regions have different areas and stray magnetic field strengths. An exception is the position at 4 mm displacement where magnetic field distribution is close to the case without magnetization deviation (Figure 9b).

Since the described theory [21] includes the model with uniformly distributed stray magnetic fields on the surface of the MI element in contrast to our case, it cannot be fully applied there. The results of the magnetic model with the GMI measurements can be interpreted as follows. The maximum MI ratio depends on Hx component. If the total field contribution of the Hx component is zero, then the maximum MI ratio does not change. For example, at the position OX = 4 mm there are two regions of magnetic field with opposite directions along the OX axis (Figure 9b) and the MI ratio curves just shift due to the Hy stray magnetic field component without decrease of the maximum MI ratio. As the magnetic composite approaches the MI element, one of these regions is expanded in area and magnetic field strength in this region increases. This leads to the rise in the total field contribution of the Hx component, thus the maximum of the MI ratio decreases. At the position OX = 0 mm the maximum value of the Hx component of the stray magnetic field is weaker than in position OX: 1 and 2 mm, however it has one direction throughout the MI element length (Figure 9b) and largest decrease in maximum MI ratio is observed (Figure 6a). If there is no Hx component of magnetic field in this position, no change in the maximum of the MI ratio will be observed.

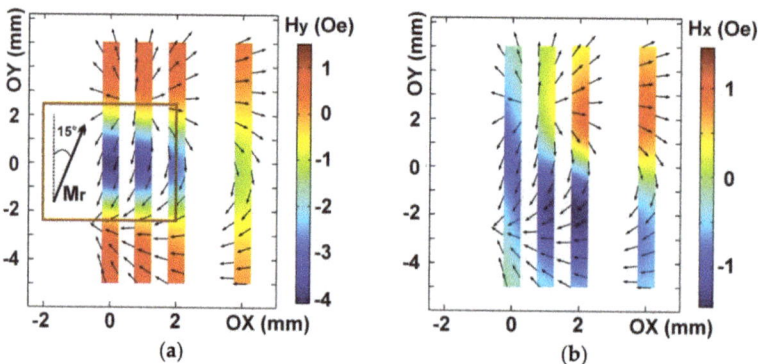

Figure 9. Stray magnetic field distribution on the surface of the GMI element at the different positions of the magnetic insert, the magnetic insert magnetization is directed at an angle 15° related to the OY axis: Hy stray magnetic field component along the long side of the GMI element (**a**); Hx stray magnetic field component along the short side of the GMI element (**b**). Mr—the magnetic insert magnetization direction in the XY plane.

4. Discussion

In the previous section, we experimentally demonstrated the validity of proposed approach for detection of the position of magnetic composite by MI magnetic field sensitive element and developed a simple model for understanding of contributions of the stray fields of magnetic particles of the composite at its different positions.

Now we would like to discuss some possible directions for future applications of MI sensors for thrombus state evaluation and thrombosis treatment. As it was mentioned in the introduction, there is a visible progress in the therapy with thrombolytic medication, which includes targeted drug delivery to a blood clot using MPs [22]. Figure 10a in schematic way describes the normal blood flow in a vein and Figure 10b part shows the region affected by thrombosis with irregular blood flow. The therapy involving the magnetic nanoparticles allows significantly increased local concentration of the thrombolytic drug immobilized onto their surface by directing them into the thrombosis region by the gradient magnetic field. It is also important to mention that such a procedure results in significant reduction of the time between the intravenous magnetic particles injection and their actuation time. The last means a greater number of the particles not localized by the microphages before their arrival to the actuation point. The main advantage of the whole procedure is a reduction of the side effects.

However, it is very difficult to evaluate the efficiency of the drug delivery without direct measurement of the number of magnetic particles at the therapy action point. Therefore, we propose to evaluate the amount of the particles there by the measurements of the superposition of their stray fields by magnetic field MI sensor (Figure 10c). Here all results described in the previous section become useful. As soon as the concentration of magnetic particles delivered by applied gradient magnetic field becomes non-zero at the point of therapy the MI sensor can define both the blood clot geometry and particles concentration. To make this possible the scanning procedure must be considered, i.e., one should measure the MI sensor responses at different positions of the device. Exactly as it was demonstrated above in the model experiments with filled composite samples. As it is critically important to determine the local concentration of magnetic particles in the thrombus region during the therapy, after definition of a central point of the blood clot the MI measurements can be periodically made at fixed position corresponding to a center.

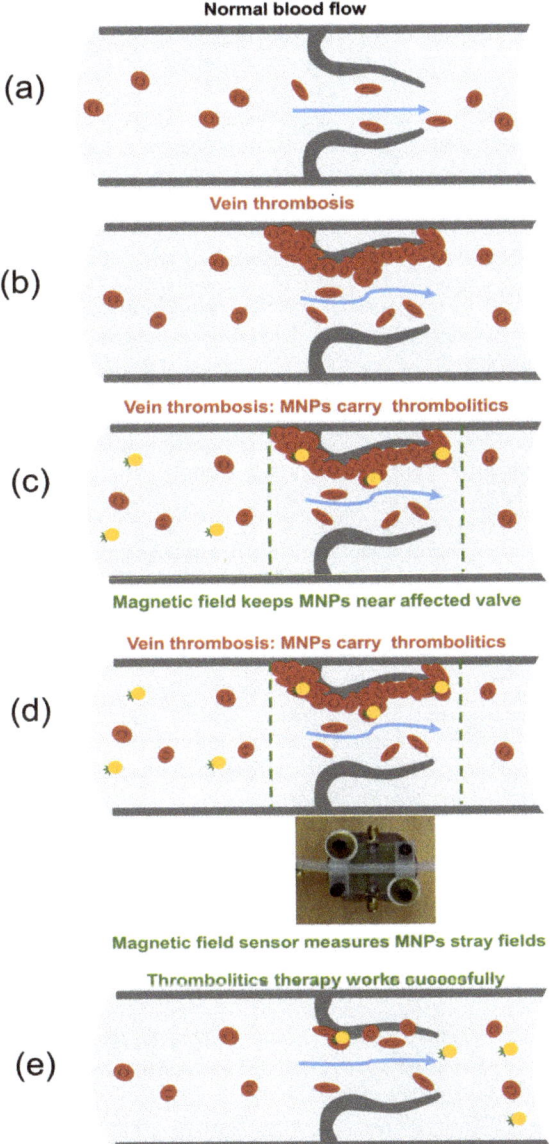

Figure 10. Schematic description of the normal blood flow in a vein (**a**) and blood flow near thrombosis affected vessel (**b**). Intravenous injection of magnetic particles carried on thrombotic drug and directed toward the therapy zone (indicated by vertical dashed lines) by gradient magnetic field (**c**). Magnetic field sensor evaluates the stray fields of magnetic particles accumulating in the therapy zone in a series of displacements measurements, which can be extended by periodic measurements of the response for evaluation of the particles concentration near the blood clot center (**d**). End of the therapy and particles removal (**e**).

Once the therapy has ended and magnetic particles are removed from the blood vessel previously hosting the blood clot, the magnetic field sensor can confirm that the stray field strength (concentration of magnetic particles) becomes very small. On one hand, we should mention the difference between large magnetic particles used for model experiments here and magnetic nanoparticles of the iron oxide usually designed for biomedical applications [5,11,39]. On the other hand, the only difference here would be the smaller value of the stray fields created by nanoparticles. This would require the higher sensitivity of the magnetic field sensor and its operation in shielded environments but such conditions are achievable and become standards of present day applications.

The idea of the addition of the iron oxide MPs to the blood stream is widely discussed in the literature and it is under current investigation for many purposes of theranostics [10,40–42]. There are many ways to make magnetic particles useful from magnetic imaging and drug delivery through hyperthermia toward regenerative medicine [39,43–46]. In this particular study, we just discussed one of possible directions of such applications combining a previously proposed way of the treatment of thrombosis affected vessel [22] with the help of magnetic particles delivering thrombosis drugs. The better way of the treatment control can be achieved by the MI sensor implication. The future will answer the question to what extent this particular approach is realistic and to what extent it can be used for the detection of clusters (agglomerates) of artificially introduced magnetic nanoparticles formed because of specific uptake or magnetic hyperthermia/thermal ablation. It seems to be difficult but possible. The tendency of incorporation of biosensors is increasing [20] and demanding additional efforts for the development of compact analytical devices for therapies and diagnostics.

In addition, it should be mentioned that appearance of nonuniform distribution of the magnetic deposits inside the living system is a quite typical situation. It can be connected with internal biological structures [47] or reflect of the changes of morphology related to the development of particular disease [48,49]. In any case, the technique to define the morphology of such cluster (agglomerate) is highly desired for correct diagnostics.

The idea to use MI effect for the detection of the stray field of magnetic nanoparticles was introduced and proven in 2003 [50] and its validity was confirmed by many studies [51–53]. The pico-Tesla MI sensor based on amorphous wire for biomagnetic measurements without magnetic shielding at room temperature was described in 2012 [54]. However, thin film based prototype is the most compatible and requested for existing production techniques, especially for the case of magnetic label or clusters detection [17]. Despite the fact that thin film based MI sensitive elements were intensively studied in the last 10 years and very high sensitivities of the order of 300%/Oe were obtained [55], promising approaches are under development. For example, a new type of MI multilayered structures with a high MI sensitivity has been proposed recently [56]. The multilayer structure consists of a highly conductive central layer and two outer ferromagnetic layers located below and above the conductive layer. The upper layer is a periodic structure, of N-multilayered elements and N + 1-regions, in which multilayered elements are absent (the upper layer is profiled). It is shown that for a profiled structure with a decrease in the deviation angle of the effective magnetic anisotropy axis from the transverse direction, the magnetic permeability of the upper layer increases, which leads to an increase in the skin effect and an increase in the MI effect. The progress in the development of the MNPs with lower size distribution and improved properties is continuous [18,44,45]. Both increase of the magnetic field detector sensitivity and enhanced quality of magnetic carriers contribute to the hope that proposed model experiments become a reality soon.

5. Conclusions

Magnetic multilayered [FeNi (100 nm)/Cu (3 nm)]$_5$/Cu (500 nm)/[Cu (3 nm)/[FeNi (100 nm)]$_5$ MI sensitive element was prepared by magnetron sputtering deposition in the external magnetic field in order to create well defined induced magnetic anisotropy. Magnetic properties, magnetization process peculiarities and magnetoimpedance effect were carefully analyzed. For model experiments non-ferromagnetic polymer tube was used to design the sample mimicking thrombus in a blood vessel. Cylinder sample was made from epoxy resin with a magnetic insert placed in the center part-filled composite type material containing 30% weight concentration of commercial magnetic iron oxide microparticles. The detection of the position of the magnetic composite mimicking thrombus in the blood vessel was carried out by the measurements of the stray magnetic fields of iron oxide microparticles using magnetoimpedance sensitive element. Changes of the MI ratio in the presence of composite can be characterized by the shift and the decrease of the maximum value of the MI. The magnetic field sensitivity of the longitudinal MI effect strongly depends on the value of the external field. The longitudinal component of the field can shift working point from the sensitive range of the field and transverse component can decrease sensitivity. We were able to detect the position of the magnetic composite sample mimicking thrombus in blood vessels. Comsol modeling of stray fields was successfully used for the analysis of the obtained experimental results. Although the model is rather simple, it provides good understanding of the origin the MI sensitivity in the proposed configuration. We describe a possible use of MI sensors for biomedical applications in the field of thrombus state evaluation and therapy.

Author Contributions: Conceptualization, G.V.K. and A.P.S.; methodology, A.V.S., G.Y.M., A.P.S. and G.V.K.; software, G.Y.M.; validation, A.V.S., V.N.L. and G.Y.M.; formal analysis, A.P.S.; investigation, V.N.L., A.V.S., G.Y.M.; data curation, V.N.L., A.V.S., G.Y.M.; writing—original draft preparation, G.Y.M. and G.V.K.; writing—review and editing, G.V.K., A.V.S., A.P.S., G.Y.M.; visualization, G.Y.M. and G.V.K.; supervision, G.V.K. All authors have read and agreed to the published version of the manuscript.

Funding: This research was in part founded by Russian Science Foundation, grant number 18-19-00090 and in part by the Ministry of Science and Higher Education of the Russian Federation (project No. FEUZ -2020-0051) and Proyecto Elkartek AVANSITE of the Basque Government.

Institutional Review Board Statement: This work did not involve humans or animals and therefore it did not require the Institutional Review Board Statement and approval.

Informed Consent Statement: Not applicable.

Data Availability Statement: Data available from the corresponding author upon reasonable request.

Acknowledgments: Selected studies were performed at SGIKER services of the Basque Country University UPV-EHU. We thank I. Orue and S. Fernandez Armas for special support.

Conflicts of Interest: The authors declare no conflict of interest.

References

1. Prinz, G.A. Magnetoelectronics applications. *J. Magn. Magn. Mater.* **1999**, *200*, 57–68. [CrossRef]
2. Zamani Kouhpanji, M.R.; Stadler, B.J.H. A guideline for Effectively Synthesizing and Characterizing Magnetic Nanoparticles for Advancing Nanobiotechnology: A Review. *Sensors* **2020**, *20*, 2554. [CrossRef]
3. Bukreev, D.A.; Moiseev, A.A.; Derevyanko, M.S.; Semirov, A.V. High-Frequency Electric Properties of Amorphous Soft Magnetic Cobalt-Based Alloys in the Region of Transition to the Paramagnetic State. *Russ. Phys. J.* **2015**, *58*, 141–145. [CrossRef]
4. Ivanov, A.O.; Camp, P.J. Effects of interactions on magnetization relaxation dynamics in ferrofluids. *Phys. Review E* **2020**, *102*, 032610. [CrossRef]
5. Wang, L.; Yan, Y.; Wang, M.; Yang, H.; Zhou, Z.; Peng, C.; Yang, S. An integrated nanoplatform for theranostics via multifunctional core-shell ferrite nanocubes. *J. Mater. Chem. B* **2016**, *4*, 1908–1914. [CrossRef]
6. Um, J.; Zhang, Y.; Zhou, W.; Zamani Kouhpanji, M.R.; Radu, C.; Franklin, R.R.; Stadler, B.J.H. Magnetic Nanowire Biolabels Using Ferromagnetic Resonance Identification. *ACS Appl. Nano Mater.* **2021**, *4*, 3557–3564. [CrossRef]
7. Svalov, A.V.; Arkhipov, A.V.; Andreev, S.V.; Neznakhin, D.S.; Larrañaga, A.; Kurlyandskaya, G.V. Modified field dependence of the magnetocaloric effect in Gd powder obtained by ball milling. *Mater. Lett.* **2021**, *284*, 128921. [CrossRef]
8. Sciancalepore, C.; Gualtieri, A.F.; Scardi, P.; Flor, A.; Allia, P.; Tiberto, P.; Barrera, G.; Messori, M.; Bondioli, F. Structural characterization and functional correlation of Fe_3O_4 nanocrystals obtained using 2-ethyl-1,3-hexanediol as innovative reactive solvent in non-hydrolytic sol-gel synthesis. *Mater. Chem. Phys.* **2018**, *207*, 337–349. [CrossRef]
9. Hergt, R.; Hiergeist, R.; Zeisberger, M.; Schüler, D.; Heyen, U.; Hilger, I.; Kaiser, W.A. Magnetic properties of bacterial magnetosomes as potential diagnostic and therapeutic tools. *J. Magn. Magn. Mater.* **2005**, *293*, 80–86. [CrossRef]

10. Grossman, J.H.; McNeil, S.E. Nanotechnology in cancer medicine. *Phys. Today* **2012**, *65*, 38–42. [CrossRef]
11. Safronov, A.P.; Beketov, I.V.; Komogortsev, S.V.; Kurlyandskaya, G.V.; Medvedev, A.I.; Leiman, D.V.; Larranaga, A.; Bhagat, S.M. Spherical magnetic nanoparticles fabricated by laser target evaporation. *AIP Adv.* **2013**, *3*, 052135. [CrossRef]
12. Khawja Ansari, S.A.M.; Ficiara, E.; Ruffinatti, F.A.; Stura, I.; Argenziano, M.; Abollino, O.; Cavalli, R.; Guiot, C.; D'Agata, F. Magnetic Iron Oxide Nanoparticles: Synthesis, Characterization and Functionalization for Biomedical Applications in the Central Nervous System. *Materials* **2019**, *12*, 465. [CrossRef]
13. Agra, K.; Mori, T.J.A.; Dorneles, L.S.; Escobar, V.M.; Silva, U.C.; Chesmana, C.; Bohna, F.; Corrêa, M.A. Dynamic magnetic behavior in non-magnetostrictive multilayered films grown on glass and flexible substrates. *J. Magn. Magn. Mater.* **2014**, *355*, 136–141. [CrossRef]
14. Lee, W.; Joo, S.; Kim, S.U.; Rhie, K.; Hong, J.; Shin, K.-H.; Kim, K.H. Magnetic bead counter using a micro-Hall sensor for biological applications. *Appl. Phys. Lett.* **2009**, *94*, 153903. [CrossRef]
15. Yuvchenko, A.A.; Lepalovskii, V.N.; Vas'kovskii, V.O.; Safronov, A.P.; Volchkov, S.O.; Kurlyandskaya, G.V. Magnetic impedance of structured film meanders in the presence of magnetic micro- and nanoparticles. *Tech. Phys.* **2014**, *59*, 230–236. [CrossRef]
16. Baselt, D.R.; Lee, G.U.; Natesan, M.; Metzger, S.W.; Sheehan, P.E.; Colton, R. A biosensor based on magnetoresistance technology. *Biosens. Bioelectron.* **1998**, *13*, 731–739. [CrossRef]
17. Cardoso, S.; Leitao, D.C.; Dias, T.M.; Valadeiro, J.; Silva, M.D.; Chicharo, A.; Silverio, V.; Gaspar, J.; Freitas, P.P. Challenges and trends in magnetic sensor integration with microfluidics for biomedical applications. *J. Phys. D Appl. Phys.* **2017**, *50*, 213001. [CrossRef]
18. Kurlyandskaya, G.V.; Portnov, D.S.; Beketov, I.V.; Larrañaga, A.; Safronov, A.P.; Orue, I.; Medvedev, A.I.; Chlenova, A.A.; Sanchez-Ilarduya, M.B.; Martinez-Amesti, A.; et al. Nanostructured materials for magnetic biosensing Nanostructured materials for magnetic biosensing. *Biochim. Biophys. Acta (BBA) Gen. Subj.* **2017**, *1861*, 1494–1506. [CrossRef] [PubMed]
19. Morikawa, T.; Nishibe, Y.; Yamadera, H.; Nonomura, Y.; Takeuchi, M.; Taga, Y. Giant magneto-impedance effect in layered thin films. *IEEE Trans. Magn.* **1997**, *33*, 4367–4372. [CrossRef]
20. Llandro, J.; Palfreyman, J.J.; Ionescu, A.; Barnes, C.H.W. Magnetic biosensor technologies for medical applications: A review. *Med. Biol. Eng. Comput.* **2010**, *48*, 977–998. [CrossRef] [PubMed]
21. Buznikov, N.A.; Safronov, A.P.; Orue, I.; Golubeva, E.V.; Lepalovskij, V.N.; Svalov, A.V.; Chlenova, A.A.; Kurlyandskaya, G.V. Modelling of magnetoimpedance response of thin film sensitive element in the presence of ferrogel: Next step toward development of biosensor for in tissue embedded magnetic nanoparticles detection. *Bios. Bioelectr.* **2018**, *117*, 366–372. [CrossRef] [PubMed]
22. Prilepskii, A.Y.; Fakhardo, A.F.; Drozdov, A.S.; Vinogradov, V.V.; Dudanov, I.P.; Shtil, A.A.; Bel'tyukov, P.P.; Shibeko, A.M.; Koltsova, E.M.; Nechipurenko, D.Y.; et al. Urokinase-Conjugated Magnetite Nanoparticles as a Promising Drug Delivery System for Targeted Thrombolysis: Synthesis and Preclinical Evaluation. *ACS Appl. Mater. Interfaces* **2018**, *10*, 36764–36775. [CrossRef]
23. Kurlyandskaya, G.V.; Elbaile, L.; Alves, F.; Ahamada, B.; Barrue, R.; Svalov, A.V.; Vas'kovskiy, V.O. Domain structure and magnetization process of a giant magnetoimpedance geometry FeNi/Cu/FeNi(Cu)FeNi/Cu/FeNi sensitive element. *J. Phys. Condens. Matter* **2004**, *16*, 6561–6568. [CrossRef]
24. Cheng, S.F.; Lubitz, P.; Zheng, Y.; Edelstein, A.S. Effects of spacer layer on growth, stress and magnetic properties of sputtered permalloy film. *J. Magn. Magn. Mater* **2004**, *282*, 109–113. [CrossRef]
25. Corrêa, M.A.; Viegas, A.D.C.; Da Silva, R.B.; De Andrade, A.M.H.; Sommer, R.L. Magnetoimpedance of single and multilayered FeCuNbSiB films in frequencies up to 1.8 GHz. *J. Appl. Phys* **2007**, *101*, 043905. [CrossRef]
26. Saito, N.; Fujiwara, H.; Sugita, Y. A new type magnetic domain in negative magnetostriction Ni-Fe films. *J. Phys. Soc. Jpn.* **1964**, *19*, 1116–1125. [CrossRef]
27. Svalov, A.V.; Kurlyandskaya, G.V.; Hammer, H.; Savin, P.A.; Tutynina, O.I. Modification of the "transcritical" state in NiFeCuMo films produced by RF sputtering. *Tech. Phys.* **2004**, *49*, 868–871. [CrossRef]
28. Coïsson, M.; Vinal, F.; Tiberto, P.; Celegato, F. Magnetic properties of FeSiB thin films displaying stripe domains. *J. Magn. Magn. Mater.* **2009**, *321*, 806–809. [CrossRef]
29. Buznikov, N.A.; Svalov, A.V.; Kurlyandskaya, G.V. Influence of the Parameters of Permalloy-Based Multilayer Film Structures on the Sensitivity of Magnetic Impedance Effect. *Phys. Met. Metallogr.* **2021**, *122*, 223–229. [CrossRef]
30. Hubert, A.; Schäfer, R. *Magnetic Domains*; Springer: Berlin, Germany, 1998.
31. Barabesi, L.; Greco, L. A Note on the exact computation of the Student t, Snedecor F and sample correlation coefficient distribution functions. *J. R. Stat. Soc. Ser. D* **2002**, *51*, 105–110.
32. Rikken, R.S.M.; Engelkamp, H.; Nolte, R.J.M.; Maan, J.C.; van Hest, J.C.M.; Wilson, D.A.; Christianen, P.C.M. Shaping polymersomes into predictable morphologies via out-of-equilibrium self-assembly. *Nat. Commun.* **2016**, *7*, 12606. [CrossRef] [PubMed]
33. Bender, P.; Günther, A.; Tschöpe, A.; Birringer, R. Synthesis and characterization of uniaxial ferrogels with Ni nanorods as magnetic phase. *J. Magn. Magn. Mater.* **2011**, *323*, 2055–2063. [CrossRef]
34. Alzola, N.V.; Kurlyandskaya, G.V.; Larrañaga, A.; Svalov, A.V. Structural Peculiarities and Magnetic Properties of FeNi Films and FeNi/Ti-Based Magnetic Nanostructures. *IEEE Trans. Magn.* **2012**, *48*, 1605–1608. [CrossRef]
35. O'Handley, R.C. *Modern Magnetic Materials*; John Wiley & Sons: New York, USA, 1972; p. 740.

36. Vas'kovskii, V.O.; Savin, P.A.; Volchkov, S.O.; Lepalovskii, V.N.; Bukreev, D.A.; Buchkevich, A.A. Nanostructuring Effects in Soft Magnetic Films and Film Elements with Magnetic Impedance. *Tech. Phys.* **2013**, *58*, 105–110. [CrossRef]
37. Kurlyandskaya, G.V.; Bebenin, N.G.; Vas'kovskiy, V.O. Giant magnetic impedance of wires with a thin magnetic coating. *Phys. Met. Metallogr.* **2011**, *111*, 133–154. [CrossRef]
38. Semirov, A.V.; Moiseev, A.A.; Bukreev, D.A.; Kovaleva, N.P.; Vasyukhno, N.V.; Nemirova, V.A. Asymmetric magnetoimpedance of a magnetically soft wire. *Phys. Met. Metallogr.* **2017**, *118*, 535–540. [CrossRef]
39. Spizzo, F.; Sgarbossa, P.; Sieni, E.; Semenzato, A.; Dughiero, F.; Forzan, M.; Bertani, R.; Del Bianco, L. Synthesis of ferrofluids made of iron oxide nanoflowers: Interplay between carrier fluid and magnetic properties. *Nanomaterials* **2017**, *7*, 373. [CrossRef]
40. Kaczmarek, K.; Hornowski, T.; Dobosz, B.; Józefczak, A. Influence of magnetic nanoparticles on the focused ultrasound hyperthermia. *Materials* **2018**, *11*, 1607. [CrossRef]
41. Pavlov, A.M.; De Geest, B.G.; Louage, B.; Lybaert, L.; De Koker, S.; Koudelka, Z.; Sapelkin, A.; Sukhorukov, G.B. Magnetically engineered microcapsules as intracellular anchors for remote control over cellular mobility. *Adv. Mater.* **2013**, *25*, 6945–6950. [CrossRef]
42. Novoselova, I.P.; Safronov, A.P.; Samatov, O.M.; Beketov, I.V.; Medvedev, A.I.; Kurlyandskaya, G.V. Water based suspensions of iron oxide obtained by laser target evaporation for biomedical applications. *J. Magn. Magn. Mater.* **2016**, *415*, 35–38. [CrossRef]
43. Vallejo-Fernandez, G.; Whear, O.; Roca, A.G.; Hussain, S.; Timmis, J.; Patel, V.; O'Grady, K. Mechanisms of hyperthermia in magnetic nanoparticles. *J. Phys. D Appl. Phys.* **2013**, *46*, 312001. [CrossRef]
44. Barrera, G.; Coisson, M.; Celegato, F.; Martino, L.; Priyanka Tiwari, P.; Verma, R.; Kane, S.N.; Mazaleyrat, F.; Tiberto, P. Specific loss power of Co/Li/Zn-mixed ferrite powders for magnetic hyperthermia. *Sensors* **2020**, *20*, 2151. [CrossRef] [PubMed]
45. Zverev, V.I.; Pyatakov, A.P.; Shtil, A.A.; Tishin, A.M. Novel applications of magnetic materials and technologies for medicine. *J. Magn. Magn. Mater.* **2018**, *459*, 182–186. [CrossRef]
46. Li, Y.; Huang, G.; Zhang, X.; Li, B.; Chen, Y.; Lu, T.; Lu, T.J.; Xu, F. Magnetic hydrogels and their potential biomedical applications. *Adv. Funct. Matters* **2013**, *3*, 660–672. [CrossRef]
47. Kurlyandskaya, G.V.; Novoselova, I.P.; Schupletsova, V.V.; Andrade, R.; Dunec, N.A.; Litvinova, L.S.; Safronov, A.P.; Yurova, K.A.; Kulesh, N.A.; Dzyuman, A.N.; et al. Nanoparticles for magnetic biosensing systems. *J. Magn. Magn. Mater.* **2017**, *431*, 249–254. [CrossRef]
48. Bulk, M.; van der Weerd, L.; Breimer, W.; Lebedev, N.; Webb, A.; Goeman, J.J.; Ward, R.J.; Huber, M.; Oosterkamp, T.H.; Bossoni, L. Quantitative comparison of different iron forms in the temporal cortex of Alzheimer patients and control subjects. *Sci. Rep.* **2018**, *8*, 6898. [CrossRef] [PubMed]
49. Brar, S.; Henderson, D.; Schenck, J.; Zimmerman, E.A. Iron accumulation in the substantia nigra of patients with Alzheimer disease and Parkinsonism. *Arch Neurol.* **2009**, *66*, 371–374. [CrossRef] [PubMed]
50. Kurlyandskaya, G.V.; Sánchez, M.L.; Hernando, B.; Prida, V.M.; Gorria, P.; Tejedor, M. Giant-magnetoimpedance-based sensitive element as a model for biosensors. *Appl. Phys. Lett.* **2003**, *82*, 3053–3056.
51. Blanc-Béguin, F.; Nabily, S.; Gieraltowski, J.; Turzo, A.; Querellou, S.; Salaun, P.Y. Cytotoxicity and GMI bio-sensor detection of maghemite nanoparticles internalized into cells. *J. Magn. Magn. Mater.* **2009**, *321*, 192–197. [CrossRef]
52. Yang, Z.; Wang, H.H.; Dong, X.W.; Yan, H.L.; Lei, C.; Luo, Y.S. Giant magnetoimpedance based immunoassay for cardiac biomarker myoglobin. *Anal. Methods* **2017**, *9*, 3636–3642. [CrossRef]
53. Chlenova, A.A.; Buznikov, N.A.; Safronov, A.P.; Golubeva, E.V.; Lepalovskii, V.N.; Melnikov, G.Y.; Kurlyandskaya, G.V. Detecting the total stray fields of ferrogel nanoparticles using a prototype magnetoimpedance sensor: Modeling and experiment. *Bull. Russ. Acad. Sci. Phys.* **2019**, *83*, 906–908.
54. Uchiyama, T.; Mohri, K.; Honkura, Y.; Panina, L.V. Recent advances of pico-Tesla resolution magneto-impedance sensor based on amorphous wire CMOS IC MI Sensor. *IEEE Trans. Magn.* **2012**, *48*, 3833–3839. [CrossRef]
55. García-Arribas, A.; Fernández, E.; Svalov, A.; Kurlyandskaya, G.V.; Barandiaran, J.M. Thin-film magneto-impedance structures with very large sensitivity. *J. Magn. Magn. Mater.* **2016**, *400*, 321–326. [CrossRef]
56. Buznikov, N.A.; Kurlyandskaya, G.V. Magnetoimpedance of Periodic Partly Profiled Multilayer Film Structures. *Fizika Metallov i Metallovedenie* **2021**, *122*, 755–760. [CrossRef]

MDPI
St. Alban-Anlage 66
4052 Basel
Switzerland
Tel. +41 61 683 77 34
Fax +41 61 302 89 18
www.mdpi.com

Sensors Editorial Office
E-mail: sensors@mdpi.com
www.mdpi.com/journal/sensors

www.ingramcontent.com/pod-product-compliance
Lightning Source LLC
LaVergne TN
LVHW070705100526
838202LV00013B/1033